A New Tooth Geometry and Its Applications

An attempt to unify gear theories

Dr. Sho Honda

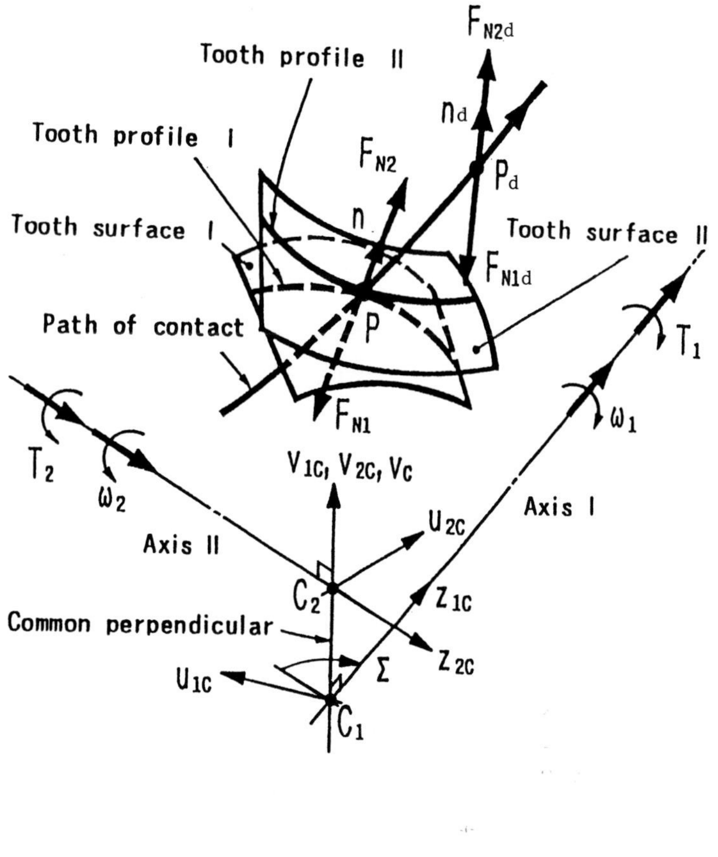

A New Tooth Geometry and Its Applications

- An attempt to unify gear theories -

The First Pulished June 11, 2016

Sho Honda

Published by

SOEISHA Co., Ltd.／SANSEIDO BOOKSTORE

1-1 Kanda Jimbocho, Chiyoda-ku, Tokyo 101-0051

Tel：03-3291-2295 Fax：03-3292-7687

Printed by

NIHON PRINTING CO., LTD.

Printed in Japan

ISBN978-4-88142-979-2 C3053

Preface

The first job when the author began working at Toyota Motor Corporation was to develop a rack and pinion for steering gears with inconstant rotational motion (Chapter 7). The steering systems at that time were manual, so that the steering force became heavier with the rotation of the wheel. The new rack and pinion was intended to reduce the steering force by varying the gear ratio. Gear pairs with a variable ratio had been known as oval gears, but the profile curves had not been clarified theoretically. Therefore, he tried to repeat drawings to understand how the profile could be obtained, developed a calculation method and succeeded in producing variable ratio steering gears. Through this development, he was possessed by the charm of gears.

Afterwards, the author was engaged in designing almost all kinds of gears for automobiles and machine tools, from cylindrical to hypoid gears, and in making experiments to solve noise problems of gears. During these works, it worried him that every kind of gears had its own calculation method in the design procedure and that the calculated dynamic increment of tooth load hardly explained the results of experiment in noise and vibration. That must be, he thought, because the present tooth geometry was made for every kind of gears independently and was effective only under a static condition (constant gear ratio), which was the reason why the theory of dynamic loads was discussed by replacing a gear pair with a mass-spring system having no relation to the tooth geometry (the fundamental requirement for contact).

Although a gear pair realizes the constant rotational motion on average determined by a gear ratio, an individual pair of profiles makes inconstant rotational motion caused by errors and deflection. Therefore, if a tooth geometry is based on the inconstant rotational motion and is made under coordinate systems common to all kinds of gears, it makes it possible to design all kinds of gears by common calculating equations defined in the common coordinate systems and to deal with the dynamic rotational motion, which is the basic idea of the new tooth geometry.

The new tooth geometry may be unfamiliar to those who are accustomed to the present one because of its coordinate systems, the definition of the angles such as pressure and spiral angles and so on. In fact, at present, only the engineers and researchers around the author understand it. However, once you read this new tooth geometry, he believes that you will understand the design procedure and the dynamic rotational motion of almost all kinds of gears at the same time. For example, hypoid gears can be designed more easily and work more quietly through the new method. It would be his great pleasure if you would read and apply it to your work.

March, 2016.
Sho Honda

Contents

Preface

1. Introduction 1

2. Basic theory 4
 2.1 Basic coordinate systems 4
 2.2 Fundamental requirement for contact and ratio of angular velocities 13
 2.3 Path of contact and its tooth profiles 16
 2.4 Tooth surface in the vicinity of a point of contact and limit of action 20
 2.5 Equations of motion and bearing loads 27
 2.6 Summary and references 29

3. Design method of tooth surfaces for quieter gear pairs 31
 3.1 Conditions for no variation of bearing loads 31
 3.2 Equations of path of contact 36
 3.3 Limit of action and selection of path of contact 45
 3.4 Equivalent rack 56
 3.5 Surface of action, conjugate tooth surfaces, design reference bodies of revolution and tooth traces 63
 3.6 Contact ratios 70
 3.7 Design method for tooth surfaces of hypoid gears 73
 3.8 Design examples of tooth surfaces of hypoid gears 84
 3.9 Summary and references 92

4. Involute helicoid and its conjugate tooth surface 94
 4.1 Definition of coordinate systems O_2, O_{q2}, O_1 and O_{q1} 94
 4.2 Equations of involute helicoid and a pair of involute gears in point contact 95
 4.3 Surface of action and conjugate tooth surface of an involute helicoid and a family of involute gear pairs 98
 4.4 Design method and example of face gears having involute helicoids for gear I (pinion) 106
 4.5 Design example of face gears having an involute spur gear for gear I (pinion) 115

4．6　Summary and references　118

5．Variation of tooth bearing and engagement error caused by assembly errors　120
5．1　Movement of a point of contact and engagement error　120
5．2　Calculated examples of variation of tooth bearing and engagement error　125
5．3　Summary　131

6．Rotational motion of gears and dynamic increment of tooth load　132
6．1　Path of contact with its common contact normal at each point of a gear pair having modified tooth surfaces Ⅱ and tooth surfaces Ⅰ under no loads　132
6．2　Paths of contact and their common contact normal at each point of a gear pair under load　141
6．3　Equations of motion of gears　151
6．4　Dynamic increment of tooth load and variation of bearing loads　160
6．5　Dynamic increment of tooth load of a gear pair having a single flank error composed of symmetrical convex curves　169
6．6　Verification by experiment through helical gear pairs having a single flank error composed of symmetrical convex curves　175
6．7　Summary and references　183

7．Design of a pair of tooth surfaces transmitting inconstant rotational motion　185
7．1　Tooth surfaces Ⅰ and Ⅱ and a tooth surface of a rack transmitting inconstant rotational motion under a given center distance　185
7．2　Tooth surface of a rack driven by a screw with constant lead and its mating tooth surface of gear Ⅰ (pinion) transmitting inconstant rotational motion　192
7．3　Design examples　196
7．4　Summary and references　200

8．Summary　202

Postscript

1. Introduction

Gear pairs have been widely used and researched for a long time and many excellent books and papers on gear theory are published until now [(1)~(11)], therefore it is said that gear theory has been fully researched and has no new knowledge to be added. In spite of that rumor, the reasons why the author intends to publish another book on this subject are that the present theory has following problems.

(1) It is not a system common to all kinds of gears. Cylindrical, worm, crossed helical, bevel and hypoid gears have their own theoretical systems which prevent design methods from being in common with all kinds of gears.

(2) Moreover, the present tooth geometry is effective only in spaces without the inertial force, so that it is useless for analyzing dynamic rotational motion of a gear pair with errors and deflection. Therefore, the present vibration models have no relation to the tooth geometry (especially the fundamental requirement for contact) and the accuracy is not sufficient to estimate the dynamic increment of tooth load.

(3) As computer methods by which tooth surfaces are expressed numerically have been developed remarkably, when one tooth surface is given numerically, the other can be obtained by the generating roll of machine and the conditions of contact can be analyzed. However, although a pair of tooth surfaces should be determined based on the required performances having no relation to the cutting methods, yet these computer methods can calculate only the tooth surfaces which are realized by a given cutting machine and cutter. Moreover, as the defects in the theory mentioned above are unsolved, it is still remained unclear what kinds of tooth surfaces should be given to realize the required performances and how the dynamic increment of tooth load of the gear pair having errors and deflection can be estimated.

Therefore, a new tooth geometry is proposed which is based on the results of my research and experience for about 40 years.

In Chapter 1. *Introduction*, the problems of the present theories and the outline of the new one are described.

In Chapter 2. *Basic theory*, eight coordinate systems common to all kinds of gears are defined, in which a point and the inclination angle of a normal through the point are expressed by five variables and the fundamental requirement for contact is introduced. A path of contact, the inclination angle of a common contact normal at each point and a pair of tooth profiles are expressed as functions of the angle of rotation and limits of action are discussed. In addition, equations of motion and bearing loads are obtained. Based on these coordinate systems and the equations of motion, the new tooth geometry becomes common to all kinds of gears and makes it possible to deal with the dynamic rotational motion of gears at the same time.

In Chapter 3. *Design method of tooth surfaces for quieter gear pairs*, the basic theory mentioned above is applied to designing tooth surfaces of gears for power transmission. At first, it is clarified that the path of contact satisfying the conditions for no variation of bearing loads is a straight line which coincides with the common contact normal and the selection methods of the path of contact which realizes a pair of smooth tooth surfaces are discussed. The concepts of equivalent rack and plane surface of action common to all kinds of gears having the path of contact mentioned above are clarified and equations of conjugate tooth surfaces corresponding to the plane surface of action are obtained. Moreover, the effective plane surface of action is determined and the contact ratios common to all kinds of gears are introduced. Based on this design method, the pitch surface, the surface of action, the pressure angle, the spiral angle and the contact ratio are redefined as definitions common to all kinds of gears and finally a design example of hypoid gears is shown and compared with one designed by Wildhaber's method. As present tooth surfaces of bevel and hypoid gears can be regarded as approximate ones corresponding to those having the plane surface of action, the plane surface of action is effective for analyzing contact conditions of the present tooth surfaces simply and theoretically.

In Chapter 4. *Involute helicoid and its conjugate tooth surface*, instead of giving the plane surface of action to realize no variation of bearing loads, an involute helicoid and its conjugate tooth surface having the same path of contact as that on the plane surface of action are discussed. Equations of an involute helicoid, its surface of action and its conjugate tooth surface are introduced and design examples of face gears are shown.

In Chapter 5. *Variation of tooth bearing and engagement error caused by assembly errors*, they are discussed based on the design examples in Chapter 4.

In Chapter 6. *Rotational motion of gears and dynamic increment of tooth load*, the paths of contact under no loads and under load of a gear pair having involute helicoids for one member which are modified slightly and convexly and the conjugate tooth surfaces of the involute helicoids for the other are obtained, the equations of motion are solved and the dynamic increment of tooth load and the variation of bearing load are obtained. The dynamic increment of tooth load of the gear pair having a single flank error composed of symmetrical convex curves which is practically important is algebraically obtained and verified by experiment through helical gear pairs. The dynamic increment of tooth load obtained here makes it possible to estimate the performances of noise and vibration in the design stage of gear pairs.

In Chapter 7. *Design of a pair of tooth surfaces transmitting inconstant rotational motion*, a surface of action and its corresponding tooth surfaces of a rack and pinion transmitting inconstant rotational motion are analyzed and it is shown that the new tooth geometry can be applied to designing a gear pair transmitting inconstant rotational motion.

In Chapter 8. *Summary*, the entire contents are summarized.

References

(1) 成瀬政男, 歯車, (1934), 岩波書店.
(2) Wildhaber, E., "Basic Relationship of Hypoid Gears Ⅰ－Ⅷ", American Machinist, Feb. 14, 28, March 14, June 6, 20, July 18, Aug. 1, 15, 1946.
(3) Buckingham, E., *Analytical Mechanics of Gears*, (1949), McGraw-Hill Book Co.
(4) 渡辺茂, 歯車歯形論, (1949), コロナ社.
(5) 成瀬政男編, 歯車の研究, (1960), 養賢堂.
(6) 仙波正荘, 歯車第１巻, 日刊工業新聞社.
(7) Dudley, D.W., *Gear Handbook*, (1962), McGraw-Hill Book Co.
(8) 歯車便覧 (1962), 日刊工業新聞社.
(9) Litvin, F.L., Gutman, Y., "Methods of Synthesis and Analysis for Hypoid Gear-Drives of "Formate" and "Helixform" (Part1-3)", "A Method of Local Synthesis of Gears Grounded on the Connections between the Principal and Geodetic Curvatures of Surfaces", Trans. ASME,

Vol.103 (January 1981), 83-125.
(10) Harris, S.L., "Dynamic Loads on the Teeth of Spur Gears", Proceedings of the Institution of Mechanical Engineers, 172 (1958), 87-112.
(11) 歌川正博，平歯車の動荷重，日本機械学会誌，61-470 (1958)，296.

2. Basic theory

2. 1 Basic coordinate systems (21), (23), (25)

Figure 2.1-1 shows the basic concept of a new tooth geometry, which is represented by a point of contact P, its common contact normal **n**, a path of contact, a pair of tooth profiles I and II, its tooth surfaces I and II and normal components $\mathbf{F}_{N1}$ and $\mathbf{F}_{N2}$ of concentrated loads in basic coordinate systems. At an instant, a gear pair rotates at angular velocities ω_1 and ω_2 (positive in the direction in Fig.2.1-1) and transmits constant input and output torques T_1 and T_2 (positive in the same direction as ω_1 and ω_2) around axes I and II respectively. The tooth surfaces I and II make contact at P and the normal component $\mathbf{F}_{N2}$ and the reaction one $\mathbf{F}_{N1}$ ($= -\mathbf{F}_{N2}$) of the load act on the tooth surfaces II and I respectively. The common contact normal **n** at P is a unit vector and represents a directed line of action of $\mathbf{F}_{N2}$ at the same time.

When the gear pair rotates by an angle chosen at will, the point of contact moves to P_d, the angular velocities vary to ω_{1d} and ω_{2d}, the normal components to $\mathbf{F}_{N1d}$ and $\mathbf{F}_{N2d}$, the common contact normal to $\mathbf{n}_d$ and PP_d draws the path of contact whose common contact normal varies from **n** to $\mathbf{n}_d$. Transforming the path of contact PP_d with the common contact normals **n** and $\mathbf{n}_d$ into the coordinate systems which rotate with each gear, the tooth profiles I and II are obtained as three-dimensional curves, which transmit the same rotational motion as the tooth surfaces I and II and represent the traces (tooth bearings) of the concentrated loads on the tooth surfaces. These tooth profiles I and II are those of the new theory, which are the three-dimensional curves with the normal (or plane element) at each point. Therefore, in order to analyze the conditions of contact in the vicinity of a point of contact on the surfaces I and II and the dynamic rotational motion of the gear pair, it is sufficient to use the tooth profiles I and II instead of the surfaces I and II. In addition, when the tooth profiles I and II are given, any pair of surfaces which includes the tooth profiles I and II and does not interfere with each other can be the tooth surfaces I and II and the conjugate pair of tooth surfaces are just one of them.

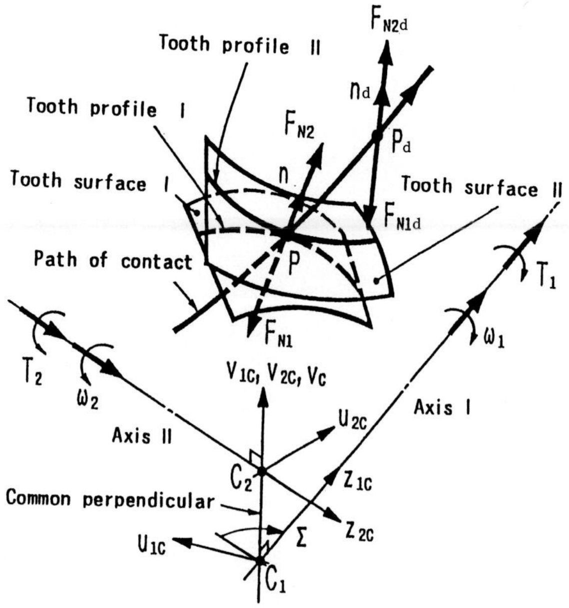

Fig.2.1-1 Path of contact and its contact normals in the coordinate systems C_1 and C_2

The concentrated load means the resultant load of the distributed loads which form an oval of contact on a tooth surface. The point of contact means the point of action of the concentrated load and includes deflection according to the concentrated load. Moreover, each pair of surfaces is the same one having the same pitch, so that they draw the same path of contact (including the deflection) according to the concentrated load. In the case of multiple tooth pair mesh, the concentrated loads according to their own share stand every one pitch on the path of contact.

2．1．1　Coordinate systems C_1, C_2, C_{q1} and C_{q2}

There are eight basic coordinate systems, six of which are defined in this section 2.1 and the other two, which rotate with each gear, are defined in 2.3.2. One of the characteristics of the new tooth geometry is that a point always has a normal (a plane element) and how the point and the inclination angle of the normal are expressed in the six coordinate systems is shown in this section.

(1) Coordinate systems C_1 and C_2

Figure 2.1-1 shows the relationship of the two axes and coordinate systems C_1 and C_2. The two axes I and II, the shaft angle Σ, the offset E (C_1C_2) and the angular velocities ω_1 and ω_2 ($\omega_1 \geqq \omega_2$) at an instant are given. When the positive direction of $\omega_2 \times \omega_1$ is given to the common perpendicular of the two axes, the directed common perpendicular $\mathbf{v}_c$ intersects the axes I and II at C_1 and C_2 where C_2 is assumed to be located above C_1 on the axis $\mathbf{v}_c$ in this tooth geometry.

A coordinate system C_2 (u_{2c}, v_{2c}, z_{2c}) is defined as follows, where the point C_2 is the origin, the axis II is the axis z_{2c} whose direction is the same as that of ω_2, the directed common perpendicular $\mathbf{v}_c$ is the axis v_{2c} and the axis u_{2c} is set perpendicular to both the axes z_{2c} and v_{2c} such that the coordinate system C_2 forms a right-handed one.

A coordinate system C_1 (u_{1c}, v_{1c}, z_{1c}) is defined in the same way, where the point C_1 is the origin, the axis I is the axis z_{1c} whose direction is the same as that of ω_1, the directed common perpendicular $\mathbf{v}_c$ is the axis v_{1c} and the axis u_{1c} is set perpendicular to both the axes z_{1c} and v_{1c} such that the coordinate system C_1 forms a right-handed one.

The relation between the coordinate systems C_1 and C_2 is as follows.

$$\begin{aligned} u_{1c} &= -u_{2c}\cos\Sigma - z_{2c}\sin\Sigma \\ v_{1c} &= v_{2c} + E \\ z_{1c} &= u_{2c}\sin\Sigma - z_{2c}\cos\Sigma \end{aligned} \quad \text{or} \quad \left.\begin{aligned} u_{2c} &= -u_{1c}\cos\Sigma + z_{1c}\sin\Sigma \\ v_{2c} &= v_{1c} - E \\ z_{2c} &= -u_{1c}\sin\Sigma - z_{1c}\cos\Sigma \end{aligned}\right\} \qquad (2.1\text{-}1)$$

(2) Coordinate systems C_{q1} and C_{q2}

Figure 2.1-2 shows a point of contact P chosen at will, the normal components F_{N1} and F_{N2} of the concentrated loads and the common contact normal $\mathbf{n}$ in the coordinate systems C_1, C_2, C_{q1} and C_{q2}. The normal component F_{N2} is positive in the direction of $\mathbf{n}$ and the components of F_{N2} in the directions of the q_{2c} and z_{2c} are F_{q2} and F_{z2} respectively. The planes which include the common contact normal $\mathbf{n}$ (the normal components F_{N2} and F_{N1}) and are parallel to the axis I or II are defined as planes of action G_1 and G_2 respectively. Therefore, the common contact normal $\mathbf{n}$ (the normal component F_{N2} or F_{N1}) is on the intersection of the planes of action G_1 and G_2. The cylinders which have the axis I or II and are tangent to the plane of action G_1 or G_2 are defined as base cylinders whose radii are R_{b1} and R_{b2} respectively.

A coordinate system C_{q2} (q_{2c}, v_{q2c}, z_{2c}) is defined by the coordinate system C_2 as follows. It has the origin C_2 and the axis z_{2c} in common and is obtained by rotating the coordinate system C_2 by an angle χ_2 (positive in the direction in Fig. 2.1-2) around the axis z_{2c} until the plane $v_{2c} = 0$ becomes parallel to the plane of action G_2 and the component of the common contact normal **n** in the direction of the axis u_{2c} becomes positive, where the axis u_{2c} is called axis q_{2c} and the axis v_{2c} is called axis v_{q2c}. The plane of action G_2 is expressed by $v_{q2c} = -R_{b2}$ in the coordinate system C_{q2} and is the plane which inclines to the plane $v_{2c} = 0$ by χ_2 and is tangent to the base cylinder with the radius R_{b2} in the coordinate system C_2.

The relation between the coordinate systems C_2 and C_{q2} is as follows because the axis z_{2c} is common.

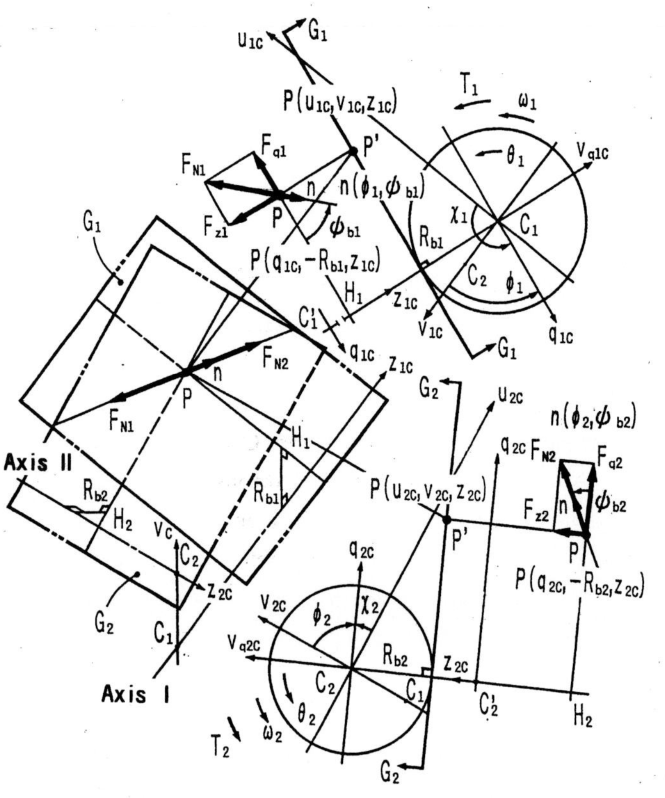

Fig.2.1-2 Relationship among coordinate systems C_1, C_{q1}, C_2 and C_{q2}

$$u_{2c} = q_{2c}\cos\chi_2 - v_{q2c}\sin\chi_2$$
$$v_{2c} = q_{2c}\sin\chi_2 + v_{q2c}\cos\chi_2$$

The plane of action G_2 is expressed by $v_{q2c} = -R_{b2}$, so that the following equations are obtained.

$$\left.\begin{array}{lcl} u_{2c} = q_{2c}\cos\chi_2 + R_{b2}\sin\chi_2 & & q_{2c} = u_{2c}\cos\chi_2 + v_{2c}\sin\chi_2 \\ v_{2c} = q_{2c}\sin\chi_2 - R_{b2}\cos\chi_2 & \text{or} & R_{b2} = u_{2c}\sin\chi_2 - v_{2c}\cos\chi_2 \\ z_{2c} = z_{2c} & & z_{2c} = z_{2c} \end{array}\right\} \quad (2.1\text{-}2)$$

In the same way, a coordinate system C_{q1} $(q_{1c}, v_{q1c}, z_{1c})$, the plane of action G_1 and the radius of base cylinder R_{b1} are defined. The relation between the coordinate systems C_1 and C_{q1} is as follows because the axis z_{1c} is common.

$$\left.\begin{array}{lcl} u_{1c} = q_{1c}\cos\chi_1 + R_{b1}\sin\chi_1 & & q_{1c} = u_{1c}\cos\chi_1 + v_{1c}\sin\chi_1 \\ v_{1c} = q_{1c}\sin\chi_1 - R_{b1}\cos\chi_1 & \text{or} & R_{b1} = u_{1c}\sin\chi_1 - v_{1c}\cos\chi_1 \\ z_{1c} = z_{1c} & & z_{1c} = z_{1c} \end{array}\right\} \quad (2.1\text{-}3)$$

(3) Expression of the inclination angle of the common contact normal n in the coordinate systems C_1 and C_2

The common contact normal **n** is on the plane of action G_2 and the component in the direction of the axis q_{2c} is positive, so that the inclination angle from the axis q_{2c} can be represented by ψ_{b2} (positive in the direction in Fig. 2. 1-2) on the plane G_2. This means that the inclination angle of **n** in the coordinate system C_2 is represented by a combination of the inclination angle ϕ_2 (the complementary angle of χ_2) of the plane of action G_2 from the directed common perpendicular v_c and ψ_{b2}, which is defined as $n(\phi_2=\pi/2-\chi_2, \psi_{b2}; C_2)$. In the same way, the inclination angle of the common contact normal **n** in the coordinate system C_1 is represented by $n(\phi_1=\pi/2-\chi_1, \psi_{b1}; C_1)$. The relation between $n(\phi_1=\pi/2-\chi_1, \psi_{b1}; C_1)$ and $n(\phi_2=\pi/2-\chi_2, \psi_{b2}; C_2)$ is obtained as follows, because **n** is on the intersection of the planes of action G_1 and G_2 (Fig. 2. 1-2).

The components (direction cosine) of **n** in the direction of each axis of the coordinate system C_2 are expressed as follows, where $|n|=1$.

$$\left.\begin{aligned} L_{u2c} &= \cos\psi_{b2}\sin\phi_2 && \text{(component in the direction of } u_{2c}) \\ L_{v2c} &= \cos\psi_{b2}\cos\phi_2 && \text{(component in the direction of } v_{2c}) \\ L_{z2c} &= \sin\psi_{b2} && \text{(component in the direction of } z_{2c}) \end{aligned}\right\} \quad (2.1\text{-}4)$$

Using Eq. (2. 1-1), the components in the direction of each axis of the coordinate system C_1 are expressed as follows.

$$\begin{aligned} L_{u1c} &= -L_{u2c}\cos\Sigma - L_{z2c}\sin\Sigma && \text{(component in the direction of } u_{1c}) \\ L_{v1c} &= L_{v2c} && \text{(component in the direction of } v_{1c}) \\ L_{z1c} &= L_{u2c}\sin\Sigma - L_{z2c}\cos\Sigma && \text{(component in the direction of } z_{1c}) \end{aligned}$$

Therefore,

$$\left.\begin{aligned} \tan\phi_1 &= L_{u1c}/L_{v1c} = -\tan\phi_2\cos\Sigma - \tan\psi_{b2}\sin\Sigma/\cos\phi_2 \\ \sin\psi_{b1} &= L_{z1c}/|n| = \cos\psi_{b2}\sin\phi_2\sin\Sigma - \sin\psi_{b2}\cos\Sigma \\ &\text{where } \phi_1 = \pi/2-\chi_1, \quad \phi_2 = \pi/2-\chi_2 \end{aligned}\right\} \quad (2.1\text{-}5)$$

In the same way, the components of **n** in the direction of each axis of the coordinate system C_1 are expressed as follows.

$$\left.\begin{aligned} L_{u1c} &= \cos\psi_{b1}\sin\phi_1 && \text{(component in the direction of } u_{1c}) \\ L_{v1c} &= \cos\psi_{b1}\cos\phi_1 && \text{(component in the direction of } v_{1c}) \\ L_{z1c} &= \sin\psi_{b1} && \text{(component in the direction of } z_{1c}) \end{aligned}\right\} \quad (2.1\text{-}6)$$

Using Eq. (2. 1-1), Eq. (2. 1-6) is transformed into the coordinate system C_2 and arranged,

$$\left.\begin{aligned} \tan\phi_2 &= L_{u2c}/L_{v2c} = -\tan\phi_1\cos\Sigma + \tan\psi_{b1}\sin\Sigma/\cos\phi_1 \\ \sin\psi_{b2} &= L_{z2c}/|n| = -(\cos\psi_{b1}\sin\phi_1\sin\Sigma + \sin\psi_{b1}\cos\Sigma) \end{aligned}\right\} \quad (2.1\text{-}7)$$

2.1.2 Instantaneous axis S and coordinate systems C_S and C_{qS}

(1) Instantaneous axis S

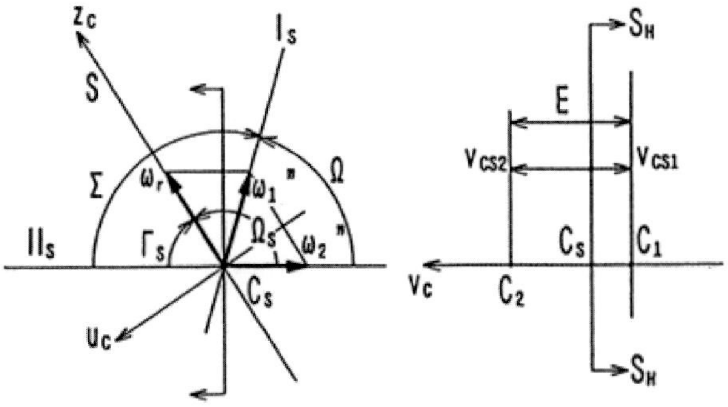

Fig.2.1-3(a) Instantaneous axis S and coordinate system C_S

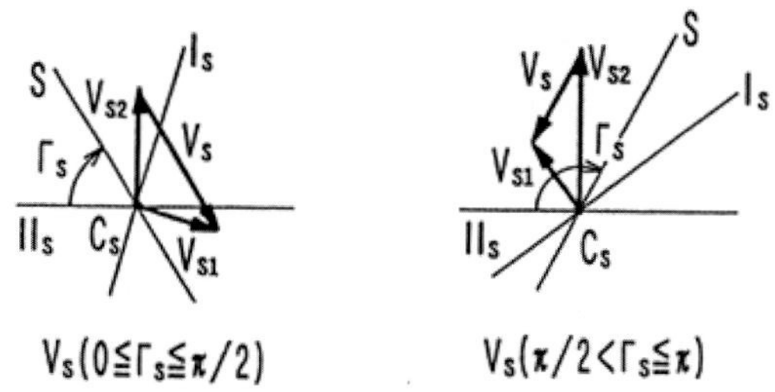

Fig.2.1-3(b) Relative velocity $\mathbf{V}_S$ at point C_S

Figure 2.1-3(a) shows the relationship between an instantaneous axis S and a coordinate system C_S, where C_1 and C_2 are the origins of the coordinate systems C_1 and C_2. When the relative velocity is $\omega_r = \omega_1 - \omega_2$ whose axis is the instantaneous axis S, the plane which includes the instantaneous axis S and is perpendicular to the common perpendicular v_c is S_H and the intersection of the plane S_H and the common perpendicular v_c is C_S, the instantaneous axis is a line through the intersection C_S on the plane S_H, so that the location of the instantaneous axis S is determined as follows.

When the projections of the axes I (ω_1) and II (ω_2) on to the plane S_H are I_S (ω_1") and II_S (ω_2") respectively and the angle of I_S to II_S seen from positive to negative in the direction of the common perpendicular v_c is Ω, because of the definition of $\omega_2 \times \omega_1$, I_S exists in the region of $0 \leqq \Omega \leqq \pi$ (positive in the counterclockwise direction) to II_S. When the angle of the instantaneous axis S (ω_r) to II_S on the plane S_H is Ω_S (positive in the counterclockwise direction), because of the definition of the instantaneous axis ($\omega_r = \omega_1 - \omega_2$), the components perpendicular to the instantaneous axis S of ω_1" and ω_2" must be equal, so that the angle Ω_S (positive in the direction in Fig. 2.1-3(a)) is obtained as follows.

$$\sin\Omega_S/\sin(\Omega_S-\Omega) = \omega_1/\omega_2 \quad \text{or} \quad \sin\Gamma_S/\sin(\Sigma-\Gamma_S) = \omega_1/\omega_2 \tag{2.1-8}$$

where $\Sigma = \pi - \Omega$ (shaft angle), $\Gamma_S = \pi - \Omega_S$

The location of C_S on the common perpendicular v_c is obtained as follows. Figure 2.1-3(b) shows the relative velocity V_S at C_S. Because it is assumed that C_1 is below C_2 on the common perpendicular v_c and $\omega_1 \geqq \omega_2$, C_S is below C_2. When the peripheral velocities of gears I and II at C_S are V_{S1} and V_{S2} respectively, the relative velocity $V_S = V_{S1} - V_{S2}$ is on the instantaneous axis S due to the definition, so that the components perpendicular to the instantaneous axis S of V_{S1} and V_{S2} which are on the plane S_H must be equal. Therefore, the relative velocity $V_S = V_{S1} - V_{S2}$ at C_S on the plane S_H varies according to the location (Γ_S) of the instantaneous axis S as shown in Fig. 2.1-3(b) and C_2C_S is obtained as follows.

$$C_2C_S = E\tan\Gamma_S/\{\tan(\Sigma-\Gamma_S)+\tan\Gamma_S\} \qquad (2.1\text{-}9)$$

Equation (2.1-9) is effective in the region of $0 \leqq \Gamma_S \leqq \pi$, where C_S is above C_1 when $0 \leqq \Gamma_S \leqq \pi/2$ and below C_1 when $\pi/2 < \Gamma_S \leqq \pi$ according to the location (Γ_S) of the instantaneous axis S.

(2) Coordinate systems C_S and C_{qS}

As the instantaneous axis S is determined in the coordinate systems C_1 and C_2 by Eqs. (2.1-8) and (2.1-9), a coordinate system C_S (u_c, v_c, z_c) is defined as shown in Fig. 2.1-3(a), where the point C_S is the origin, the directed common perpendicular $\mathbf{v}_c$ is the axis v_c, the instantaneous axis S is the axis z_c (positive in the direction of ω_r) and the axis u_c is set perpendicular to both the axes z_c and v_c such that the coordinate system C_S forms a right-handed one. Therefore the coordinate system C_S varies according to the ratio of angular velocities. When the points C_1 and C_2 are expressed in the coordinate system C_S by C_1 $(0, v_{cs1}, 0\ ;\ C_S)$ and C_2 $(0, v_{cs2}, 0\ ;\ C_S)$, v_{cs1} and v_{cs2} are obtained as follows.

$$\left.\begin{aligned} v_{cs2} &= C_SC_2 = E\tan\Gamma_S/\{\tan(\Sigma-\Gamma_S)+\tan\Gamma_S\} \\ v_{cs1} &= C_SC_1 = v_{cs2} - E = -E\tan(\Sigma-\Gamma_S)/\{\tan(\Sigma-\Gamma_S)+\tan\Gamma_S\} \end{aligned}\right\} \qquad (2.1\text{-}10)$$

As C_2 is always above C_S on the axis v_c, the relations among the coordinate systems C_S, C_1 and C_2 are expressed by v_{cs1}, v_{cs2}, Σ and Γ_S as follows.

$$\begin{aligned} u_c &= -u_{2c}\cos\Gamma_S - z_{2c}\sin\Gamma_S \\ v_c &= v_{2c} + v_{cs2} \\ z_c &= u_{2c}\sin\Gamma_S - z_{2c}\cos\Gamma_S \end{aligned} \quad \text{or} \quad \left.\begin{aligned} u_{2c} &= -u_c\cos\Gamma_S + z_c\sin\Gamma_S \\ v_{2c} &= v_c - v_{cs2} \\ z_{2c} &= -u_c\sin\Gamma_S - z_c\cos\Gamma_S \end{aligned}\right\} \qquad (2.1\text{-}11)$$

$$\begin{aligned} u_c &= u_{1c}\cos(\Sigma-\Gamma_S) - z_{1c}\sin(\Sigma-\Gamma_S) \\ v_c &= v_{1c} + v_{cs1} \\ z_c &= u_{1c}\sin(\Sigma-\Gamma_S) + z_{1c}\cos(\Sigma-\Gamma_S) \end{aligned} \quad \text{or} \quad \left.\begin{aligned} u_{1c} &= u_c\cos(\Sigma-\Gamma_S) + z_c\sin(\Sigma-\Gamma_S) \\ v_{1c} &= v_c - v_{cs1} \\ z_{1c} &= -u_c\sin(\Sigma-\Gamma_S) + z_c\cos(\Sigma-\Gamma_S) \end{aligned}\right\} \qquad (2.1\text{-}12)$$

Figure 2.1-4 shows the relation among the coordinate systems C_S, C_1 and C_2 and that among the reference planes S_H ($v_c = 0$), S_S and S_p in the coordinate system C_S. When the common contact normal $\mathbf{n}$ intersects the plane S_H ($v_c = 0$) at P_H, the plane perpendicular to the axis z_c at P_H is S_S and the plane perpendicular to the axis u_c at P_H is S_p.

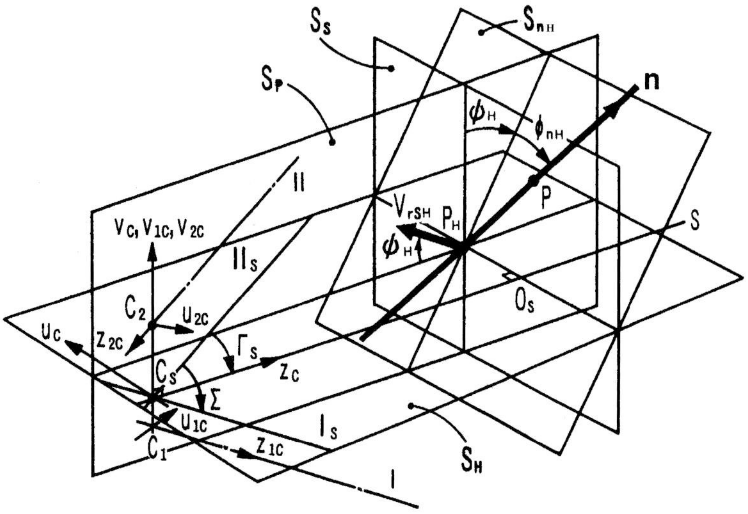

Fig.2.1-4 Relationship among coordinate systems C_1, C_2 and C_S with planes S_H, S_p, S_S and S_{nH}

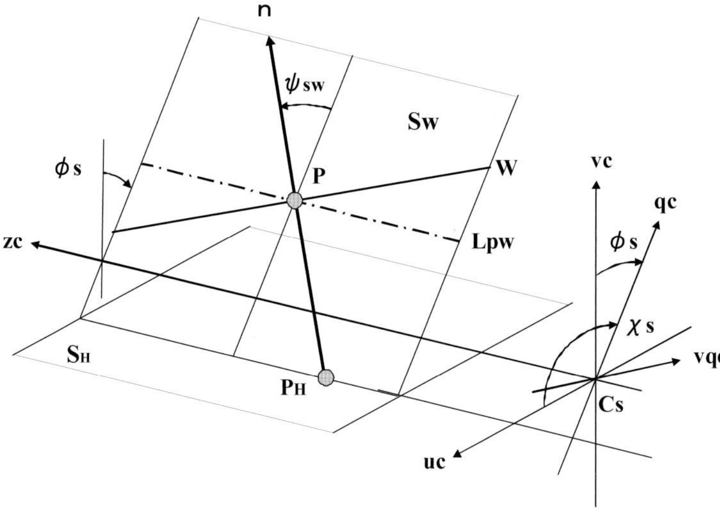

Fig.2.1-5 Relationship between coordinate systems C_S and C_{qS}

Figure 2.1-5 shows the relation between coordinate systems C_S and C_{qS}. In the same way as the coordinate systems C_{q1} and C_{q2} are defined to the coordinate systems C_1 and C_2, the plane which includes the common contact normal **n** and is parallel to the instantaneous axis S is defined as a plane of action S_W and the cylinder whose axis is the instantaneous one and which is tangent to the plane S_W is defined as a base cylinder whose radius is R_{bC}. A coordinate system C_{qS} (q_c, v_{qc} = $-R_{bc}$, z_c ; C_{qS}) is defined by the coordinate system C_S(u_c, v_c, z_c ; C_S) as follows, which has the origin C_S and the axis z_c in common and is obtained by rotating the coordinate system C_S by an angle χ_S (positive in the direction in Fig.2.1-5) around the axis z_c until the plane v_c = 0 becomes parallel to the plane of action S_W and the component of the common contact normal **n** in the direction of the u_c axis becomes positive, where the axis u_c is called axis q_c and the axis v_c is called axis v_{qc}. The plane of action S_W is expressed by $v_{qc} = -R_{bC}$ in the coordinate system C_{qS} and is the plane which inclines to the plane $v_c = 0$ by χ_S and is tangent to the base cylinder with the radius R_{bC} in the coordinate system C_S.

The relation between the coordinate systems C_S and C_{qS} is as follows because the axis z_c is common.

$$\left.\begin{aligned} u_c &= q_c\cos\chi_S + R_{bc}\sin\chi_S \\ v_c &= q_c\sin\chi_S - R_{bc}\cos\chi_S \\ z_c &= z_c \end{aligned}\quad \text{or} \quad \begin{aligned} q_c &= u_c\cos\chi_S + v_c\sin\chi_S \\ R_{bc} &= u_c\sin\chi_S - v_c\cos\chi_S \\ z_c &= z_c \end{aligned}\right\} \qquad (2.1\text{-}13)$$

(3) Expression of the inclination angle of the common contact normal n by the plane of action S_W in the coordinate system C_S

Figure 2.1-5 shows the plane of action S_W. The common contact normal **n** is on the plane of action S_W and the component of the axis q_c is positive because of the definition, so that the inclination angle of **n** from the axis q_c can be represented by ψ_{SW} (positive in the direction in Fig.2.1-5) on the plane S_W. This means that the inclination angle of the common contact normal **n** in the coordinate system C_S is represented by a combination of the inclination angle ϕ_S (the complementary angle of χ_S) of the plane of action S_W from the directed common perpendicular v_c and ψ_{SW}, which is defined as $n(\phi_S = \pi/2 - \chi_S, \psi_{SW}; C_S)$. (Note 2.1-1).

The relation between $n(\phi_S, \psi_{SW}; C_S)$ and $n(\phi_2 = \pi/2 - \chi_2, \psi_{b2}; C_2)$ is obtained as follows, because **n** is on the intersection of the planes of action S_W and G_2. When the components of **n** in the direction of each axis in the coordinate system C_2 expressed by Eq.(2.1-4) are transformed into the coordinate system C_S by Eq.(2.1-11), the components L_{uC},

L_{vC} and L_{zC} in the direction of each axis of the coordinate system C_S are as follows.

$$\left.\begin{aligned} L_{uC} &= -L_{u2c}\cos\Gamma_S - L_{z2c}\sin\Gamma_S = -(\cos\phi_{b2}\sin\phi_2\cos\Gamma_S + \sin\phi_{b2}\sin\Gamma_S) \\ L_{vC} &= L_{v2c} = \cos\phi_{b2}\cos\phi_2 \\ L_{zC} &= L_{u2c}\sin\Gamma_S - L_{z2c}\cos\Gamma_S = \cos\phi_{b2}\sin\phi_2\sin\Gamma_S - \sin\phi_{b2}\cos\Gamma_S \end{aligned}\right\} \quad (2.1\text{-}14)$$

Therefore,

$$\left.\begin{aligned} \tan\phi_S &= L_{uC}/L_{vC} = -\tan\phi_2\cos\Gamma_S - \tan\phi_{b2}\sin\Gamma_S/\cos\phi_2 \\ \sin\phi_{SW} &= L_{zC}/|\mathbf{n}| = \cos\phi_{b2}\sin\phi_2\sin\Gamma_S - \sin\phi_{b2}\cos\Gamma_S \end{aligned}\right\} \quad (2.1\text{-}15)$$

In the same way, using Eqs. (2.1-6) and (2.1-12), the relation between $\mathbf{n}(\phi_S, \phi_{SW}; C_S)$ and $\mathbf{n}(\phi_1 = \pi/2 - \chi_1, \phi_{b1}; C_1)$ is obtained as follows.

$$\left.\begin{aligned} \tan\phi_S &= L_{uC}/L_{vC} = \tan\phi_1\cos(\Sigma-\Gamma_S) - \tan\phi_{b1}\sin(\Sigma-\Gamma_S)/\cos\phi_1 \\ \sin\phi_{SW} &= L_{zC}/|\mathbf{n}| = \cos\phi_{b1}\sin\phi_1\sin(\Sigma-\Gamma_S) + \sin\phi_{b1}\cos(\Sigma-\Gamma_S) \end{aligned}\right\} \quad (2.1\text{-}16)$$

In the same way, the transformations from $\mathbf{n}(\phi_S, \phi_{SW}; C_S)$ to $\mathbf{n}(\phi_2, \phi_{b2}; C_2)$ and $\mathbf{n}(\phi_1, \phi_{b1}; C_1)$ are as follows. The components of $\mathbf{n}$ in the direction of each axis in the coordinate system C_S are expressed as follows.

$$\left.\begin{aligned} L_{uc} &= \cos\phi_{SW}\sin\phi_S && \text{(component of } \mathbf{n} \text{ in the direction of } u_c) \\ L_{vc} &= \cos\phi_{SW}\cos\phi_S && \text{(component of } \mathbf{n} \text{ in the direction of } v_c) \\ L_{zc} &= \sin\phi_{SW} && \text{(component of } \mathbf{n} \text{ in the direction of } z_c) \end{aligned}\right\} \quad (2.1\text{-}17)$$

Transforming Eqs. (2.1-17) into the coordinate systems C_2 and C_1 through Eqs. (2.1-11) and (2.1-12), the following equations are obtained.

$$\left.\begin{aligned} \tan\phi_2 &= L_{u2c}/L_{v2c} = -\tan\phi_S\cos\Gamma_S + \tan\phi_{SW}\sin\Gamma_S/\cos\phi_S \\ \sin\phi_{b2} &= L_{z2c}/|\mathbf{n}| = -\cos\phi_{SW}\sin\phi_S\sin\Gamma_S - \sin\phi_{SW}\cos\Gamma_S \end{aligned}\right\} \quad (2.1\text{-}18)$$

$$\left.\begin{aligned} \tan\phi_1 &= L_{u1c}/L_{v1c} = \tan\phi_S\cos(\Sigma-\Gamma_S) + \tan\phi_{SW}\sin(\Sigma-\Gamma_S)/\cos\phi_S \\ \sin\phi_{b1} &= L_{z1c}/|\mathbf{n}| = -\cos\phi_{SW}\sin\phi_S\sin(\Sigma-\Gamma_S) + \sin\phi_{SW}\cos(\Sigma-\Gamma_S) \end{aligned}\right\} \quad (2.1\text{-}19)$$

(4) Expression of the inclination angle of the common contact normal $\mathbf{n}$ by the plane S_{nH} in the coordinate system C_S

Figure 2.1-4 shows the plane S_{nH} which includes the common contact normal $\mathbf{n}$, is parallel to the axis u_c and inclines from the plane S_S by an angle ϕ_H. The inclination angle of $\mathbf{n}$ can be represented by a combination of an angle ϕ_{nH} of $\mathbf{n}$ on the plane S_{nH} from the plane S_p and the angle ϕ_H as $\mathbf{n}(\phi_H, \phi_{nH}; C_S)$, where ϕ_H and ϕ_{nH} are positive in the direction in Fig. 2.1-4. (Note 2.1-2).

Using Eq. (2.1-14), $\mathbf{n}(\phi_H, \phi_{nH}; C_S)$ is obtained from $\mathbf{n}(\phi_2, \phi_{b2}; C_2)$ as follows.

$$\left.\begin{aligned} \sin\phi_{nH} &= -L_{uc}/|\mathbf{n}| = \cos\phi_{b2}\sin\phi_2\cos\Gamma_S + \sin\phi_{b2}\sin\Gamma_S \\ \tan\phi_H &= L_{zc}/L_{vc} = \tan\phi_2\sin\Gamma_S - \tan\phi_{b2}\cos\Gamma_S/\cos\phi_2 \end{aligned}\right\} \quad (2.1\text{-}20)$$

In the same way, using Eqs. (2.1-6) and (2.1-12), it is obtained from $n(\phi_1, \varphi_{b1}; C_1)$ as follows.

$$\left.\begin{aligned} \sin\phi_{nH} &= -L_{uc}/|\mathbf{n}| = -\cos\varphi_{b1}\sin\phi_1\cos(\Sigma-\Gamma_S)+\sin\varphi_{b1}\sin(\Sigma-\Gamma_S) \\ \tan\varphi_H &= L_{zc}/L_{vc} = \tan\phi_{10}\sin(\Sigma-\Gamma_S)+\tan\varphi_{b10}\cos(\Sigma-\Gamma_S)/\cos\phi_{10} \end{aligned}\right\} \quad (2.1\text{-}21)$$

In the same way, using Eq. (2.1-17), it is obtained from $n(\phi_S, \varphi_{SW}; C_S)$ as follows.

$$\left.\begin{aligned} \sin\phi_{nH} &= -L_{uc}/|\mathbf{n}| = -\cos\varphi_{SW}\sin\phi_S \\ \tan\varphi_H &= L_{zc}/L_{vc} = \tan\varphi_{SW}/\cos\phi_S \end{aligned}\right\} \quad (2.1\text{-}22)$$

Conversely, the transformations from $n(\varphi_H, \phi_{nH}; C_S)$ to $n(\phi_S, \varphi_{SW}; C_S)$, $n(\phi_2, \varphi_{b2}; C_2)$ and $n(\phi_1, \varphi_{b1}; C_1)$ are as follows. The components of **n** in the direction of each axis in the coordinate system C_S are expressed by $n(\varphi_H, \phi_{nH}; C_S)$ as follows.

$$\left.\begin{aligned} L_{uc} &= -\sin\phi_{nH} && \text{(component of } \mathbf{n} \text{ in the direction of } u_c) \\ L_{vc} &= \cos\phi_{nH}\cos\varphi_H && \text{(component of } \mathbf{n} \text{ in the direction of } v_c) \\ L_{zc} &= \cos\phi_{nH}\sin\varphi_H && \text{(component of } \mathbf{n} \text{ in the direction of } z_c) \end{aligned}\right\} \quad (2.1\text{-}23)$$

$n(\phi_S, \varphi_{SW}; C_S)$ is expressed as follows.

$$\left.\begin{aligned} \tan\phi_S &= L_{uc}/L_{vc} = -\tan\phi_{nH}/\cos\varphi_H \\ \sin\varphi_{SW} &= L_{zc}/|\mathbf{n}| = \cos\phi_{nH}\sin\varphi_H \end{aligned}\right\} \quad (2.1\text{-}24)$$

Using Eqs. (2.1-11) and (2.1-12), $n(\phi_2, \varphi_{b2}; C_2)$ and $n(\phi_1, \varphi_{b1}; C_1)$ are obtained as follows.

$$\left.\begin{aligned} \tan\phi_2 &= L_{u2c}/L_{v2c} = \tan\phi_{nH}\cos\Gamma_S/\cos\varphi_H+\tan\varphi_H\sin\Gamma_S \\ \sin\varphi_{b2} &= L_{z2c}/|\mathbf{n}| = \sin\phi_{nH}\sin\Gamma_S-\cos\phi_{nH}\sin\varphi_H\cos\Gamma_S \end{aligned}\right\} \quad (2.1\text{-}25)$$

$$\left.\begin{aligned} \tan\phi_1 &= L_{u1c}/L_{v1c} = -\tan\phi_{nH}\cos(\Sigma-\Gamma_S)/\cos\varphi_H+\tan\varphi_H\sin(\Sigma-\Gamma_S) \\ \sin\varphi_{b1} &= L_{z1c}/|\mathbf{n}| = \sin\phi_{nH}\sin(\Sigma-\Gamma_S)+\cos\phi_{nH}\sin\varphi_H\cos(\Sigma-\Gamma_S) \end{aligned}\right\} \quad (2.1\text{-}26)$$

The plane S_{nH} defined here is effective for all kinds of gears. However in the case of crossed helical, worm and cylindrical gears where the common contact normal **n** intersects the axis v_c, another S_{nH} defined as the plane which includes the axis v_c is more familiar because it is nearer to the present normal plane. This is discussed further in 3.2.

(Note 2.1-1) : the positive direction of ϕ_S is the reverse to that in reference (35), the paper in MPT2009.

(Note 2.1-2) : the positive direction of ϕ_{nH} corresponds to that of the pressure angles of the present bevel and hypoid gears.

2．2　Fundamental requirement for contact and ratio of angular velocities [23), (25]

In this section, the fundamental requirement for contact and the ratio of angular velocities are discussed when a point of contact P chosen at will and its common contact normal **n** are given as shown in Fig. 2. 1-1.

2．2．1　Relative velocity and a cylinder formed by relative velocities

A point of contact P chosen at will and the inclination angle of its common contact normal **n** are given as follows in the coordinate systems C_2, C_{q2}, C_1, C_{q1}, C_S and C_{qS}.

$P(u_{2c}, v_{2c}, z_{2c};\ C_2)$ or
$P(q_{2c}, -R_{b2}, z_{2c};\ C_{q2})$
$n(\phi_2, \psi_{b2};\ C_2)$
where $\phi_2 = \pi/2 - \chi_2$

$P(u_{1c}, v_{1c}, z_{1c};\ C_1)$ or
$P(q_{1c}, -R_{b1}, z_{1C};\ C_{q1})$
$n(\phi_1, \psi_{b1};\ C_1)$
where $\phi_1 = \pi/2 - \chi_1$

$P(u_c, v_c, z_c;\ C_S)$ or
$P(q_c, -R_{bC}, z_c;\ C_{qS})$
$n(\phi_S, \psi_{SW};\ C_S)$
where $\phi_S = \pi/2 - \chi_S$

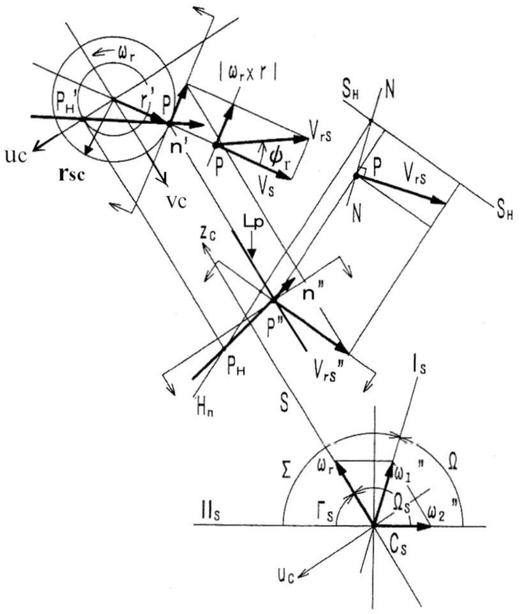

Fig.2.2-1 Relationship between relative velocity $\mathbf{V}_{rS}$ and contact normal **n**

When position vectors of a point P from the origins C_1, C_2 and C_S are $\mathbf{r}_1$, $\mathbf{r}_2$ and $\mathbf{r}$ respectively, the peripheral velocities $\mathbf{V}_1$ and $\mathbf{V}_2$ and the relative velocity $\mathbf{V}_{rS}$ are expressed as follows.

(1) Peripheral velocities $\mathbf{V}_1$ and $\mathbf{V}_2$

$$\mathbf{V}_1 = \omega_1 \times \mathbf{r}_1 = \omega_1 \times (C_1C_S + \mathbf{r}), \qquad \mathbf{V}_2 = \omega_2 \times \mathbf{r}_2 = \omega_2 \times (C_2C_S + \mathbf{r}) \qquad (2.2\text{-}1)$$

where
$$|\mathbf{V}_1| = \sqrt{(u_{1C}{}^2 + v_{1C}{}^2)} \cdot \omega_1 = \sqrt{(q_{1C}{}^2 + R_{b1}{}^2)} \cdot \omega_1$$
$$|\mathbf{V}_2| = \sqrt{(u_{2C}{}^2 + v_{2C}{}^2)} \cdot \omega_2 = \sqrt{(q_{2C}{}^2 + R_{b2}{}^2)} \cdot \omega_2$$

(2) Relative velocity $\mathbf{V}_{rS}$

Figure 2. 2-1 shows the relation between the common contact normal **n** and the relative velocity $\mathbf{V}_{rS}$ at a point of contact P chosen at will, in which the sign (' or ") indicates the orthographic projection of a vector or a point toward the reference planes. The relative velocity $\mathbf{V}_{rS}$ at P is expressed as follows, using Eqs. (2. 1-8), (2. 1-9) and (2. 2-1).

$$V_{rS} = V_1 - V_2 = \omega_r \times r + V_S \quad (2.2\text{-}2)$$

where
$$\omega_r = \omega_1 - \omega_2$$
$$\omega_r = \omega_2 \sin\Sigma/\sin(\Sigma-\Gamma_S) = \omega_1 \sin\Sigma/\sin\Gamma_S$$
$$V_S = \omega_1 \times C_1C_S - \omega_2 \times C_2C_S$$
$$V_S = \omega_2 E\sin\Gamma_S = \omega_1 E\sin(\Sigma-\Gamma_S)$$
$$|\omega_r \times r| = \sqrt{(u_c^2+v_c^2)}\,\omega_2 \sin\Sigma/\sin(\Sigma-\Gamma_S)$$
$$V_{rS} = \omega_2\sqrt{\{(E\sin\Gamma_S)^2+(u_c^2+v_c^2)\sin^2\Sigma/\sin^2(\Sigma-\Gamma_S)\}}$$

Equation (2.2-2) means that the relative velocity V_{rS} is represented by the motion of a screw around the axis z_c and does not depend on the position on the ordinate z_c. In other words, the relative velocity V_{rS} is on a plane tangent to a cylinder with the radius r_{SC} whose axis is the instantaneous axis S and all the V_{rS} s on a line element L_P of the cylinder are equal. The inclination angle ϕ_r of V_{rS} from V_S (L_P) on this tangent plane is as follows.

$$\left.\begin{aligned} \tan\phi_r &= |\omega_r \times r|/|V_S| = \sqrt{(u_c^2+v_c^2)}\sin\Sigma/\{E\sin\Gamma_S\sin(\Sigma-\Gamma_S)\} \\ r_{SC} &= \sqrt{(u_c^2+v_c^2)} = E\sin\Gamma_S\tan\phi_r\sin(\Sigma-\Gamma_S)/\sin\Sigma \end{aligned}\right\} \quad (2.2\text{-}3)$$

2. 2. 2 Normal velocity V_n and fundamental requirement for contact

(1) Normal velocity

The relative velocity V_{rS} can be expressed by the normal components V_{n1} and V_{n2} in the direction of n and the components V_{rs1} and V_{rs2} perpendicular to n of the peripheral velocities V_1 and V_2 respectively as follows, where the common contact normal n is on the intersection of the planes G_1 and G_2.

$$V_{rS} = V_1 - V_2 = (V_{n1} - V_{n2}) + (V_{rs1} - V_{rs2}) \quad (2.2\text{-}4)$$

where $|V_{n1}| = R_{b1}\omega_1\cos\phi_{b1}$, $|V_{n2}| = R_{b2}\omega_2\cos\phi_{b2}$

(2) Fundamental requirement for contact

At the point of contact P, the common contact normal n is perpendicular to the relative velocity V_{rS}, so that the fundamental requirement for contact is obtained as follows, using Eq. (2.2-4).

$$\begin{aligned} V_{rS}\cdot n &= \{(V_{n1} - V_{n2}) + (V_{rs1} - V_{rs2})\}\cdot n \\ &= (V_{n1}-V_{n2})\cdot n = 0 \end{aligned} \quad (2.2\text{-}5)$$

Therefore,

$$R_{b1}\omega_1\cos\phi_{b1} = R_{b2}\omega_2\cos\phi_{b2} \quad (2.2\text{-}6)$$

Equation (2.2-6) is the fundamental requirement for contact expressed in the coordinate systems C_1, C_2, C_{q1} and C_{q2}.

2. 2. 3 Another expression of fundamental requirement for contact

When the plane N is perpendicular to the relative velocity V_{rs} in Fig. 2. 2-1, the common contact normal **n** is a directed line on the plane N at P and the intersection H_n of the planes N and S_H is a line which intersects the instantaneous axis S generally. The intersection point P_H of the common contact normal **n** and the plane S_H is on the line H_n and varies according to the location of the two axes (kinds of gears), Σ and E as follows.

(1) $\Sigma=0$ or π or $E=0$ (in the cases of cylindrical and bevel gears)

$V_S=0$ means that V_{rs} is just the peripheral velocity around the instantaneous axis S, so that the plane N includes the instantaneous axis S, the line H_n coincides with the instantaneous axis S and the common contact normal **n** always passes the instantaneous axis S, which means that the point P_H is on the instantaneous axis S and the common contact normal **n** in these kinds of gears is a directed line through a point P_H chosen at will on the instantaneous axis S.

(2) Others (in the cases of hypoid, crossed helical and worm gears)

A point of contact P chosen at will determines its own relative velocity V_{rs}, plane N and line H_n. The common contact normal **n** is a directed line through a point P_H on the line H_n and does not generally intersect the instantaneous axis S.

When the intersection of the common contact normal **n** at a point of contact P and the plane S_H ($v_c=0$) is P_H (u_{cH}, 0, z_{cH} ; C_S), P_H is also a point of contact by Eq. (2. 2-2) as follows, where $\mathbf{r}_H$ is the position vector from C_S to P_H and V_{rSH} is the relative velocity at P_H.

$$\begin{aligned} V_{rSH}\cdot \mathbf{n} &= (\omega_r\times \mathbf{r}_H+V_S)\cdot \mathbf{n} \\ &= (\omega_r\times(\mathbf{r}-|PP_H|\cdot \mathbf{n})+V_S)\cdot \mathbf{n} \\ &= (\omega_r\times \mathbf{r}+V_S)\cdot \mathbf{n}-(\omega_r\times|PP_H|\cdot \mathbf{n})\cdot \mathbf{n}=0 \end{aligned} \qquad (2.2\text{-}7)$$

Actually, Eq. (2. 2-7) is effective in the case not only where P_H is on the plane S_H, but also where P_H is any point on **n**.

The relative velocity V_{rSH} at P_H is on the plane S_p as shown Fig. 2. 1-4, whose inclination angle from V_S is represented by ψ_H. The plane S_{nH} is perpendicular to the relative velocity V_{rSH}, so that any directed line on the plane S_{nH} through P_H can be a common contact normal. Therefore, the fundamental requirement for contact at P is that the common contact normal **n** exists on the plane S_{nH}, which means that the inclination angle n (ψ_H, ϕ_{nH} ; C_S) has the following ψ_H obtained by Eq. (2. 2-3).

$$\tan\psi_H = u_{cH}\sin\Sigma/\{E\sin\Gamma_S\sin(\Sigma-\Gamma_S)\} \qquad (2.2\text{-}8)$$

when $\Sigma=0$ or π or $E=0$, $u_{cH}=0$

Conversely, if there exists the relation shown by Eq. (2. 2-8) between an intersection P_H (u_{cH}, 0, z_{cH} ; C_S) of a normal **n** and the plane S_H and its inclination angle n (ψ_H, ϕ_{nH} ; C_S) of **n**, then the normal **n** is the common contact one. Therefore, Eq. (2. 2-8) is another expression of the fundamental requirement for contact.

The most popular expression of the fundamental requirement for contact is shown by Eq. (2.2-5). However, Eq. (2.2-6) is convenient for algebraical analysis in the case where the ratio of angular velocities varies and Eq. (2.2-8) is useful to examine whether a normal n through P is the common contact one or not.

2.2.4 Ratio of angular velocities (gear ratio)

Transforming Eq. (2.2-6) of the fundamental requirement for contact, the ratio of angular velocities is obtained as follows.

$$i = \omega_1/\omega_2 = R_{b2}\cos\phi_{b2}/(R_{b1}\cos\phi_{b1}) \quad (2.2\text{-}9)$$

When a point of contact P chosen at will and the inclination angle of the common contact normal n are given by five independent variables, the ratio of angular velocities i is determined by Eq. (2.2-9) as a dependent variable. Conversely, it is necessary to pay attention that when the ratio of angular velocities i is given as an independent one, one of the five variables of a point of contact P and the inclination angle of the common contact normal n becomes a dependent one.

2.3 Path of contact and its tooth profiles (21), (25)

In this section, it is discussed how a path of contact PP_d with its common contact normals n and n_d shown in Fig. 2.1-1 is expressed.

2.3.1 Equations of path of contact with its common contact normal at each point

Figure 2.3-1 shows a point of contact P with its common contact normal n on a path of contact and its tangent plane W which is indicated by the intersection line w with the plane of action G_2 of gear II at an angle of rotation θ_2 of gear II in the coordinate systems C_2 and C_{q2}. When gear II rotates by a small angle of rotation $\Delta\theta_2$, the point of contact P is supposed to move to P_d, the common contact normal n to n_d and the tangent plane W (w) to W_d(w_d). The plane of action including n_d through P_d is G_{2d}, which intersects the tangent plane W_d along the intersection line w_d, where W_d and w_d are not shown here.

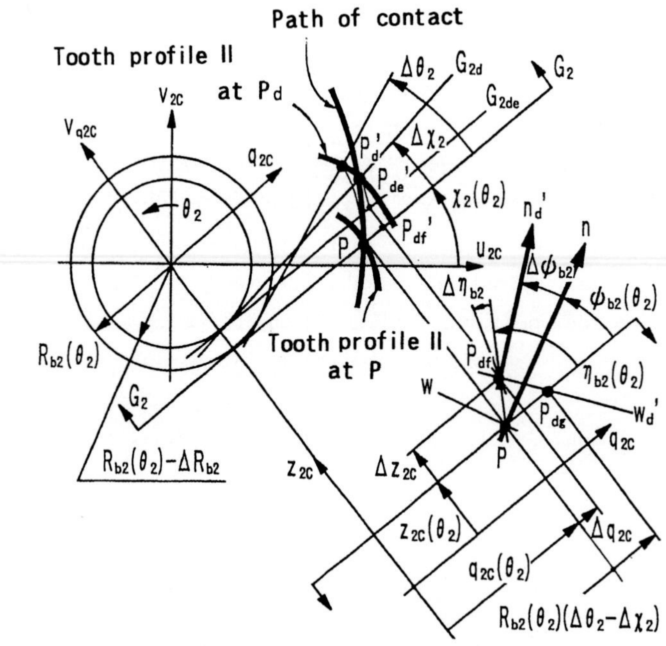

Fig.2.3-1 Introduction of equations of path of contact and its common contact normal

A point of contact P chosen at will and its common contact normal **n** are expressed as functions of θ_2 in the coordinate systems C_2 and C_{q2} as follows, where the angle of rotation θ_2 is positive in the direction on Fig. 2. 3-1.

$$P\{u_{2c}(\theta_2), v_{2c}(\theta_2), z_{2c}(\theta_2);\ C_2\}, \qquad P\{q_{2c}(\theta_2), -R_{b2}(\theta_2), z_{2c}(\theta_2);\ C_{q2}\}$$
$$n\{\phi_2(\theta_2), \phi_{b2}(\theta_2);\ C_2\}, \quad \text{where } \phi_2(\theta_2) = \pi/2 - \chi_2(\theta_2)$$

In the same way, a point of contact P_d and its common contact normal $\mathbf{n}_d$ are expressed as follows.

$$P_d\{q_{2c}(\theta_2)+\Delta q_{2c}, -R_{b2}(\theta_2)+\Delta R_{b2}, z_{2c}(\theta_2)+\Delta z_{2c};\ C_{q2}\}$$
$$n_d\{\pi/2-\chi_2(\theta_2)-\Delta\chi_2, \phi_{b2}(\theta_2)+\Delta\phi_{b2};\ C_2\}$$

When gear II rotates back by $\Delta\chi_2$ for G_{2d} to be parallel to the plane of action G_2, the G_{2d} moves to G_{2de}, the point P_d to P_{de} whose projection to the plane G_2 is P_{df} and w_d and $\mathbf{n}_d$ to w_d' and $\mathbf{n}_d'$ through P_{df} respectively which are the projections of w_d and $\mathbf{n}_d$ to the plane G_2. w_d' intersects the plane of rotation through the point P at P_{dg}.

w_d' moves from w on the plane of action G_2 by $(\Delta\theta_2 - \Delta\chi_2)$ and inclines to w at P_{df} by $\Delta\phi_{b2}$, so that the displacement PP_{dg} of w_d' from w in the direction of the axis q_{2c} is expressed as follows.

$$PP_{dg} = \{R_{b2}(\theta_2)-\Delta R_{b2}/2\}(\Delta\theta_2-\Delta\chi_2)+\Delta z_{2c}\Delta\phi_{b2}/\cos^2\phi_{b2}(\theta_2)$$
$$= R_{b2}(\theta_2)(\Delta\theta_2-\Delta\chi_2)$$

When the inclination angle of the tangent of the path of contact at P projected on the plane G_2 is defined as $\eta_{b2}(\theta_2)$, then

$$\tan\eta_{b2}(\theta_2) = (dz_{2c}/d\theta_2)/(dq_{2c}/d\theta_2) \tag{2. 3-1}$$

A small displacement Δz_{2c} on the plane G_2 by $\Delta\theta_2$ is expressed by Eq. (2. 3-1) as follows, using the relation shown in Fig. 2. 3-1.

$$\Delta z_{2c}[\tan\{\phi_{b2}(\theta_2)+\Delta\phi_{b2}\}+1/\tan\{\eta_{b2}(\theta_2)+\Delta\eta_{b2}\}] = R_{b2}(\theta_2)(\Delta\theta_2-\Delta\chi_2)$$

By neglecting infinitesimal of the second order,

$$\Delta z_{2c} = R_{b2}(\theta_2)(\Delta\theta_2-\Delta\chi_2)/\{\tan\phi_{b2}(\theta_2)+1/\tan\eta_{b2}(\theta_2)\}$$

Using Eq. (2. 3-1), Δq_{2c} is expressed as follows.

$$\Delta q_{2c} = R_{b2}(\theta_2)(\Delta\theta_2-\Delta\chi_2)/\{\tan\phi_{b2}(\theta_2)\tan\eta_{b2}(\theta_2)+1\}$$

ΔR_{b2}, $\Delta\chi_2$, $\Delta\phi_{b2}$ and $\Delta\eta_{b2}$ are functions of θ_2, so that they can be expressed by $\Delta\theta_2$ formally as follows.

$$\triangle R_{b2} = (dR_{b2}/d\theta_2)\triangle\theta_2, \qquad \triangle\eta_{b2} = (d\eta_{b2}/d\theta_2)\triangle\theta_2$$
$$\triangle\chi_2 = (d\chi_2/d\theta_2)\triangle\theta_2, \qquad \triangle\phi_{b2} = (d\phi_{b2}/d\theta_2)\triangle\theta_2$$

By integrating from 0 to θ_2,

$$\left.\begin{aligned}
q_{2c}(\theta_2) &= \int_0^{\theta 2} [R_{b2}(\theta_2)(1-d\chi_2/d\theta_2)/\{\tan\phi_{b2}(\theta_2)\tan\eta_{b2}(\theta_2)+1\}]d\theta_2 + q_{2c}(0) \\
R_{b2}(\theta_2) &= \int_0^{\theta 2} (dR_{b2}/d\theta_2)d\theta_2 + R_{b2}(0) \\
z_{2c}(\theta_2) &= \int_0^{\theta 2} [R_{b2}(\theta_2)(1-d\chi_2/d\theta_2)/\{\tan\phi_{b2}(\theta_2)+1/\tan\eta_{b2}(\theta_2)\}]d\theta_2 + z_{2c}(0) \\
\eta_{b2}(\theta_2) &= \int_0^{\theta 2} (d\eta_{b2}/d\theta_2)d\theta_2 + \eta_{b2}(0) \\
\chi_2(\theta_2) &= \int_0^{\theta 2} (d\chi_2/d\theta_2)d\theta_2 + \chi_2(0) = \pi/2 - \phi_2(\theta_2) \\
\phi_{b2}(\theta_2) &= \int_0^{\theta 2} (d\phi_{b2}/d\theta_2)d\theta_2 + \phi_{b2}(0)
\end{aligned}\right\} \tag{2.3-2}$$

The integral constants mean the position of the point of contact P_0, the inclination angles of the common contact normal $\mathbf{n}_0$ and the inclination angle η_{b20} of the tangent of the path of contact projected on the plane G_2, when $\theta_2 = 0$.

Equation (2.3-2) represents a path of contact with its common contact normal at each point as functions of θ_2 in the coordinate system C_{q2}, which is determined by 10 variables, namely, $P_0\{q_{2c}(0), -R_{b2}(0), z_{2c}(0)\}$, $\eta_{b2}(0)$, $n_0\{\pi/2-\chi_2(0), \phi_{b2}(0)\}$, $dR_{b2}/d\theta_2$, $d\eta_{b2}/d\theta_2$, $d\chi_2/d\theta_2$ and $d\phi_{b2}/d\theta_2$. The ratio $i(\theta_2)$ of angular velocities at each point given by Eq. (2.3-2) is obtained by Eq. (2.2-9), so that Eq. (2.3-2) represents a path of contact with its common contact normal at each point whose ratio of angular velocities varies according to the angle of rotation θ_2. In the ordinary design of gears, in which the given ratio i_0 of angular velocities is constant, Eq. (2.3-2) is constrained by two equations $i(0) = i_0$ and $i(\theta_2) = i_0$, so that it has eight independent variables.

A point of contact P is transformed into the coordinate system $C_2(u_{2c}, v_{2c}, z_{2c})$ using Eq. (2.1-2), then

$$\left.\begin{aligned}
u_{2c}(\theta_2) &= q_{2c}(\theta_2)\cos\chi_2(\theta_2) + R_{b2}(\theta_2)\sin\chi_2(\theta_2) \\
v_{2c}(\theta_2) &= q_{2c}(\theta_2)\sin\chi_2(\theta_2) - R_{b2}(\theta_2)\cos\chi_2(\theta_2) \\
z_{2c}(\theta_2) &= z_{2c}(\theta_2)
\end{aligned}\right\} \tag{2.3-3}$$

In the same way, using Eqs. (2.1-1), (2.1-3) and (2.1-5), a point of contact $P\{u_{1c}(\theta_2), v_{1c}(\theta_2), z_{1c}(\theta_2); C_1\}$ and the inclination angle $n\{\phi_1(\theta_2), \phi_{b1}(\theta_2); C_1\}$ of its common contact normal $\mathbf{n}$ are expressed as functions of θ_2 in the coordinate systems C_1 and C_{q1} as follows.

$$\left.\begin{aligned} u_{1c}(\theta_2) &= q_{1c}(\theta_2)\cos\chi_1(\theta_2) + R_{b1}(\theta_2)\sin\chi_1(\theta_2) \\ v_{1c}(\theta_2) &= q_{1c}(\theta_2)\sin\chi_1(\theta_2) - R_{b1}(\theta_2)\cos\chi_1(\theta_2) \\ z_{1c}(\theta_2) &= z_{1c}(\theta_2) \\ \chi_1(\theta_2) &= \pi/2 - \phi_1(\theta_2) \end{aligned}\right\} \quad (2.3\text{-}4)$$

In the same way, using Eqs. (2.1-11), (2.1-13) and (2.1-15), a point of contact $P\{u_c(\theta_2), v_c(\theta_2), z_c(\theta_2); C_S\}$ and the inclination angle $n\{\phi_S(\theta_2), \phi_{SW}(\theta_2); C_S\}$ of its common contact normal n are expressed as functions of θ_2 in the coordinate systems C_S and C_{qS} as follows.

$$\left.\begin{aligned} u_c(\theta_2) &= q_c(\theta_2)\cos\chi_S(\theta_2) + R_{bC}(\theta_2)\sin\chi_S(\theta_2) \\ v_c(\theta_2) &= q_c(\theta_2)\sin\chi_S(\theta_2) - R_{bC}(\theta_2)\cos\chi_S(\theta_2) \\ z_c(\theta_2) &= z_c(\theta_2) \\ \chi_S(\theta_2) &= \pi/2 - \phi_S(\theta_2) \end{aligned}\right\} \quad (2.3\text{-}5)$$

2. 3. 2 Equations of tooth profile

(1) Equations of tooth profile II

Figure 2.3-2 shows a tooth profile II in the coordinate system $C_{r2}(u_{r2c}, v_{r2c}, z_{r2c})$ which is defined to the coordinate system C_2 as follows. It has the origin C_2 and the axis z_{2c} in common and rotates around the axis z_{2c} with gear II whose angle of rotation is θ_2, where the axis u_{r2c} coincides with the axis u_{2c} when $\theta_2 = 0$.

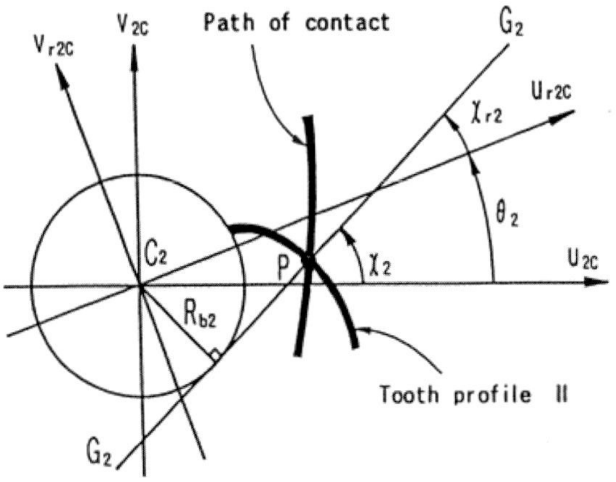

Fig.2.3-2 Tooth profile II in the coordinate system C_{r2}

As a point of contact P and the inclination angle of its common contact normal n are given by Eq. (2.3-2), a point $P(u_{r2c}, v_{r2c}, z_{r2c}; C_{r2})$ and the inclination angle of its normal $n(\phi_{r2}, \phi_{b2}; C_{r2})$ are expressed in the coordinate system C_{r2} as follows.

$$\left.\begin{aligned} \chi_{r2}(\theta_2) &= \chi_2(\theta_2) - \theta_2 = \pi/2 - \phi_2(\theta_2) - \theta_2 \\ \phi_{r2}(\theta_2) &= \phi_2(\theta_2) + \theta_2 \\ u_{r2c}(\theta_2) &= q_{2c}(\theta_2)\cos\chi_{r2} + R_{b2}(\theta_2)\sin\chi_{r2} \\ v_{r2c}(\theta_2) &= q_{2c}(\theta_2)\sin\chi_{r2} - R_{b2}(\theta_2)\cos\chi_{r2} \\ z_{r2c}(\theta_2) &= z_{2c}(\theta_2) \end{aligned}\right\} \quad (2.3\text{-}6)$$

(2) Equations of tooth profile I

In the same way, the coordinate system $C_{r1}(u_{r1c}, v_{r1c}, z_{r1c})$ is defined to the coordinate system C_1 as follows. It has the origin C_1 and the axis z_{1c} in common and rotates around the axis z_{1c} with gear I whose angle of rotation is θ_1, where the axis u_{r1c} coincides with the axis u_{1c} when $\theta_1 = 0$. A point $P(u_{r1c}, v_{r1c}, z_{r1c}; C_{r1})$ and the inclination angle of its normal $\mathbf{n}(\phi_{r1}, \phi_{b1}; C_{r1})$ are expressed using Eqs. (2.2-9) and (2.3-4) as follows.

$$\left.\begin{aligned}
\theta_1 &= \int_0^{\theta 2} i(\theta_2)d\theta_2 \\
\chi_{r1}(\theta_2) &= \chi_1(\theta_2)-\theta_1 = \pi/2 - \phi_1(\theta_2)-\theta_1 \\
\phi_{r1}(\theta_2) &= \phi_1(\theta_2)+\theta_1 \\
u_{r1c}(\theta_2) &= q_{1c}(\theta_2)\cos\chi_{r1} + R_{b1}(\theta_2)\sin\chi_{r1} \\
v_{r1c}(\theta_2) &= q_{1c}(\theta_2)\sin\chi_{r1} - R_{b1}(\theta_2)\cos\chi_{r1} \\
z_{r1c}(\theta_2) &= z_{1c}(\theta_2)
\end{aligned}\right\} \qquad (2.3\text{-}7)$$

Equations (2.3-6) and (2.3-7) mean a pair of tooth profiles and their normals whose ratio of angular velocities varies when they make contact at each point.

2.4 Tooth surface in the vicinity of a point of contact and limit of action (24), (26)

2.4.1 Infinitesimal surface of action and its corresponding tooth surfaces

Figure 2.4-1(a) shows a line of contact PP_w in the vicinity of a point of contact P chosen at will when P and its common contact normal $\mathbf{n}$ are given on a path of contact $\mathbf{h}$, where the tangent plane at P is W and $\mathbf{n}(\phi_H, \phi_{nH}; C_S)$ intersects the plane S_H at $P_H(u_{cH}, 0, z_{cH}; C_S)$. When another point of contact P_w on the plane W except P in the vicinity of P and its common contact normal $\mathbf{n}_w(\phi_w, \phi_{nw}; C_S)$ are given, $\phi_w = \phi_H$ and $\phi_{nw} = \phi_{nH}$ are obtained because $\mathbf{n}_w$ is parallel to $\mathbf{n}$. Therefore, when $\mathbf{n}_w$ intersects the plane S_H at $P_{Hw}(u_{cHw}, 0, z_{cHw}; C_S)$, the following relation between u_{cHw} and ϕ_H is obtained from Eq. (2.2-8).

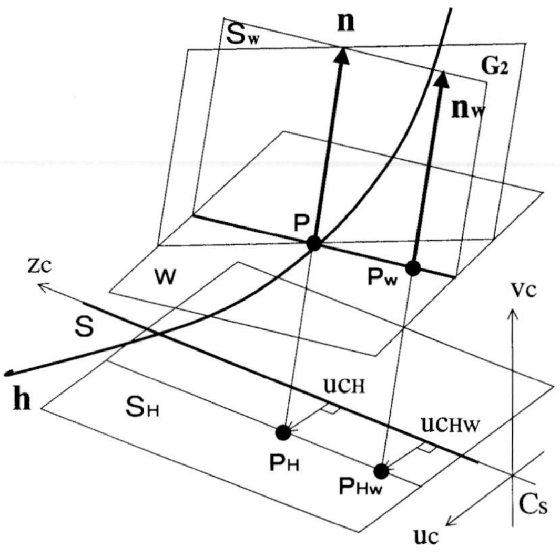

Fig.2.4-1(a) Instantaneous surface of action S_W at P

$$u_{cHW} = E\tan\phi_H \sin(\Sigma-\Gamma_s)\sin\Gamma_s/\sin\Sigma = u_{cH} \qquad (2.4\text{-}1)$$

Equation (2.4-1) means that $P_H P_{Hw}$ is parallel to the instantaneous axis S. Conversely, when any point P_{Hw} is put on the line parallel to the instantaneous axis S through P_H and a normal $\mathbf{n}_w$ parallel to $\mathbf{n}$ is set through P_{Hw}, $\mathbf{n}_w$ becomes the common contact normal from Eq. (2.4-1). Therefore, the line of contact PP_w is the intersection of the tangent plane W and the plane S_w including $\mathbf{n}$ and parallel to the instantaneous axis S.

Figure 2.4-1(b) shows an infinitesimal surface of action $\triangle PP_dP_{wd}$ and the corresponding infinitesimal tooth surface $\triangle P_{r2f}P_dP_{wd}$ which are made in the vicinity of P by an infinitesimal angle of rotation of gear II. By the infinitesimal angle of rotation of

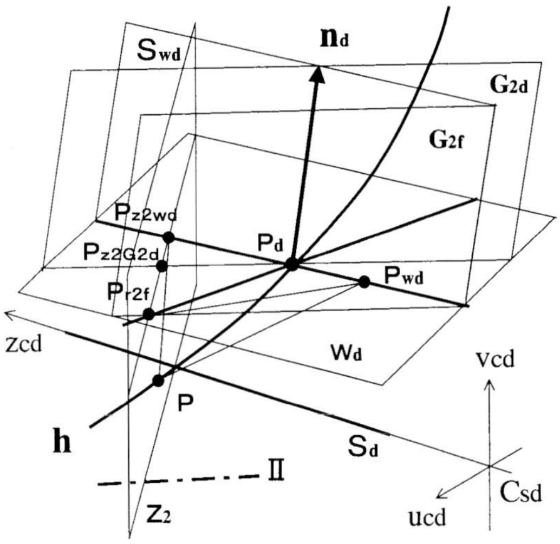

Fig.2.4-1(b) Surface of action and generated tooth surface II

gear II, the point of contact P moves to P_d, the common contact normal $\mathbf{n}$ to $\mathbf{n}_d$, the instantaneous axis S to S_d, the tangent plane W to W_d, the plane S_w to S_{wd} and the plane of action G_2 to G_{2d}, while the point of contact P moves to P_{r2f} on the tangent plane W_d and the plane of action through P_{r2f} is shown by G_{2f}. When a point P_{wd} is chosen at will on the line of contact (the intersection of the planes S_{wd} and W_d), P_dP_{wd} represents the line of contact, $P_{r2f}P_d$ the tooth profile II, $\triangle PP_dP_{wd}$ the infinitesimal surface of action and $\triangle P_{r2f}P_dP_{wd}$ the corresponding infinitesimal tooth surface II (strictly speaking its tangent plane), where the line of contact P_dP_{wd} intersects the plane of rotation Z_2 through P regarding the axis II at P_{Z2wd}.

In the same way, the corresponding infinitesimal tooth surface I ($\triangle P_{r1f}P_dP_{wd}$) and the tooth profile I ($P_{r1f}P_d$) which intersects the tooth profile II ($P_{r2f}P_d$) at P_d are obtained, though they are not shown here.

When a path of contact with its common contact normal at each point is given, the envelope of the infinitesimal surface of action $\triangle PP_dP_{wd}$ along the path of contact makes a belt-like surface of action according to the location of the instantaneous axis S, which generates the corresponding pair of belt-like tooth surfaces I and II.

2．4．2 Generation of infinitesimal tooth surfaces (point contact and line contact)

A tooth surface II which has a given path of contact $\mathbf{h}$ with its contact normal $\mathbf{n}$ at each point is determined independently of gear I by Eqs. (2.3-2) and (2.3-6). Therefore, when a shaft angle Σ_{Ct2} and an offset E_{Ct2} of a generating gear axis C_{t2} for tooth surface II are chosen at will, its instantaneous axis (a generating gear ratio) is determined and an infinitesimal tooth surface II can be generated around the path of contact $\mathbf{h}$ according to a given generating tooth surface which has the tangent plane W. The generating gear axis C_{t2} can be chosen at will in itself but it is chosen to be perpendicular to the axis v_{2c} in this theory to be dealt with simply. An infinitesimal tooth surface I can be generated in the same way by selecting a generating axis C_{t1} and a generating tooth surface.

When the generating gear axes C_{t1} and C_{t2} are chosen as axes perpendicular to the common perpendicular except for gear axes II and I, the instantaneous axis S_1 determined by the gear axis I and the generating gear axis C_{t1} is different from the instantaneous axis S_2 determined by the gear axis II and the generating gear axis C_{t2}. Therefore, the plane S_w including the common contact normal $\mathbf{n}$ and parallel to the instantaneous axis becomes two different planes S_{w1} and S_{w2} whose intersection is $\mathbf{n}$, the line of contact PP_w becomes two different lines PP_{w1} and PP_{w2} which intersect at P and the generated infinitesimal tooth surfaces I and II make point contact at P.

When the generating gear axes C_{t1} and C_{t2} coincide with gear axes II and I respectively, the generating instantaneous axes S_1 and S_2 coincide with the instantaneous axis S. Therefore the plane S_w and the line of contact PP_w which is the intersection of the planes S_w and W become the same ones, so that the generated infinitesimal tooth surfaces II and I are conjugate ones having the line of contact PP_w in common.

Depending on the choice of the generating gear axes C_{t2} and C_{t1}, innumerable infinitesimal tooth surfaces with the same tangent plane W are generated. Therefore, the two most practical cases are discussed in this theory as follows.

(1) Tooth surfaces in line contact : The instantaneous axis of a generating gear and a generated one is chosen so as to coincide with the instantaneous axis S of gears I and II, which means that generating gear axes C_{t2} and C_{t1} are chosen generally to coincide with the mating gear axes I and II respectively. The ordinary hypoid gears are cut in this way, so that they have tooth surfaces in line contact. In cylindrical and bevel gears, any generating gear can be used which has the instantaneous axis S in common with a generated one.

(2) Tooth surfaces in point contact : The instantaneous axis of a generating gear and a generated one is chosen not to coincide with the instantaneous axis S of gears I and II, which means in this theory that a generating gear axis C_{t2} is chosen to be parallel to the axis of gear II to have the plane of action G_2 in common and a generating gear axis C_{t1} is chosen to be parallel to the axis of gear I to have the plane of action G_1 in common. The ordinary crossed helical and conical gears are cut in this way. Because the generating gear axis is parallel to each gear axis, the plane S_w including the common contact normal **n** and parallel to the generating instantaneous axis coincides with the plane of action G_1 or G_2, which means that the line of contact PP_{w1} or PP_{w2} is the intersection of the tangent plane W and the plane of action G_1 or G_2 respectively.

2. 4. 3 Limit of action

When a path of contact with its common contact normal at each point is given, a belt-like surface of action of a generating gear and gear I or II is determined and a corresponding tooth surface is generated mentioned above, the limit of action of which can be defined as follows.

"Limit of action is a certain point of contact on a tooth surface which has no displacement or continues to make contact on the tooth surface during an infinitesimal angle of rotation."

When in Fig. 2. 4-1(b), $\triangle P_{r2f} P_d P_{wd}$ is an infinitesimal tooth surface generated through an infinitesimal angle of rotation $\triangle\theta_2$, the requirement for limit of action at P_{r2f} is that there exists a point P_{wd} $(u_{r2C} + \triangle u_{r2Cwd}, v_{r2C} + \triangle v_{r2Cwd}, z_{r2c} + \triangle z_{r2Cwd} ; C_{r2})$ on the line of contact which makes the infinitesimal displacement $P_{r2f} P_{wd}$ from P_{r2f} $(u_{r2C}, v_{r2C}, z_{r2c} ; C_{r2})$ equal 0. Therefore, the requirement for limit of action at P_{r2f} is obtained as follows.

$$\triangle u_{r2Cwd} = \triangle v_{r2Cwd} = \triangle z_{r2cwd} = 0 \qquad (2.4\text{-}2)$$

2．4 Tooth surface in the vicinity of a point of contact and limit of action

A point of contact P_d, its common contact normal $\mathbf{n}_d$, another point of contact P_{wd} and its common contact normal $\mathbf{n}_{wd}$ are expressed in the coordinate systems C_2 and C_{q2} as follows, where $\mathbf{n}_{wd}$ is parallel to $\mathbf{n}_d$ because the line of contact P_dP_{wd} is the intersection of the tangent plane W_d and the plane S_{wd}.

$$P_d(q_{2c}+\Delta q_{2c}, -R_{b2}+\Delta v_{q2c}, z_{2c}+\Delta z_{2c};\ C_{q2})$$
$$\mathbf{n}_d\{\pi/2-(\chi_2+\Delta\chi_2),\ \phi_{b2}+\Delta\phi_{b2};\ C_2\}$$
$$P_{wd}(q_{2c}+\Delta q_{2cwd}, -R_{b2}+\Delta v_{q2cwd}, z_{2c}+\Delta z_{2cwd};C_{q2})$$
$$\mathbf{n}_{wd}=\mathbf{n}_d$$

Therefore, using Eq. (2.3-2), Δz_{2cwd} is obtained as follows.

$$\Delta z_{2cwd}=\{R_{b2}(1-d\chi_2/d\theta_2)\tan\eta_{b2wd}/(\tan\phi_{b2}\tan\eta_{b2wd}+1)\}\Delta\theta_2$$

where η_{b2wd} is the inclination angle of PP_{wd} on the plane of action.

From Eq. (2.4-2) which means $\Delta z_{r2cwd}=\Delta z_{2cwd}=0$, the following relation is obtained.

$$\eta_{b2wd}=0\ \ (\text{where } 1-d\chi_2/d\theta_2\neq 0)$$

Namely, $\eta_{b2wd}=0$ means that the point of contact P_{wd} coincides with P_{Z2wd}. The point of contact P_{z2wd} and its common contact normal $\mathbf{n}_{z2wd}$ are expressed in the coordinate systems C_{r2}, C_2 and C_{q2} as follows, where subscript z means that the infinitesimal displacement is on the plane of rotation.

$$P_{z2wd}(u_{r2C}+\Delta u_{r2CZ}, v_{r2C}+\Delta v_{r2CZ}, z_{r2c};\ C_{r2})$$
$$P_{z2wd}(q_{2c}+\Delta q_{2cZ}, -R_{b2}+\Delta v_{q2cZ}, z_{2c};\ C_{q2})$$
$$\Delta q_{2cZ}=R_{b2}(\Delta\theta_2-\Delta\chi_2)\ \ (\because \eta_{b2wd}=0)$$
$$\Delta v_{q2cZ}=-(dR_{b2Z}/d\theta_2)\Delta\theta_2$$
$$\mathbf{n}_{z2wd}\{\pi/2-(\chi_2+\Delta\chi_2),\ \phi_{b2}+\Delta\phi_{b2};\ C_2\}=\mathbf{n}_d$$

Using Eq. (2.3-6), $\Delta u_{r2CZ}=\Delta v_{r2CZ}=0$ are transformed as follows.

$$\Delta u_{r2CZ}=\{q_{2c}(1-d\chi_2/d\theta_2)+dR_{b2Z}/d\theta_2\}\cdot\Delta\theta_2\sin\chi_{r2}=0$$
$$\Delta v_{r2CZ}=-\{q_{2c}(1-d\chi_2/d\theta_2)+dR_{b2Z}/d\theta_2\}\cdot\Delta\theta_2\cos\chi_{r2}=0$$

Therefore, the requirement for limit of action is obtained as follows.

$$\eta_{b2}=0,\ \ q_{2c}(1-d\chi_2/d\theta_2)+dR_{b2Z}/d\theta_2=0 \qquad (2.4\text{-}3)$$

Conversely, when any point of contact P on the path of contact $\mathbf{h}$ satisfies Eq. (2.4-3), $\Delta u_{r2Cwd}=\Delta v_{r2Cwd}=\Delta z_{r2cwd}=0$ is realized, so that Eq. (2.4-3) is the necessary and sufficient condition for limit of action at $P(=P_{r2f})$ on the tooth surface II.

When the point of contact P_{z2wd} satisfies Eq. (2.4-3), the radius $R_{2Z}=R_2+\Delta R_{2Z}$ at P_{z2wd} has the following relation.

$$(R_2+\triangle R_{2Z})^2=(q_{2c}+\triangle q_{2cZ})^2+(R_{b2}+\triangle R_{b2Z})^2$$
$$\triangle R_{2Z}R_2=\triangle q_{2cZ}q_{2c}+\triangle R_{b2Z}R_{b2}$$
$$=q_{2c}\{R_{b2}(1-d\chi_2/d\theta_2)\}\triangle\theta_2+R_{b2}(dR_{b2Z}/d\theta_2)\triangle\theta_2$$
$$=\{q_{2c}(1-d\chi_2/d\theta_2)+dR_{b2Z}/d\theta_2\}R_{b2}\triangle\theta_2=0$$
$$\therefore\ dR_{2Z}/d\theta_2=\{q_{2c}(1-d\chi_2/d\theta_2)+dR_{b2Z}/d\theta_2\}R_{b2}/R_2=0 \qquad (2.4\text{-}4)$$

Equation (2.4-4) means that Eq. (2.4-3) is the condition under which the path of contact PP_{Z2wd} on the plane of rotation is tangent to a cylinder having the axis II at P, so that Eq. (2.4-4) is more convenient to obtain the limit of action numerically in the case of complex surfaces of action.

In the same way, the limit of action of gear I is obtained as follows.

$$\eta_{b1}=0,\ q_{1c}(1-d\chi_1/d\theta_1)+dR_{b1Z}/d\theta_1=0 \qquad (2.4\text{-}5)$$

The limits of action are obtained as the solutions θ_2 and θ_1 of Eqs. (2.4-3) and (2.4-5).

When the axis of a generating gear is parallel to that of a generated one, the plane S_{wd} coincides with G_{2d} and $P_{Z2wd}=P_{Z2G2d}$ is obtained. Because PP_{Z2G2d} is the path of contact on the plane of rotation, from Eq. (2.3-2) the following relations are obtained.

$$\triangle v_{q2cZ}=\triangle v_{q2C}=-(dR_{b2}/d\theta_2)\triangle\theta_2$$
$$P_{Z2G2d}(q_{2c}+\triangle q_{2cZ},\ -R_{b2}+\triangle v_{q2C},\ z_{2c};\ C_{q2})$$

Therefore, Eqs. (2.4-3) and (2.4-5) are transformed as follows.

$$\eta_{b2}=0,\ q_{2c}(1-d\chi_2/d\theta_2)+dR_{b2}/d\theta_2=0 \qquad (2.4\text{-}6)$$
$$\eta_{b1}=0,\ q_{1c}(1-d\chi_1/d\theta_1)+dR_{b1}/d\theta_1=0 \qquad (2.4\text{-}7)$$

Equation (2.4-3) means $P_{r2f}P_{Z2wd}=0$ in Fig. 2.4-1(b). Generally speaking, whether a point (for example P_{r2f}) given on a tooth profile (or a path of contact) is the limit of action or not depends on the shape of the intersection of the surface of action given by a generating gear and the plane of rotation through the point.

Varying the plane of rotation, the locus of the limit of action draws the limit line of action. Eq. (2.4-4) means that the limit of action is the foot point of the perpendicular which drops from the gear axis toward the intersection of the surface of action and the plane of rotation, so that the limit line of action is the orthographic projection of the gear axis to the surface of action.

2.4.4 Radii of curvature of infinitesimal tooth surfaces

(1) Inclination angles of line of contact on infinitesimal tooth surfaces

Figure 2.4-2 shows the tangent plane W_d on Fig. 2.4-1(b) seen from the positive direction of the common contact normal $\mathbf{n}_d$, where the plane S_{wd} and the planes of action G_{2d} and G_{1d} intersect at right angles with the tangent plane W_d whose intersection is $\mathbf{n}_d$. By rotating the plane of action G_2 by $\triangle\chi_2$ so as to be parallel to the G_{2d}, the plane of action G_{2p} is defined and the coordinate system C_{q2d} is obtained by rotating the coordinate system C_{q2}

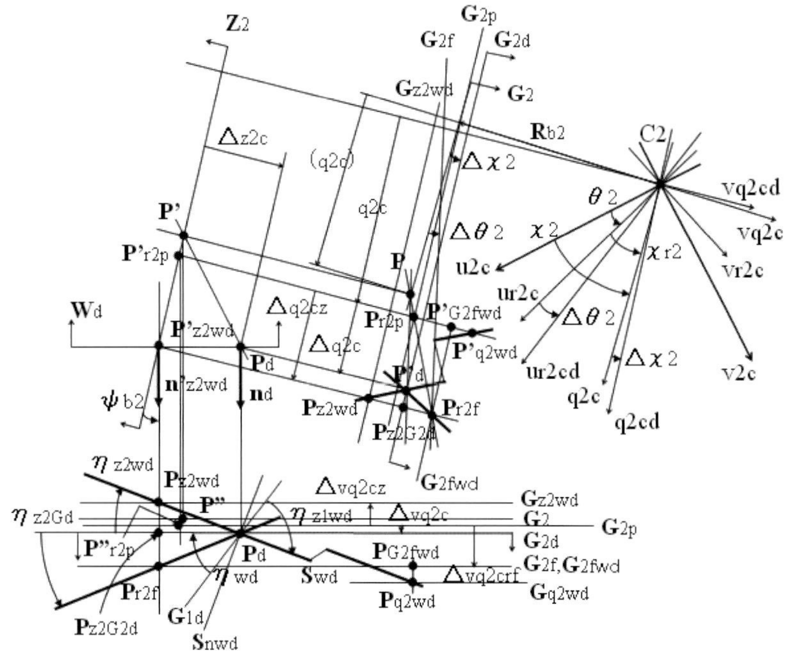

Fig.2.4-2 Path of contact PP_d, tooth curve $P_{r2f}\,P_d$ and line of contact $P_{z2wd}\,P_d\,P_{q2wd}$ on planes W_d, G_{2d} and Z_2

by $\triangle\chi_2$. The intersection P_{r2P} of the plane G_{2p} and PP_{r2f} represents the position of P in the coordinate system C_{q2d}, so that the infinitesimal displacements in the coordinate system C_{q2d} are defined from the point P_{r2P} as shown in Fig. 2. 4-2. The planes of action G_{2d} and G_{Z2wd} are parallel ($\mathbf{n}_d = \mathbf{n}_{z2wd}$) and the planes W_d and Z_2 intersect at right angles with the plane G_{2d}, therefore the intersection $P_{Z2wd}P_{Z2G2d}$ of the planes W_d and Z_2 intersects at right angles with the plane of action $G_{Z2wd}(G_{2d})$, where $P_{Z2wd}P_{Z2G2d}$ is positive in the direction of the axis v_{q2cd}. The point P_{r2f}, its normal $\mathbf{n}_{r2f}$ and the intersection P_{Z2G2d} of the tooth profile $P_{r2f}P_{Z2wd}$ on the plane of rotation and the plane of action G_{2d} are expressed as follows.

$$
\begin{aligned}
&P_{r2f}(q_{2c}+\triangle q_{2crf},\ -R_{b2}+\triangle v_{q2crf},\ z_{2c};\ C_{q2d})\\
&\qquad \triangle q_{2crf} = R_{b2}(\triangle\theta_2-\triangle\chi_2) = \triangle q_{2cZ}\\
&\qquad \triangle v_{q2crf} = q_{2c}(\triangle\theta_2-\triangle\chi_2)\\
&\mathbf{n}_{r2f}\{\pi/2-(\chi_2+\triangle\theta_2),\ \psi_{b2};\ C_2\}\\
&P_{Z2G2d}(q_{2c}+\triangle q_{2cZ},\ -R_{b2}+\triangle v_{q2C},\ z_{2c};\ C_{q2d})\\
&\qquad \triangle v_{q2C} = -(dR_{b2}/d\theta_2)\triangle\theta_2
\end{aligned}
$$

Therefore, using the right triangle $P_dP_{Z2G2d}P_{Z2wd}$, η_{Z2wd} is obtained as follows, where η_{Z2wd} is positive in the clockwise direction seen from the positive direction of $\mathbf{n}_d$ as shown in Fig. 2. 4-2.

$$P_dP_{Z2G2d} = -\triangle z_{2c}/\cos\phi_{b2}$$
$$P_{Z2G2d}P_{Z2wd} = \triangle v_{q2cZ} - \triangle v_{q2c}$$
$$\tan\eta_{Z2wd} = P_{Z2G2d}P_{Z2wd}/P_dP_{Z2G2d}$$
$$= (dR_{b2Z}/d\theta_2 - dR_{b2}/d\theta_2)\cos\phi_{b2}$$
$$/\{R_{b2}(1-d\chi_2/d\theta_2)\tan\eta_{b2}/(\tan\eta_{b2}\tan\phi_{b2}+1)\} \quad (2.4\text{-}8)$$

In the same way, using the right triangle $P_dP_{Z2G2d}P_{r2f}$, η_{Z2Gd} is obtained as follows, where η_{Z2Gd} is positive in the clockwise direction seen from the positive direction of $\mathbf{n}_d$ (the negative direction is shown in Fig. 2.4-2).

$$P_{Z2G2d}P_{r2f} = \triangle v_{q2crf} - \triangle v_{q2c}$$
$$\tan\eta_{Z2Gd} = P_{Z2G2d}P_{r2f}/P_dP_{Z2G2d}$$
$$= -\{q_{2c}(1-d\chi_2/d\theta_2) + dR_{b2}/d\theta_2\}\cos\phi_{b2}$$
$$/\{R_{b2}(1-d\chi_2/d\theta_2)\tan\eta_{b2}/(\tan\eta_{b2}\tan\phi_{b2}+1)\} \quad (2.4\text{-}9)$$

η_{Z1wd} and η_{Z1Gd} on the infinitesimal tooth surface I are obtained in the same way.

$$\tan\eta_{Z1wd} = (dR_{b1Z}/d\theta_1 - dR_{b1}/d\theta_1)\cos\phi_{b1}$$
$$/\{R_{b1}(1-d\chi_1/d\theta_1)\tan\eta_{b1}/(\tan\eta_{b1}\tan\phi_{b1}+1)\} \quad (2.4\text{-}10)$$
$$\tan\eta_{Z1Gd} = -\{q_{1c}(1-d\chi_1/d\theta_1) + dR_{b1}/d\theta_1\}\cos\phi_{b1}$$
$$/\{R_{b1}(1-d\chi_1/d\theta_1)\tan\eta_{b1}/(\tan\eta_{b1}\tan\phi_{b1}+1)\} \quad (2.4\text{-}11)$$

Therefore, the angle η_{wd} made by the planes of action G_{2d} and G_{1d} is as follows.

$$\eta_{wd} = \eta_{Z1wd} - \eta_{Z2wd} \quad (2.4\text{-}12)$$

(2) Radii of curvature of infinitesimal tooth surfaces

When the path of contact with its common contact normal at each point is given, $\triangle P_{r2f}P_{Z2wd}P_d$ can represent an infinitesimal tooth surface in the vicinity of the point P_{r2f}, where the point P_{wd} on the line of contact (Fig. 2.4-1(b)) coincides with P_{Z2wd}, $P_{r2f}P_{Z2wd}$ means the tooth profile on the plane of rotation and $P_{Z2wd}P_d$ means the line of contact.

When the line of contact $P_{Z2wd}P_d$ can be assumed to be a straight line in the vicinity of P_d, the radius of curvature ρ_{n2W} along the line of contact becomes a principal one of the tooth surface, because it is infinity.

The radius of curvature ρ_{Z2} in the plane of rotation and the radius of curvature ρ_{nZ2} in the plane including the common contact normal $\mathbf{n}_{z2wd}$ $(=\mathbf{n}_d)$ of the tooth profile $P_{r2f}P_{Z2wd}$ in the plane of rotation are obtained as follows.

$$\left.\begin{aligned}
\rho_{Z2} &= P_{Z2wd}P_{r2f}/(\triangle\theta_2-\triangle\chi_2)\\
&= (\triangle v_{q2crf}-\triangle v_{q2cZ})/(\triangle\theta_2-\triangle\chi_2)\\
&= q_{2c}+dR_{b2Z}/d\theta_2/(1-d\chi_2/d\theta_2)\\
&\quad\text{(when a generating gear axis coincides with the mating gear axis)}\\
\rho_{Z2} &= P_{Z2G2d}P_{r2f}/(\triangle\theta_2-\triangle\chi_2)\\
&= (\triangle v_{q2crf}-\triangle v_{q2c})/(\triangle\theta_2-\triangle\chi_2)\\
&= q_{2c}+dR_{b2}/d\theta_2/(1-d\chi_2/d\theta_2)\\
&\quad\text{(when a generating gear axis is parallel to the generated gear axis)}\\
\rho_{nZ2} &= \rho_{Z2}/\cos\phi_{b2}
\end{aligned}\right\}\quad (2.4\text{-}13)$$

Therefore, the other principal radius of curvature ρ_{n2} perpendicular to the line of contact is obtained by Euler's formula as follows.

$$\begin{aligned}
1/\rho_{nZ2} &= \cos^2\eta_{Z2wd}/\rho_{n2}+\sin^2\eta_{Z2wd}/\rho_{n2W}\\
\rho_{n2} &= \rho_{Z2}\cos^2\eta_{Z2wd}/\cos\phi_{b2}
\end{aligned}\quad (2.4\text{-}14)$$

In the same way, the principal radius of curvature ρ_{n1} perpendicular to the line of contact of the infinitesimal tooth surface I is obtained as follows.

$$\left.\begin{aligned}
\rho_{Z1} &= q_{1C}+(dR_{b1Z}/d\theta_1)/(1-d\chi_1/d\theta_1)\\
&\quad\text{(when a generating gear axis coincides with the mating gear axis)}\\
\rho_{Z1} &= q_{1c}+dR_{b1}/d\theta_1/(1-d\chi_1/d\theta_1)\\
&\quad\text{(when a generating gear axis is parallel to the generated gear axis)}\\
\rho_{nZ1} &= \rho_{Z1}/\cos\phi_{b1}\\
\rho_{n1} &= \rho_{Z1}\cos^2\eta_{Z1wd}/\cos\phi_{b1}
\end{aligned}\right\}\quad (2.4\text{-}15)$$

The radii of curvature obtained here are based on the assumption that the infinitesimal tooth surfaces can be approximated to be cylindrical surfaces, when a path of contact with its common contact normal at each point is given and a straight line of contact is assumed. There are innumerable tooth surfaces which have the same path of contact with its common contact normal at each point, therefore the radius of curvature of the tooth surface can not be determined only from the given path of contact with its common contact normal at each point. This is discussed more precisely in 3.5.6.

2. 5 Equations of motion and bearing loads [22], [25]

Figure 2.5-1 shows the relation among the normal component F_{N2} of the concentrated load at the point of contact P shown in Fig. 2.1-1 and bearing loads B_{z2}, B_{vq2f}, B_{vq2r}, B_{q2f} and B_{q2r} of gear II in the coordinate systems C_{q2} and C_2. The point of contact P and its common contact normal **n** are given by Eq. (2.3-2) and the friction force is negligibly small because of sufficient lubrication. Gear II has a shaft with sufficiently large stiffness and is rigidly supported by bearings b_{2a}, b_{2f} and b_{2r} in the axial and radial directions.

Gear II rotates around the axis II, receives the normal component F_{N2} from gear I and transmits the input (or output) torque T_2, so that the equation of motion and the bearing loads of gear II are obtained in the coordinate system C_{q2} as follows.

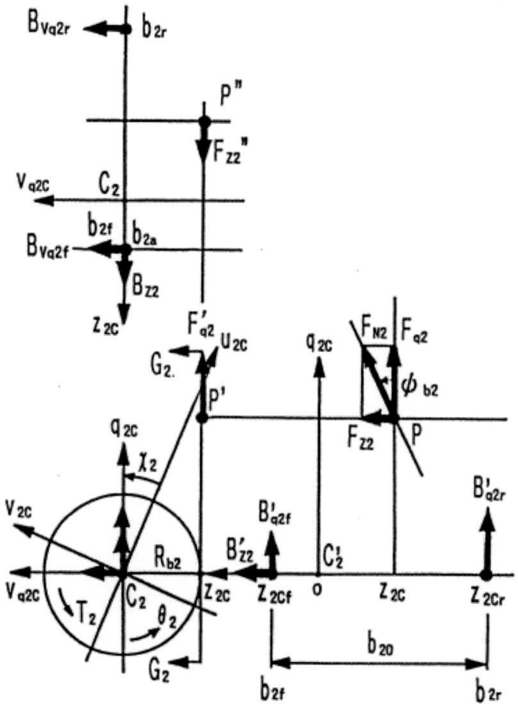

Fig.2.5-1 Tooth load and bearing loads of gear Ⅱ

$$
\left.
\begin{aligned}
J_2(d^2\theta_2/dt^2) &= F_{q2}R_{b2}(\theta_2)+T_2\\
B_{z2} &= -F_{z2} = -F_{q2}\tan\phi_{b2}(\theta_2)\\
B_{vq2r}b_{20} &= F_{z2}R_{b2}(\theta_2)\\
B_{q2f}+B_{q2r} &= -F_{q2}\\
B_{q2f}b_{20} &= -F_{q2}\{z_{2c}(\theta_2)-z_{2cr}\}+F_{q2}q_{2c}(\theta_2)\tan\phi_{b2}(\theta_2)\\
B_{vq2f}+B_{vq2r} &= 0
\end{aligned}
\right\}
\qquad (2.5\text{-}1)
$$

where

J_2 : moment of inertia of gear Ⅱ

θ_2 : angle of rotation of gear Ⅱ

T_2 : input (or output) torque of gear Ⅱ

F_{q2} and F_{z2} : components of F_{N2} in the directions of the axes q_{2c} and z_{2c}

B_{z2} : load on bearing b_{2a} in the direction of the axis z_{2c}

B_{q2f} and B_{q2r} : loads on bearings b_{2f} and b_{2r} in the direction of the axis q_{2c}

B_{vq2f} and B_{vq2r} : loads on bearings b_{2f} and b_{2r} in the direction of the axis v_{q2c}

z_{2cf} and z_{2cr} : locations of bearings b_{2f} and b_{2r} on the axis z_{2c}

b_{20} : distance between bearings b_{2f} and b_{2r} ($z_{2cf} - z_{2cr} > 0$)

and the bearing loads are positive in the positive direction of each axis of the coordinate system C_{q2}.

In the same way, those of gear Ⅰ are obtained as follows.

$$
\left.\begin{aligned}
J_1(d^2\theta_1/dt^2) &= F_{q1}R_{b1}(\theta_2)+T_1\\
B_{z1} &= -F_{z1} = -F_{q1}\tan\phi_{b1}(\theta_2)\\
B_{vq1r}b_{10} &= F_{z1}R_{b1}(\theta_2)\\
B_{q1f}+B_{q1r} &= -F_{q1}\\
B_{q1f}b_{10} &= -F_{q1}\{z_{1c}(\theta_2)-z_{1cr}\}+F_{q1}q_{1c}(\theta_2)\tan\phi_{b1}(\theta_2)\\
B_{vq1f}+B_{vq1r} &= 0
\end{aligned}\right\} \qquad (2.5\text{-}2)
$$

where J_1 : moment of inertia of gear I

θ_1 : angle of rotation of gear I

T_1 : input (or output) torque of gear I

F_{q1} and F_{z1} : components of F_{N1} in the directions of the axes q_{1c} and z_{1c}

B_{z1} : load on bearing b_{1a} in the direction of the axis z_{1c}

B_{q1f} and B_{q1r} : loads on bearings b_{1f} and b_{1r} in the direction of the axis q_{1c}

B_{vq1f} and B_{vq1r} : loads on bearings b_{1f} and b_{1r} in the direction of the axis v_{q1c}

z_{1cf} and z_{1cr} : locations of bearings b_{1f} and b_{1r} on the axis z_{1c}

b_{10} : distance between bearings b_{1f} and b_{1r} $(z_{1cf}-z_{1cr}>0)$

and the bearing loads are positive in the positive direction of each axis of the coordinate system C_{q1}.

Using the equations of motion of gears I and II, the law of action and reaction and the fundamental requirement for contact (2.2-6), the rotational motion of a gear pair is expressed by the following simultaneous equations.

$$
\left.\begin{aligned}
J_1(d^2\theta_1/dt^2) &= F_{q1}R_{b1}(\theta_2)+T_1\\
J_2(d^2\theta_2/dt^2) &= F_{q2}R_{b2}(\theta_2)+T_2\\
-F_{q1}/\cos\phi_{b1}(\theta_2) &= F_{q2}/\cos\phi_{b2}(\theta_2)\\
R_{b1}(\theta_2)(d\theta_1/dt)\cos\phi_{b1}(\theta_2) &= R_{b2}(\theta_2)(d\theta_2/dt)\cos\phi_{b2}(\theta_2)
\end{aligned}\right\} \qquad (2.5\text{-}3)
$$

A path of contact with its common contact normal at each point is given by Eq. (2.3-2), so that Eq. (2.5-3) are simultaneous equations with unknown quantities θ_1, θ_2, F_{q1} and F_{q2} and are basic equations to describe the rotational motion of a gear pair having a given pair of tooth profiles. Because Eq. (2.5-3) is effective only in the continuous and differentiable region of a path of contact, it is necessary to adopt another method to obtain the rotational motion in the region including a non-differentiable point. For example, at the point where the number of tooth pairs in mesh changes, the path of contact becomes continuous but non-differentiable and the collision of tooth pair occurs caused by the discontinuity of the normal velocity.

2．6　Summary and references

The basic theory of the new tooth geometry which is common to almost all kinds of gears and enables dealing with the dynamic rotational motion (the equations of motion) is proposed, which is summarized as follows.

(1) When the gear axes, the common perpendicular, the instantaneous axis, the angular velocities of gears and a point of contact and its common contact normal are given, eight basic

coordinate systems C_1, C_2, C_S, C_{q1}, C_{q2}, C_{qS}, C_{r1} and C_{r2} are defined.

(2) When a point of contact P chosen at will and a common contact normal **n** at P, which mean that P has a plane element in the vicinity of P, are expressed by five independent variables, the fundamental requirement for contact is introduced from the relative velocity at P and the ratio i of angular velocities at P is determined. This means that one of the six variables which are composed of a point of contact, a common contact normal and the ratio of angular velocities becomes a dependent variable constrained from the fundamental requirement for contact.

(3) Equations of a path of contact with its common contact normal at each point are introduced as functions of the angle of rotation θ_2 of gear II.

(4) Transforming the path of contact with its common contact normal at each point into the coordinate systems C_{r1} and C_{r2} which rotate with gear I or II, the tooth profiles I and II are obtained.

(5) Using the infinitesimal surface of action and the corresponding tooth surfaces in the vicinity of the path of contact, the difference between the tooth surfaces in point contact and those in line contact are discussed and the equations of the limits of action and the radii of curvature are obtained.

(6) Equations of motion and bearing loads are obtained.

Based on the basic theory mentioned above, its applications, namely, design methods of tooth surfaces, an involute helicoid and its conjugate tooth surface, variation of a tooth bearing caused by assembly errors, the rotational motion and the dynamic loads of gears and pairs of tooth surfaces transmitting inconstant rotational motion are discussed in the following chapters.

References

(21) 本多捷, 歯面の接触と動荷重の基礎理論（第１報）, 機論, 62-600, C(1996), 3262-3268.
(22) 本多捷, 歯面の接触と動荷重の基礎理論（第２報）, 機論, 62-600, C(1996), 3269-3274.
(23) 本多捷, 歯面の接触と動荷重の基礎理論（第３報）, 機論, 62-603, C(1996), 4349-4356.
(24) 本多捷, 動力伝達用歯車の設計理論（第１報）, 機論, 70-689, C(2004), 258-265.
(25) Honda, S., "A pair of tooth surfaces without variation of bearing loads", Proceedings of the 7th ASME Int. PTG Conf., Oct. 1996, 467-476.
(26) Honda, S., "Requirement for limits of action and Wildhaber's limit normal", Proceedings of ASME DETC2003/PTG-48096, Sept. 2003.

3. Design method of tooth surfaces for quieter gear pairs

In this chapter, a design method of tooth surfaces for quieter gear pairs for power transmission is clarified as follows.

(1) The conditions for no variation of bearing loads (quietness) in 3.1,

(2) design methods of a path of contact which realize the conditions in 3.2 to 3.4 and

(3) finally, a design method and examples of tooth surfaces which realize the path of contact in 3.5 to 3.8

are discussed.

3. 1 Conditions for no variation of bearing loads [31]

3. 1. 1 Variation of bearing loads

The quietness of a running gear pair is realized under the condition that there is no variation of bearing loads. As shown in Fig. 2. 5-1, gear II on which the normal component F_{N2} of the concentrated load acts at the point of contact P is supported by the bearing reaction loads B_{z2}, B_{vq2f}, B_{vq2r}, B_{q2f} and B_{q2r} and runs in a steady state. The friction component of the concentrated load is negligibly small because of sufficient lubrication. Gear II has a shaft with sufficiently large stiffness and is rigidly supported by bearings b_{2a}, b_{2f} and b_{2r} in the axial and radial directions.

The variation of the load on gear II caused by the rotation of the gear pair can be known by the loads on bearings b_{2a}, b_{2f} and b_{2r} in the coordinate system (static) C_2. Therefore, transforming the bearing loads expressed by Eq. (2. 5-1) in the coordinate system C_{q2} into the coordinate system C_2, they are obtained as follows.

$$\left.\begin{array}{ll} B_{z2c} = B_{z2} & \text{(load on } b_{2a} \text{ in the direction of } z_{2c}) \\ B_{u2cf} = B_{q2f}\cos\chi_2 - B_{vq2f}\sin\chi_2 & \text{(load on } b_{2f} \text{ in the direction of } u_{2c}) \\ B_{v2cf} = B_{q2f}\sin\chi_2 + B_{vq2f}\cos\chi_2 & \text{(load on } b_{2f} \text{ in the direction of } v_{2c}) \\ B_{u2cr} = B_{q2r}\cos\chi_2 - B_{vq2r}\sin\chi_2 & \text{(load on } b_{2r} \text{ in the direction of } u_{2c}) \\ B_{v2cr} = B_{q2r}\sin\chi_2 + B_{vq2r}\cos\chi_2 & \text{(load on } b_{2r} \text{ in the direction of } v_{2c}) \end{array}\right\} \quad (3.1\text{-}1)$$

The variation of bearing loads are obtained by differentials of Eq. (3. 1-1) as follows.

(1) Variation of load on bearing b_{2a} in the direction of the axis z_{2c}

$$\Delta B_{z2c} = \Delta B_{z2} = -\Delta F_{z2}$$
$$\Delta F_{z2} = \Delta F_{q2}\tan\phi_{b2} + F_{q2}\Delta\phi_{b2}/\cos^2\phi_{b2}$$
$$\Delta F_{q2} = \{\Delta(J_2(d^2\theta_2/dt^2)) - F_{q2}\Delta R_{b2}\}/R_{b2}$$

(2) Variation of loads on bearing b_{2f} in the directions of the axes u_{2c} and v_{2c}

$$\Delta B_{u2cf} = \Delta B_{q2f}\cos\chi_2 - B_{q2f}\sin\chi_2\Delta\chi_2 - \Delta B_{vq2f}\sin\chi_2 - B_{vq2f}\cos\chi_2\Delta\chi_2$$
$$\Delta B_{v2cf} = \Delta B_{q2f}\sin\chi_2 + B_{q2f}\cos\chi_2\Delta\chi_2 + \Delta B_{vq2f}\cos\chi_2 - B_{vq2f}\sin\chi_2\Delta\chi_2$$

$$\Delta B_{q2f} = -[\Delta F_{q2}(z_{2c} - z_{2cr} - q_{2c}\tan\phi_{b2}) + F_{q2}\{\Delta z_{2c} - \Delta q_{2c}\tan\phi_{b2} - q_{2c}\Delta\phi_{b2}/\cos^2\phi_{b2}\}]/b_{20}$$
$$\Delta B_{vq2f} = -(\Delta F_{z2}R_{b2} + F_{z2}\Delta R_{b2})/b_{20}$$

(3) Variation of loads on bearing b_{2r} in the directions of the axes u_{2c} and v_{2c}

$$\Delta B_{u2cr} = \Delta B_{q2r}\cos\chi_2 - B_{q2r}\sin\chi_2\Delta\chi_2 - \Delta B_{vq2r}\sin\chi_2 - B_{vq2r}\cos\chi_2\Delta\chi_2$$
$$\Delta B_{v2cr} = \Delta B_{q2r}\sin\chi_2 + B_{q2r}\cos\chi_2\Delta\chi_2 + \Delta B_{vq2r}\cos\chi_2 - B_{vq2r}\sin\chi_2\Delta\chi_2$$
$$\Delta B_{q2r} = -\Delta F_{q2} - \Delta B_{q2f}$$
$$\Delta B_{vq2r} = -\Delta B_{vq2f}$$

Variation of bearing loads at an angle of rotation θ_2 chosen at will are expressed by the six differentials, namely Δq_{2c}, ΔR_{b2}, Δz_{2c}, $\Delta\chi_2$, $\Delta\phi_{b2}$ and $\Delta(d^2\theta_2/dt^2)$.

3. 1. 2 Conditions for no variation of bearing loads

When gear II rotates under constant input (output) torque, to realize no variation of bearing loads, all the differentials must be 0 at any rotation angle, namely,

$$\Delta B_{z2c} = \Delta B_{u2cf} = \Delta B_{v2cf} = \Delta B_{u2cr} = \Delta B_{v2cr} = 0$$

Therefore, the conditions for no variation of bearing loads of gear II are obtained as follows.

$$\left.\begin{array}{ll} (1) & \Delta\chi_2(\theta_2) = 0 \\ (2) & \Delta\phi_{b2}(\theta_2) = 0 \\ (3) & \Delta R_{b2}(\theta_2) = 0 \\ (4) & \Delta z_{2c}(\theta_2) = \Delta q_{2c}(\theta_2)\cdot\tan\phi_{b2}(\theta_2) \\ (5) & \Delta(d^2\theta_2/dt^2) = 0 \end{array}\right\} \quad (3.1\text{-}2)$$

From Eq. (3.1-2), the variables of Eq. (2.3-2) must satisfy the following conditions.

(1) Condition of $\Delta\chi_2(\theta_2)=0$

Because of $d\chi_2/d\theta_2 = 0$, the inclination angle $\chi_2(\theta_2)$ of the plane of action G_2 must be constant, which is given by χ_{20}.

$$\chi_2(\theta_2) = \chi_2(0) = \chi_{20} = \pi/2 - \phi_{20}$$

(2) Condition $\Delta\phi_{b2}(\theta_2)=0$

Because of $d\phi_{b2}/d\theta_2 = 0$, the inclination angle $\phi_{b2}(\theta_2)$ of the common contact normal on the plane of action G_2 must be constant, which is given by ϕ_{b20}.

$$\phi_{b2}(\theta_2) = \phi_{b2}(0) = \phi_{b20}$$

(3) Condition $\Delta R_{b2}(\theta_2)=0$

Because of $dR_{b2}/d\theta_2 = 0$, the radius $R_{b2}(\theta_2)$ of base cylinder must be constant, which is given by R_{b20}.

$$R_{b2}(\theta_2) = R_{b2}(0) = R_{b20}$$

(4) Condition $\Delta z_{2c}(\theta_2) = \Delta q_{2c}(\theta_2)\cdot\tan\phi_{b2}(\theta_2)$

Using Eq. (2.3-1) and $\phi_{b2}(\theta_2) = \phi_{b20}$, $d\eta_{b2}/d\theta_2 = 0$ is obtained, therefore,

$$\eta_{b2}(\theta_2) = \phi_{b20} = \eta_{b2}(0)$$

Condition (4) means that the inclination angle of the path of contact on the plane of action G_2($q_{2c}-z_{2c}$ plane) coincides with that of the common contact normal.

Substituting the results led from conditions (1) to (4) for Eq. (2.3-2) and transforming it into the coordinate system C_2, the path of contact with its common contact normal at each point is obtained as follows.

$$\left.\begin{aligned}
q_{2c}(\theta_2) &= R_{b20}\,\theta_2\cos^2\phi_{b20} + q_{2c}(0)\\
u_{2c}(\theta_2) &= q_{2c}(\theta_2)\cos\chi_{20} + R_{b20}\sin\chi_{20}\\
v_{2c}(\theta_2) &= q_{2c}(\theta_2)\sin\chi_{20} - R_{b20}\cos\chi_{20}\\
z_{2c}(\theta_2) &= R_{b20}\,\theta_2\cos\phi_{b20}\sin\phi_{b20} + z_{2c}(0)\\
&n(\phi_{20} = \pi/2-\chi_{20},\ \phi_{b20};\ C_2)
\end{aligned}\right\}\quad (3.1\text{-}3)$$

Equation (3.1-3) means that the path of contact is a straight line which passes through the point $P_0\{q_{2c}(0), -R_{b20}, z_{2c}(0); C_{q2}\}$, has the inclination angle $n(\phi_{20} = \pi/2-\chi_{20}, \phi_{b20}; C_2)$ and coincides with the common contact normal.

Using Eqs. (2.1-1), (2.1-3) and (2.1-5) and transforming Eq. (3.1-3) into the coordinate systems C_1 and C_{q1}, the path of contact is expressed as a straight line which passes through the point $P_0\{q_{1c}(0), -R_{b10}, z_{1c}(0); C_{q1}\}$ and has the inclination angle $n(\phi_{10} = \pi/2-\chi_{10}, \phi_{b10}; C_1)$. Therefore, the ratio of angular velocities and the angle of rotation of gear I are obtained by Eq. (2.2-9) as follows.

$$\left.\begin{aligned}
i(\theta_2) &= (d\theta_1/dt)/(d\theta_2/dt) = R_{b20}\cos\phi_{b20}/(R_{b10}\cos\phi_{b10}) = i_0\\
\theta_1 &= i_0\theta_2
\end{aligned}\right\}\quad (3.1\text{-}4)$$

Therefore, the ratio of angular velocities is constant, which is given by i_0.

Substituting $q_{1c}(\theta_1)$ and $z_{1c}(\theta_1)$ for $q_{1c}(\theta_2)$ and $z_{1c}(\theta_2)$ in Eq. (3.1-3), the path of contact with its common contact normal at each point is expressed by θ_1 and transformed into the coordinate system C_1 as follows.

$$\left.\begin{aligned}
\theta_1 &= i_0\theta_2 \quad (\text{when } \theta_2 = 0,\ \theta_1 = 0)\\
q_{1c}(\theta_1) &= R_{b10}\,\theta_1\cos^2\phi_{b10} + q_{1c}(0)\\
u_{1c}(\theta_1) &= q_{1c}(\theta_1)\cos\chi_{10} + R_{b10}\sin\chi_{10}\\
v_{1c}(\theta_1) &= q_{1c}(\theta_1)\sin\chi_{10} - R_{b10}\cos\chi_{10}\\
z_{1c}(\theta_1) &= R_{b10}\,\theta_1\cos\phi_{b10}\sin\phi_{b10} + z_{1c}(0)\\
&n(\phi_{10} = \pi/2-\chi_{10},\ \phi_{b10};\ C_1)
\end{aligned}\right\}\quad (3.1\text{-}5)$$

(5) Condition $\Delta(d^2\theta_2/dt^2) = 0$

Because $d^2\theta_2/dt^2$ is constant, it means a uniformly accelerated motion. Under the assumption that the gear pair rotates in a steady state under constant input and output torques, it means $d^2\theta_2/dt^2 = 0$ and $d\theta_2/dt$ = constant, which in turn means a uniform motion. In addition, $d\theta_1/dt$ becomes constant by Eq. (3.1-4), so that $d^2\theta_1/dt^2 = 0$ is obtained in gear I. Therefore, using Eq. (2.5-3), the following equations are obtained.

$$F_{q2}R_{b20} = -T_2, \qquad F_{q1}R_{b10} = -T_1$$

Using the law of action and reaction in Eq. (2.5-3), ϕ_{b10} and ϕ_{b20} are eliminated from Eq. (3.1-4), then

$$i_0 = F_{q2} \cdot R_{b20}/(-F_{q1} \cdot R_{b10}) = -T_2/T_1 \tag{3.1-6}$$

Equation (3.1-6) means that the ratio i_0 of angular velocities must be that of output and input torques (constant under the assumption).

Concerning gear I, no variation of bearing loads is realized in the same way.

When a gear pair rotates in a steady state under constant input and output torques, in order to realize no variation of bearing loads, the path of contact with its common contact normal at each point must satisfy the following conditions.

(a) When a point of contact is given at will in the static space (coordinate system C_2), the path of contact must be a straight line which coincides with the common contact normal and must be fixed in the static space.

(b) The ratio i_0 of angular velocities must be constant and coincide with that of output and input torques.

Conversely, when conditions (a) and (b) are satisfied, no variation of bearing loads is realized, so that they are necessary and sufficient conditions for no variation of bearing loads.

Figure 3.1-1 shows the relation among the plane of action G_{20}, the base cylinder (radius R_{b20}), the path of contact (=the common contact normal **n**) and the normal component F_{N2} of the concentrated load in the coordinate systems C_2 and C_{q2}, which realizes no variation of bearing loads. In the case of multiple tooth pair mesh, condition (a) means that the straight path of contact must coincide with both the common contact normal and the line of action of the normal component of the resultant load, while condition (b) must be satisfied.

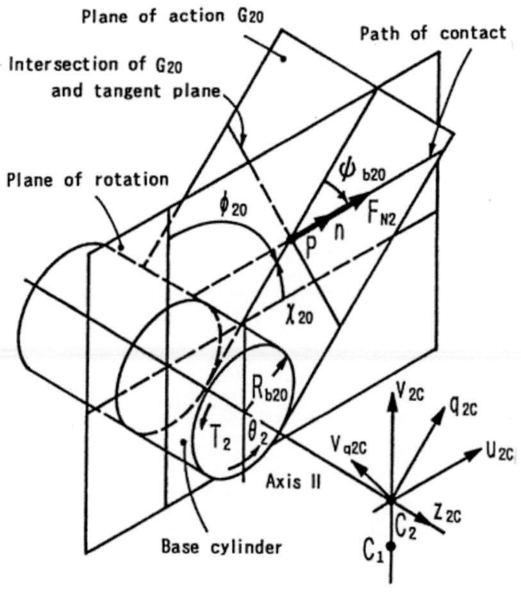

Fig.3.1-1 Path of contact without variation of bearing loads

3.1.3 Tooth profiles realizing no variation of bearing loads

Transforming Eq. (3.1-3) into the coordinate system $C_{r2}(u_{r2c}, v_{r2c}, z_{r2c})$ or $C_{r1}(u_{r1c}, v_{r1c}, z_{r1c})$, tooth profile II or I is obtained respectively. Equations of tooth profile

(2.3-6) and (2.3-7) are simplified as follows.

(1) Equations of tooth profile II and inclination angle of its normal

$$\left.\begin{aligned}&\chi_{r2} = \chi_{20} - \theta_2 = \pi/2 - \phi_{20} - \theta_2\\&u_{r2c} = q_{2c}(\theta_2)\cos\chi_{r2} + R_{b20}\sin\chi_{r2}\\&v_{r2c} = q_{2c}(\theta_2)\sin\chi_{r2} - R_{b20}\cos\chi_{r2}\\&z_{r2c} = z_{2c}(\theta_2)\\&n(\phi_{20} + \theta_2,\ \phi_{b20}\ ;\ C_{r2})\end{aligned}\right\} \tag{3.1-7}$$

(2) Equations of tooth profile I and inclination angle of its normal

$$\left.\begin{aligned}&\theta_1 = i_0\theta_2\\&\chi_{r1} = \chi_{10} - \theta_1 = \pi/2 - \phi_{10} - \theta_1\\&u_{r1c} = q_{1c}(\theta_2)\cos\chi_{r1} + R_{b10}\sin\chi_{r1}\\&v_{r1c} = q_{1c}(\theta_2)\sin\chi_{r1} - R_{b10}\cos\chi_{r1}\\&z_{r1c} = z_{1c}(\theta_2)\\&n(\phi_{10} + \theta_1,\ \phi_{b10}\ ;\ C_{r1})\end{aligned}\right\} \tag{3.1-8}$$

Using Eq. (3.1-5), they are also expressed as follows.

$$\left.\begin{aligned}&\chi_{r1} = \chi_{10} - \theta_1 = \pi/2 - \phi_{10} - \theta_1\\&u_{r1c} = q_{1c}(\theta_1)\cos\chi_{r1} + R_{b10}\sin\chi_{r1}\\&v_{r1c} = q_{1c}(\theta_1)\sin\chi_{r1} - R_{b10}\cos\chi_{r1}\\&z_{r1c} = z_{1c}(\theta_1)\\&n(\phi_{10} + \theta_1,\ \phi_{b10}\ ;\ C_{r1})\end{aligned}\right\} \tag{3.1-9}$$

Equations (3.1-7) and (3.1-9) indicate that the profiles II and I are curves on the involute helicoids corresponding to the path of contact (= the common contact normal). Any surface including the tooth profile I or II and not interfering with each other can be a tooth surface, so that involute crossed helical and conical gears realize gear pairs which make point contact along the tooth profiles I and II on the involute helicoids. Generally, this tooth profile means a tooth bearing (a trace of a point of contact on the tooth surface), therefore, for example, the tooth bearing on the tooth surface II must be along the tooth profile II in order to realize no variation of bearing loads. That is the reason why the tooth bearing has an important role to reduce noise and vibration.

3．1．4 Variation of loads caused by a running gear pair

A running gear pair in a steady state under constant input and output torques realizes no variation of bearing loads when it satisfies the five conditions given by Eq. (3.1-2). Therefore, variation of loads of the gear pair occurs when the five conditions are not satisfied and it consists of the following five items corresponding to the five conditions, where the point of contact P_0 at $\theta_2 = 0$ is a basis of variation and χ_{20}, ϕ_{b20} and $F_{q20} = -T_2/R_{b20}$ are given.

(1) Variation of load ($F_{q20}\ \triangle\chi_2$) in the direction of the axis v_{q2c} caused by the variation of angle ($\triangle\chi_2$) of the plane of action
(2) Variation of load ($F_{q20}\ \triangle\phi_{b2}$) in the direction of the axis z_{2c} caused by the variation of angle ($\triangle\phi_{b2}$) of the common contact normal on the plane of action
(3) Variation of torque ($F_{q20}\ \triangle R_{b2}$) around the axis z_{2c} caused by the variation of radius ($\triangle R_{b2}$) of base cylinder
(4) Variation of torque $\{F_{q20}\ \triangle q_{2c}(\tan\eta_{b2} - \tan\phi_{b20})\}$ around the axis v_{q2c} caused by the wrong angle ($\eta_{b2} \neq \phi_{b20}$) of the path of contact on the plane of action
(5) Variation of torque ($J_2\ \triangle(d^2\theta_2/dt^2)$) around the axis z_{2c} caused by the variation of rotation, which occurs by the deviation (errors and deflection) from the true point of contact on the path of contact which coincides with the common contact normal.

Variation of loads of a running gear pair consists of loads or torques caused by the variation of points of action or inclination angles of the static loads F_{q20} in the static space (items 1 to 4) and the torque caused by the variation of rotation (item 5). In the present theory, the dynamic increment of tooth load means only item 5, but it is necessary to consider those caused by the static load mentioned above. Moreover, the choice of the tooth bearing in item 4 is regarded as a supplementary way to compensate for assembly errors and deflection to reduce item 5, but it is the condition which determines the relation of the path of contact (tooth bearing) and the common contact normal and is an important measure to reduce the variation of the static load.

The variation of loads mentioned above is effective only in the continuous and differentiable region of the path of contact, therefore it is necessary to adopt another method to obtain the rotational motion at the continuous but non-differentiable point at which the number of tooth pairs in mesh changes, where a collision of tooth pair occurs caused by discontinuities of displacement and normal velocity and it increases item 5.

3. 2 Equations of path of contact [32]

When the two axes, the shaft angle Σ, the offset E and the ratio i_0 (constant) of angular velocities are given, the path of contact (= the common contact normal) given by Eq. (3.1-3) is called $\mathbf{g}_0$ from now on. In this section, using Eqs. (3.1-3) and (3.1-4), it is discussed how the five unknown variables ($q_{2c}(0)$, $z_{2c}(0)$, R_{b20}, ϕ_{20}, ϕ_{b20}) of the path of contact $\mathbf{g}_0$ are selected.

3. 2. 1 Selection of design point P_0 in the coordinate system C_S

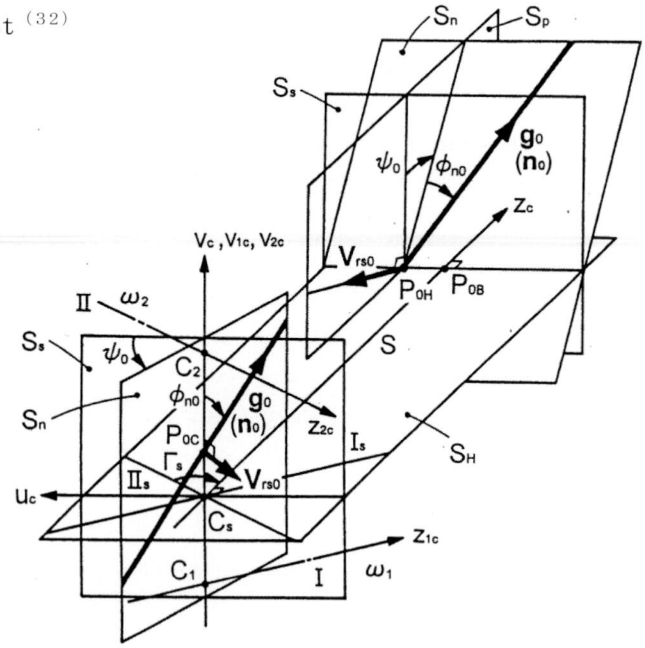

Fig.3.2-1 Selection of design point P_0, planes S_n and paths of contact $\mathbf{g}_0$

Figure 3.2-1 shows schematically a

point of contact $P_0(u_{c0}, v_{c0}, z_{c0}; C_S)$ on the path of contact $\mathbf{g}_0$ and the inclination angle $\mathbf{g}_0(\phi_0, \phi_{n0}; C_S)$ of $\mathbf{g}_0$ given by Eq. (3.1-3) in the coordinate system C_S, where $v_c = 0$ is called plane S_H, $u_c = u_{c0}$ plane S_p and $z_c = z_{c0}$ plane S_S.

In designing gears, gear pairs which have the same path of contact $\mathbf{g}_0$ have the same tooth profiles, so that it is different only which part of the tooth profile is used effectively. Therefore, when designing gear pairs, it is important where $\mathbf{g}_0$ is located in the static space (coordinate system C_S) and it has no difference substantially where a design point is put on $\mathbf{g}_0$ because it is only a point to determine the location of $\mathbf{g}_0$. Therefore, in this theory, the point of contact P_0 at $\theta_2 = 0$ in Eq. (3.1-3) is adopted as a design point P_0 $(u_{c0}, v_{c0}, z_{c0}; C_S)$ and it is given around the instantaneous axis on the plane S_H (bevel and hypoid gears) or around the common perpendicular (crossed helical, worm and cylindrical gears) from a practical point of view in the coordinate system C_S as shown in Fig. 3.2-1, namely,

(1) Hypoid gears $(\Sigma \neq 0, \pi \ \& \ E \neq 0)$: $P_0(u_{c0}, 0, z_{c0}; C_S)$

It is given on the plane S_H $(v_{c0} = 0)$ and shown by P_{0H}.

(2) Bevel gears $(\Sigma \neq 0, \pi \ \& \ E = 0)$: $P_0(0, 0, z_{c0}; C_S)$

It is given by P_{0B} on the instantaneous axis S. When $v_{c0} = 0$ is given, the intersection of the plane S_H and the plane N perpendicular to the relative velocity which is a peripheral one around the axis z_c is the instantaneous axis S (Fig. 2.2-1), so that $u_{c0} = 0$.

(3) Crossed helical and worm gears $(\Sigma \neq 0, \pi \ \& \ E \neq 0)$: $P_0(0, v_{c0}, 0; C_S)$

It is shown by P_{0C} on the common perpendicular. It is thought to be a special case of hypoid gears where $\mathbf{g}_0$ passes through the axis v_c. Therefore P_{0C} can be given on the plane S_H, but it is a little troublesome to have different P_0s on the drive and the coast sides.

(4) Cylindrical gears $(\Sigma = 0, \pi \ \& \ E \neq 0)$: $P_0(0, 0, 0; C_S)$

It is shown by C_S which is the intersection of the instantaneous axis S and the common perpendicular. When $u_{c0} = z_{c0} = 0$ is given, the intersection of the plane S_H and the plane N perpendicular to the relative velocity which is a peripheral one around the axis z_c is the instantaneous axis S (Fig. 2.2-1), so that $v_{c0} = 0$.

The domains of the variables are given from the practical point of view as follows.

$$0 \leqq u_{c0}, \quad 0 \leqq v_{c0}, \quad 0 \leqq z_{c0}$$

In this new theory, the kind of gears depends on where the design point P_0 is located on the plane S_H or the common perpendicular. In other words, the gear pair which has the design point P_0 chosen above is considered to be called hypoid, bevel, crossed helical, worm or cylindrical gears respectively. By giving the design point P_0, three of the five variables have been chosen.

3．2．2 Inclination angle of path of contact $\mathbf{g}_0$ in the coordinate system C_S

Figure 3.2-1 shows the planes S_ns which include the path of contact $\mathbf{g}_0$ and the design point P_0 chosen in 3.2.1 and are perpendicular to the relative velocity $\mathbf{V}_{rs0}$ at P_0. One of the planes S_ns coincides with the plane S_{nH} which includes $\mathbf{g}_0$ (= $\mathbf{n}$) and is parallel to the axis u_c as defined in 2.1.2(4) when P_0 is P_{0H}, P_{0B} or C_S on the plane S_H, but the other becomes different when P_0 is P_{0C} on the common perpendicular.

The path of contact $\mathbf{g}_0$ is a directed line chosen at will on the plane S_n through the design point P_0, which is positive in the direction of displacement of a point of contact when the angle of rotation θ_2 of gear II is positive. The inclination angle $g_0(\psi_0, \phi_{n0}; C_S)$ of $\mathbf{g}_0$ is expressed by a combination of the inclination angles ψ_0 of the plane S_n and the ϕ_{n0} on the plane S_n according to the kinds of gears as follows.

(1) Hypoid gears

Because the design point $P_0(u_{c0}, 0, z_{c0}; C_S)$ is P_{0H} and the relative velocity $\mathbf{V}_{rs0}$ is on the plane $S_p(u_c = u_{c0})$, the plane S_n is a plane which inclines the plane S_S $(z_c = z_{c0})$ around the axis u_c so as to be perpendicular to $\mathbf{V}_{rs0}$ at P_{0H} and whose inclination angle ψ_0 (positive in the clockwise direction around the axis u_c as shown on Fig. 3.2-1) is expressed as follows, where $E \neq 0$ and $0 < \Gamma_S < \pi$ (the inclination angle of the instantaneous axis).

$$\tan\psi_0 = u_{c0}\sin\Sigma / \{E\sin(\Sigma - \Gamma_S)\sin\Gamma_S\} \tag{3.2-1}$$

Equation (3.2-1) is the same as Eq. (2.2-8).

The inclination angle ϕ_{n0} of $\mathbf{g}_0$ on the plane S_n from the plane S_p is positive in the clockwise direction around the axis z_c as shown in Fig. 3.2-1 and the domain is $|\phi_{n0}| < \pi$.

(2) Bevel gears

Because the design point $P_0(0, 0, z_{c0}; C_S)$ is P_{0B} on the instantaneous axis S, the relative velocity $\mathbf{V}_{rs0} = 0$ and any normal through P_{0B} satisfies the fundamental requirement for contact, therefore any plane through P_{0B} can be the plane S_n. In this theory, corresponding to hypoid gears, the plane S_n is defined as a plane through P_{0B} which inclines by ψ_0 around the axis u_c to the plane S_S $(z_c = z_{c0})$, where ψ_0 can take any value.

The inclination angle ϕ_{n0} of $\mathbf{g}_0$ on the plane S_n from the plane S_p is positive in the clockwise direction around the axis z_c as shown in Fig. 3.2-1 and the domain is $|\phi_{n0}| < \pi$, corresponding to hypoid gears.

(3) Crossed helical and worm gears

Because the design point $P_0(0, v_{c0}, 0; C_S)$ is P_{0C} on the common perpendicular (the axis v_c), the relative velocity $\mathbf{V}_{rs0}$ is perpendicular to the common perpendicular and the plane S_n is a plane which includes the common perpendicular and inclines by ψ_0 around the axis v_c to the plane S_S $(z_c = 0)$, the inclination angle ψ_0 of which is positive in the clockwise direction around the axis v_c as shown in Fig. 3.2-1 and is obtained as follows, where $E \neq 0$ and $0 < \Gamma_S < \pi$.

$$\tan\psi_0 = v_{c0}\sin\Sigma / \{E\sin(\Sigma - \Gamma_S)\sin\Gamma_S\} \tag{3.2-2}$$

The inclination angle ϕ_{n0} of $\mathbf{g}_0$ on the plane S_n from the axis v_c is positive in the clockwise direction seen from the negative to the positive direction of the axis z_c (or u_c) as shown in Fig. 3.2-1 and the domain is $|\phi_{n0}| < \pi$.

(4) Cylindrical gears

Because the design point $P_0(0, 0, 0; C_S)$ is C_S, the relative velocity $\mathbf{V}_{rs0} = 0$ and any normal through C_S satisfies the fundamental requirement for contact, therefore any plane through C_S can be the plane S_n. In this theory, the following two planes through C_S are defined as the planes S_n.

(a) Corresponding to hypoid gears, one of the planes S_ns is a plane through C_S which inclines by ψ_0 around the axis u_c to the plane S_S ($z_c = 0$), which is used for tapered gears.
(b) Corresponding to crossed helical gears, the other of the planes S_ns is a plane through C_S which includes the common perpendicular (the axis v_c) and inclines by ψ_0 around the axis v_c to the plane S_S ($z_c = 0$), which is used for ordinary helical gears.

The inclination angle ϕ_{n0} of $\mathbf{g}_0$ on the plane S_n is defined in the same way as in hypoid gears or in crossed helical gears respectively.

The inclination angles ψ_0 and ϕ_{n0} correspond to the present definitions of the spiral and pressure angles, but it is necessary to know that they have some exceptions according to the kinds of gears. For example, ϕ_{n0} of cylindrical gears is the complementary angle of the present ϕ_{n0}.

In this theory, the two planes S_ns are introduced according to the kinds of gears based on the present custom, but in crossed helical and cylindrical gears, it is possible for the intersection of $\mathbf{g}_0$ chosen in the way mentioned above and the plane S_H to be a new design point and to determine a new plane S_n corresponding to hypoid gears. In this way, the design method becomes more unified, but it is a little troublesome to have different planes S_ns on the drive and the coast sides.

The path of contact $\mathbf{g}_0$ chosen above has the following independent design variables.

(a) Hypoid and bevel gears

Because the design point P_0 is chosen on the plane S_H ($v_{c0} = 0$),

Hypoid gears : 3 of u_{c0} (or ψ_0), z_{c0} and ϕ_{n0} under Eq. (3.2-1),

Bevel gears : 3 of ψ_0, z_{c0} and ϕ_{n0}, where any $\mathbf{g}_0$ through P_{0B} satisfies the fundamental requirement for contact because of $u_{c0} = v_{c0} = 0$.

(b) Crossed helical, worm and cylindrical gears

Because the design point P_0 is chosen on the common perpendicular v_c ($u_{c0} = z_{c0} = 0$),

Crossed helical and worm gears : 2 of v_{c0} (or ψ_0) and ϕ_{n0} under Eq. (3.2-2)

Cylindrical gears : 2 of ψ_0 and ϕ_{n0}, where any $\mathbf{g}_0$ through C_S satisfies the fundamental requirement for contact because of $u_{c0} = v_{c0} = z_{c0} = 0$.

3.2.3 Transformation of inclination angle of path of contact $\mathbf{g}_0$ into other coordinate systems

The inclination angle $g_0(\psi_0, \phi_{n0}; C_S)$ of $\mathbf{g}_0$ is transformed into other coordinate systems and expressed as $g_0(\phi_{20}, \psi_{b20}; C_2)$, $g_0(\phi_{10}, \psi_{b10}; C_1)$ and $g_0(\phi_{S0}, \psi_{sw0}; C_S)$ as follows.

(1) When the plane S_n does not include the common perpendicular (bevel and hypoid gears)

Figure 3.2-2 shows the path of contact $\mathbf{g}_0$ in the coordinate systems C_S, C_1 and C_2 when the plane S_n does not include the common perpendicular, and where the planes of action are

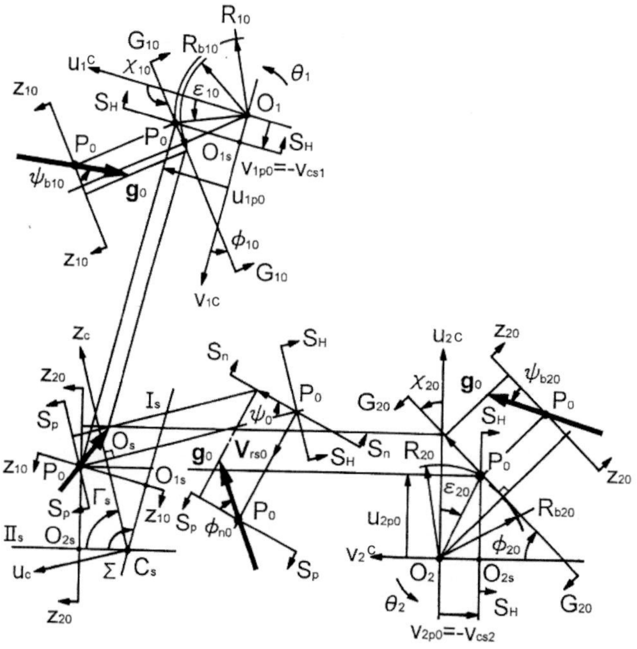

Fig.3.2-2 Relationship among inclination angles of $\mathbf{g}_0$ of bevel and hypoid gears

indicated by G_{10} and G_{20}. Equations of transformation from $g_0(\psi_0, \phi_{n0}; C_S)$ to $g_0(\phi_{20}, \psi_{b20}; C_2)$, $g_0(\phi_{10}, \psi_{b10}; C_1)$ and $g_0(\phi_{S0}, \psi_{SW0}; C_S)$ are already obtained by Eqs. (2.1-25), (2.1-26) and (2.1-24) as follows.

$$\left.\begin{aligned}
\tan\phi_{20} &= \tan\phi_{n0}\cos\Gamma_S/\cos\psi_0+\tan\psi_0\sin\Gamma_S\\
\sin\psi_{b20} &= \sin\phi_{n0}\sin\Gamma_S-\cos\phi_{n0}\sin\psi_0\cos\Gamma_S\\
\tan\phi_{10} &= -\tan\phi_{n0}\cos(\Sigma-\Gamma_S)/\cos\psi_0+\tan\psi_0\sin(\Sigma-\Gamma_S)\\
\sin\psi_{b10} &= \sin\phi_{n0}\sin(\Sigma-\Gamma_S)+\cos\phi_{n0}\sin\psi_0\cos(\Sigma-\Gamma_S)\\
\tan\phi_{S0} &= -\tan\phi_{n0}/\cos\psi_0\\
\sin\psi_{SW0} &= \cos\phi_{n0}\sin\psi_0
\end{aligned}\right\} \qquad (3.2\text{-}3)$$

(2) When the plane S_n includes the common perpendicular (crossed helical, worm and cylindrical gears)

Figure 3.2-3 shows the path of contact $\mathbf{g}_0$ in the coordinate systems C_S, C_1 and C_2 when the plane S_n includes the common perpendicular, and where the planes of action are indicated by G_{10} and G_{20}. Equations of transformation from $g_0(\psi_0, \phi_{n0}; C_S)$ to $g_0(\phi_{20}, \psi_{b20}; C_2)$, $g_0(\phi_{10}, \psi_{b10}; C_1)$ and $g_0(\phi_{S0}, \psi_{SW0}; C_S)$ are obtained as follows.

When a normal displacement on $\mathbf{g}_0$ is expressed by L_g (>0), the components L_{uc}, L_{vc} and L_{zc} of L_g in the direction of each axis of the coordinate system C_S are obtained as follows.

$$\begin{aligned}
L_{uc} &= -L_g\sin\phi_{n0}\cos\psi_0\\
L_{vc} &= L_g\cos\phi_{n0}\\
L_{zc} &= L_g\sin\phi_{n0}\sin\psi_0
\end{aligned}$$

The components L_{u2c}, L_{v2c} and L_{z2c} of L_g in the direction of each axis of the coordinate system C_2 are expressed by those in the coordinate system C_S as follows.

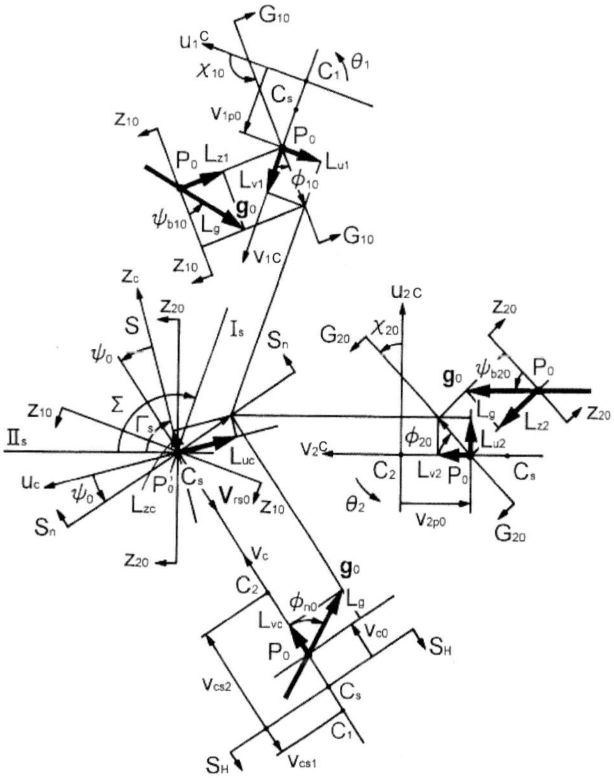

Fig.3.2-3 Relationship among inclination angles of $\mathbf{g}_0$ of worm and crossed helical gears

$$L_{u2c} = -L_{uc}\cos\Gamma_S + L_{zc}\sin\Gamma_S = L_g\sin\phi_{n0}\cos(\Gamma_S-\phi_0)$$
$$L_{v2c} = L_{vc} = L_g\cos\phi_{n0}$$
$$L_{z2c} = -L_{uc}\sin\Gamma_S - L_{zc}\cos\Gamma_S = L_g\sin\phi_{n0}\sin(\Gamma_S-\phi_0)$$

The inclination angle ϕ_{20} of the plane of action G_{20} and the inclination angle ψ_{b20} of $\mathbf{g}_0$ on the plane of action G_{20} at P_0 are obtained as follows.

$$\left.\begin{aligned}\tan\phi_{20} &= L_{u2c}/L_{v2c} = \tan\phi_{n0}\cos(\Gamma_S-\phi_0)\\ \sin\psi_{b20} &= L_{z2c}/L_g = \sin\phi_{n0}\sin(\Gamma_S-\phi_0)\end{aligned}\right\} \quad (3.2\text{-}4)$$

In the same way, the inclination angles ϕ_{10} of the plane of action G_{10} and ψ_{b10} on the plane of action G_{10} are obtained as follows, where L_{u1c}, L_{v1c} and L_{z1c} of L_g are the components in the direction of each axis of the coordinate system C_1.

$$L_{u1c} = L_{uc}\cos(\Sigma-\Gamma_S) + L_{zc}\sin(\Sigma-\Gamma_S) = -L_g\sin\phi_{n0}\cos(\Sigma-\Gamma_S+\phi_0)$$
$$L_{v1c} = L_{vc} = L_g\cos\phi_{n0}$$
$$L_{z1c} = -L_{uc}\sin(\Sigma-\Gamma_S) + L_{zc}\cos(\Sigma-\Gamma_S) = L_g\sin\phi_{n0}\sin(\Sigma-\Gamma_S+\phi_0)$$

Therefore,

$$\left.\begin{aligned} \tan\phi_{10} &= L_{u1c}/L_{v1c} = -\tan\phi_{n0}\cos(\Sigma-\Gamma_S+\phi_0) \\ \sin\phi_{b10} &= L_{z1c}/L_g = \sin\phi_{n0}\sin(\Sigma-\Gamma_S+\phi_0) \end{aligned}\right\} \tag{3.2-5}$$

In the same way,

$$\left.\begin{aligned} \tan\phi_{S0} &= L_{uc}/L_{vc} = -\tan\phi_{n0}\cos\phi_0 \\ \sin\phi_{SW0} &= L_{zc}/L_g = \sin\phi_{n0}\sin\phi_0 \end{aligned}\right\} \tag{3.2-6}$$

3. 2. 4 Equations of path of contact g_0 in the coordinate systems C_2, C_1 and C_S

(1) Equations of path of contact g_0 in the coordinate system C_2

Transforming the design point P_0 $(u_{c0}, v_{c0}, z_{c0}; C_S)$ into the coordinate systems C_2 and C_{q2}, $P_0(u_{2p0} = u_{2c}(0), v_{2p0} = v_{2c}(0), z_{2p0} = z_{2c}(0); C_2)$ and $P_0(q_{2p0} = q_{2c}(0), -R_{b20}, z_{2p0} = z_{2c}(0); C_{q2})$ are obtained as follows.

$$\left.\begin{aligned} u_{2p0} &= -u_{c0}\cos\Gamma_S + z_{c0}\sin\Gamma_S \\ v_{2p0} &= v_{c0} - v_{cs2} \\ z_{2p0} &= -u_{c0}\sin\Gamma_S - z_{c0}\cos\Gamma_S \\ v_{cs2} &= E\tan\Gamma_S/\{\tan\Gamma_S+\tan(\Sigma-\Gamma_S)\} \\ \tan\varepsilon_{20} &= v_{2p0}/u_{2p0} \\ R_{20} &= \sqrt{(u_{2p0}^2+v_{2p0}^2)}\ \text{(Radius of the design point)} \\ q_{2p0} &= u_{2p0}\cos\chi_{20} + v_{2p0}\sin\chi_{20} \\ R_{b20} &= u_{2p0}\sin\chi_{20} - v_{2p0}\cos\chi_{20} = R_{20}\cos(\phi_{20}+\varepsilon_{20}) \\ \chi_{20} &= \pi/2 - \phi_{20} \end{aligned}\right\} \tag{3.2-7}$$

Substituting Eq. (3. 2-7) for Eq. (3. 1-3), the path of contact g_0 is obtained in the coordinate system C_2, where χ_{20} and ϕ_{b20} are given by Eq. (3. 2-3) or (3. 2-4) according to the kinds of gears.

(2) Equations of path of contact g_0 in the coordinate system C_1

In the same way, transforming the design point P_0 $(u_{c0}, v_{c0}, z_{c0}; C_S)$ into the coordinate systems C_1 and C_{q1}, $P_0(u_{1p0} = u_{1c}(0), v_{1p0} = v_{1c}(0), z_{1p0} = z_{1c}(0); C_1)$ and $P_0(q_{1p0} = q_{1c}(0), -R_{b10}, z_{1p0} = z_{1c}(0); C_{q1})$ are obtained as follows.

$$\left.\begin{aligned} u_{1p0} &= u_{c0}\cos(\Sigma-\Gamma_S) + z_{c0}\sin(\Sigma-\Gamma_S) \\ v_{1p0} &= v_{c0} - v_{cs1} \\ z_{1p0} &= -u_{c0}\sin(\Sigma-\Gamma_S) + z_{c0}\cos(\Sigma-\Gamma_S) \\ v_{cs1} &= -E\tan(\Sigma-\Gamma_S)/\{\tan\Gamma_S+\tan(\Sigma-\Gamma_S)\} \\ \tan\varepsilon_{10} &= v_{1p0}/u_{1p0} \\ R_{10} &= \sqrt{(u_{1p0}^2+v_{1p0}^2)}\ \text{(Radius of the design point)} \\ q_{1p0} &= u_{1p0}\cos\chi_{10} + v_{1p0}\sin\chi_{10} \\ R_{b10} &= u_{1p0}\sin\chi_{10} - v_{1p0}\cos\chi_{10} = R_{10}\cos(\phi_{10}+\varepsilon_{10}) \\ \chi_{10} &= \pi/2 - \phi_{10} \end{aligned}\right\} \tag{3.2-8}$$

Substituting Eq. (3. 2-8) for Eq. (3. 1-5), the path of contact g_0 is obtained in the coordinate system C_1.

(3) Equations of path of contact $\mathbf{g}_0$ in the coordinate system C_s

Transforming the design point P_0 $(u_{c0}, v_{c0}, z_{c0}; C_s)$ into the coordinate system C_{qs} by Eq. (2.1-13), $P_0(q_{c0}, -R_{bc0}, z_{c0}; C_{qs})$ is obtained as follows.

$$
\begin{aligned}
q_{c0} &= u_{c0}\cos\chi_{s0} + v_{c0}\sin\chi_{s0} \\
R_{bc0} &= u_{c0}\sin\chi_{s0} - v_{c0}\cos\chi_{s0} \\
z_{c0} &= z_{c0}
\end{aligned}
$$

The inclination angle $\mathbf{g}_0(\phi_{s0}, \phi_{sw0}; C_s)$ of $\mathbf{g}_0$ is given by Eq. (3.2-3) or (3.2-6). Therefore, because a normal displacement on $\mathbf{g}_0$ is expressed by $R_{b20}\theta_2 \cos\phi_{b20}$ $(= R_{b10}\theta_1 \cos\phi_{b10})$, the path of contact and the inclination angle of $\mathbf{g}_0$ are obtained in the coordinate systems C_s and C_{qs} as follows.

$$
\left.
\begin{aligned}
q_c(\theta_2) &= R_{b20}\theta_2\cos\phi_{b20}\cos\phi_{sw0} + q_{c0} \\
u_c(\theta_2) &= q_c(\theta_2)\cos\chi_{s0} + R_{bc0}\sin\chi_{s0} \\
v_c(\theta_2) &= q_c(\theta_2)\sin\chi_{s0} - R_{bc0}\cos\chi_{s0} \\
z_c(\theta_2) &= R_{b20}\theta_2\cos\phi_{b20}\sin\phi_{sw0} + z_{c0} \\
\mathbf{g}_0(\phi_{s0} &= \pi/2 - \chi_{s0}, \ \phi_{sw0}; C_s)
\end{aligned}
\right\} \qquad (3.2\text{-}9)
$$

3. 2. 5 Geometric characteristics in the vicinity of the path of contact $\mathbf{g}_0$

Because the path of contact $\mathbf{g}_0$ satisfies Eq. (3.1-2), the variables are obtained as follows.

$$
\left.
\begin{aligned}
& d\chi_2/d\theta_2 = dR_{b2}/d\theta_2 = 0, \quad \chi_2 = \chi_{20}, \ R_{b2} = R_{b20}, \ \eta_{b2} = \phi_{b2} = \phi_{b20} \\
& d\chi_1/d\theta_1 = dR_{b1}/d\theta_1 = 0, \quad \chi_1 = \chi_{10}, \ R_{b1} = R_{b10}, \ \eta_{b1} = \phi_{b1} = \phi_{b10}
\end{aligned}
\right\} \qquad (3.2\text{-}10)
$$

Therefore, the equations obtained in 2.4 are simplified as follows.

(1) Infinitesimal surface of action

The infinitesimal surface of action S_W in the vicinity of the path of contact $\mathbf{g}_0$ on which the generating and generated gears engage is a plane which is formed by moving $\mathbf{g}_0$ in parallel to the instantaneous axis. Therefore, all the common contact normals on the S_W are parallel to $\mathbf{g}_0$.

(2) Limit of action

(a) When the generating axis coincides with the mating gear axis

Substituting Eq. (3.2-10) for Eqs. (2.4-3) and (2.4-5), the equations of limit of action are obtained as follows.

$$
\left.
\begin{aligned}
q_{2c}(\theta_2) + dR_{b2z}/d\theta_2 = 0 \\
q_{1c}(\theta_1) + dR_{b1z}/d\theta_1 = 0
\end{aligned}
\right\} \qquad (3.2\text{-}11)
$$

(b) When the generating gear axis is parallel to the generated gear axis

Substituting Eq. (3.2-10) for Eqs. (2.4-6) and (2.4-7), the equations of limit of action are obtained as follows.

$$\left.\begin{aligned} &q_{2c}(\theta_2) = 0 \\ &q_{1c}(\theta_1) = 0 \end{aligned}\right\} \qquad (3.2\text{-}12)$$

The point on $\mathbf{g}_0$ which satisfies Eq. (3.2-11) or (3.2-12) is the limit of action.

(3) Radii of curvature of infinitesimal tooth surfaces

(a) The inclination angles η_{z1w}, η_{z2w}, η_w, η_{z1G} and η_{z2G}

Substituting Eq. (3.2-10) for Eqs. (2.4-8), (2.4-9), (2.4-10), (2.4-11) and (2.4-12),

$$\left.\begin{aligned} \tan\eta_{z2w} &= (dR_{b2z}/d\theta_2)/(R_{b20}\sin\phi_{b20}) \\ \tan\eta_{z1w} &= (dR_{b1z}/d\theta_1)/(R_{b10}\sin\phi_{b10}) \\ \eta_w &= \eta_{z1w} - \eta_{z2w} \\ \tan\eta_{z2G} &= -q_{2c}/(R_{b20}\sin\phi_{b20}) \\ \tan\eta_{z1G} &= -q_{1c}/(R_{b10}\sin\phi_{b10}) \end{aligned}\right\} \qquad (3.2\text{-}13)$$

where η_{z1w} and η_{z2w} : angles of the line of contact and the planes G_1 and G_2 respectively on the tangent plane,

η_{z1G} : angle of tooth profile I and the plane G_1 on the tangent plane,

η_{z2G} : angle of tooth profile II and the plane G_2 on the tangent plane.

(b) Radii of curvature of infinitesimal tooth surfaces

Substituting Eq. (3.2-10) for Eqs. (2.4-13), (2.4-14) and (2.4-15),

$$\left.\begin{aligned} \rho_{z2} &= q_{2c}(\theta_2) + dR_{b2z}/d\theta_2 \\ &\quad \text{(when a generating gear axis coincides with the mating gear axis)} \\ &= q_{2c}(\theta_2) \\ &\quad \text{(when a generating gear axis is parallel to the generated gear axis)} \\ \rho_{nz2} &= \rho_{z2}/\cos\phi_{b20} \\ \rho_{n2} &= \rho_{nz2}\cos^2\eta_{z2w} \qquad \text{(perpendicular to the line of contact)} \end{aligned}\right\} \qquad (3.2\text{-}14)$$

$$\left.\begin{aligned} \rho_{z1} &= q_{1c}(\theta_1) + dR_{b1z}/d\theta_1 \\ &\quad \text{(when a generating gear axis coincides with the mating gear axis)} \\ &= q_{1c}(\theta_1) \\ &\quad \text{(when a generating gear axis is parallel to the generated gear axis)} \\ \rho_{nz1} &= \rho_{z1}/\cos\phi_{b10} \\ \rho_{n1} &= \rho_{nz1}\cos^2\eta_{z1w} \qquad \text{(perpendicular to the line of contact)} \end{aligned}\right\} \qquad (3.2\text{-}15)$$

$dR_{b1z}/d\theta_1$ and $dR_{b2z}/d\theta_2$ will be obtained in 3.3.7.

(4) Conditions of contact between infinitesimal tooth surfaces I and II

The radius of curvature in this theory is positive in the positive direction of the common contact normal **n** (= $\mathbf{g}_0$), so that conditions of contact between infinitesimal tooth surfaces I and II are as follows.

$$\left.\begin{array}{ll}\rho_{nz1}\rho_{nz2}<0\ ; & \text{convex-convex contact}\\ \rho_{nz1}\rho_{nz2}>0\ ; & \text{convex-concave contact}\end{array}\right\} \qquad (3.2\text{-}16)$$

3. 3 Limit of action and selection of path of contact [33]

When the design point P_0 and the path of contact $\mathbf{g}_0$ are given as in section 3.2, it is discussed how the inclination angle $\mathbf{g}_0(\phi_0, \phi_{n0}\ ;\ C_S)$ of $\mathbf{g}_0$ is selected in order to realize $\mathbf{g}_0$ which has a smooth tooth surface around P_0. ϕ_0 is already given with the selection of P_0, so that it is discussed how ϕ_{n0} is selected to the given ϕ_0 in this section.

3. 3. 1 Path of contact which has a limit of action at design point P_0

For a designed path of contact $\mathbf{g}_0$ to have a smooth tooth surface around P_0, it is sufficient that P_0 is not a limit of action. Therefore, $\mathbf{g}_0$ is necessary to incline adequately from the path of contact which has a limit of action at P_0, which is called a limit path of contact.

Figure 3.3-1 shows the relation among the infinitesimal tooth surface approximated by its tangent plane W_0, the path of contact $\mathbf{g}_0$, the surface of action S_{W0}, the plane of rotation Z_{20}, the plane of action G_{20} and the plane S_n at P_0, where the three planes S_{W0}, S_n and G_{20} intersect along $\mathbf{g}_0$ (intersection P_0P_{GSWn}). Because there are innumerable tooth surfaces having the tangent plane W_0 according to selected generating gear axes, two typical cases of generating gear axes are selected to obtain the limit path of contact.

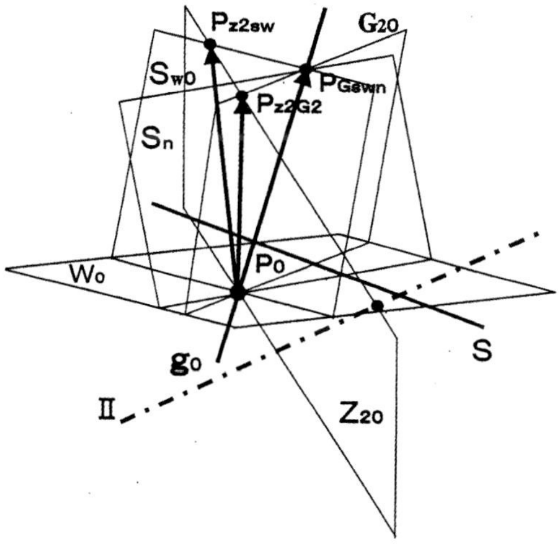

Fig.3.3-1 Relationship among planes W_0, G_{20}, S_{w0}, S_n and Z_{20} concerning $\mathbf{g}_0$

(1) When a generating gear axis coincides with the mating gear axis I (line contact)

The surface of action S_{W0} which is made by an infinitesimal tooth surface (its tangent plane W_0) of a generating gear shown in 2.4 is a plane which is formed by moving $\mathbf{g}_0$ in parallel to the instantaneous axis S because the ratio of angular velocity is constant. The limit plane of action S_{W2S} is defined as the plane which is made by inclining S_{W0} until a cylinder around the axis II makes contact at P_0 on the intersection P_0P_{Z2SW} of S_{W0} and the plane of rotation Z_{20} of gear II. The intersection of S_{W2S} and S_n is defined as the limit path of contact $\mathbf{g}_{2S}$, which is the path of contact having the limit of action at P_0 in the case of conjugate tooth surfaces.

(2) When a generating gear axis is parallel to the gear axis II (point contact)

The surface of action S_{W0} which is made by an infinitesimal tooth surface (its tangent plane W_0) of a generating gear becomes the plane of action G_{20}. The limit plane of action G_{2k} is defined as the plane which is made by inclining G_{20} until a cylinder around the axis II makes contact at P_0 on the intersection P_0P_{Z2G2} of S_{W0} and the plane of rotation Z_{20} of gear II. The intersection of G_{2k} and S_n is defined as the limit path of contact $\mathbf{g}_{2k}$, which is the path of contact having the limit of action at P_0 in the case of crossed helical and conical gears with involute helicoids for gear II.

In the same way, the limit surface of action S_{W1S}, the limit plane of action G_{1k} and the limit paths of contact $\mathbf{g}_{1S}$ and $\mathbf{g}_{1k}$ of gear I which have a limit of action at P_0 are defined.

In this theory, a pair of tooth surfaces which realize the path of contact $\mathbf{g}_0$ are limited in the case of either (1) or (2) and the limit paths of contact $\mathbf{g}_{1S}$, $\mathbf{g}_{2S}$, $\mathbf{g}_{1k}$ and $\mathbf{g}_{2k}$ are obtained as follows.

3. 3. 2 Limit paths of contact $\mathbf{g}_{1S}$ and $\mathbf{g}_{2S}$

(1) When the plane S_n does not include the common perpendicular (bevel and hypoid gears)

Figure 3.3-2 shows the relation among the intersection $\mathbf{g}_0$ ($= P_0P_{GSWn}$) of the plane S_n and the surface of action S_{W0}, the intersection $\mathbf{h}_{2Z}$ ($= P_0P_{Z2SW}$) of S_{W0} and the plane of rotation Z_{20} and the intersection $\mathbf{h}_{1Z}$ ($= P_0P_{Z1SW}$) of S_{W0} and the plane of rotation Z_{10}, where the points P_{Z1SW}, P_{GSWn} and P_{Z2SW} are on the intersection of the plane $v_{2c} = 0$ (shown by II − II) and the surface of action S_{W0}. The components of $\mathbf{h}_{2Z}$ in the direction of each axis of the coordinate system C_S are expressed by a directed segment $P_0P_{GSWn} = L_g$ as follows.

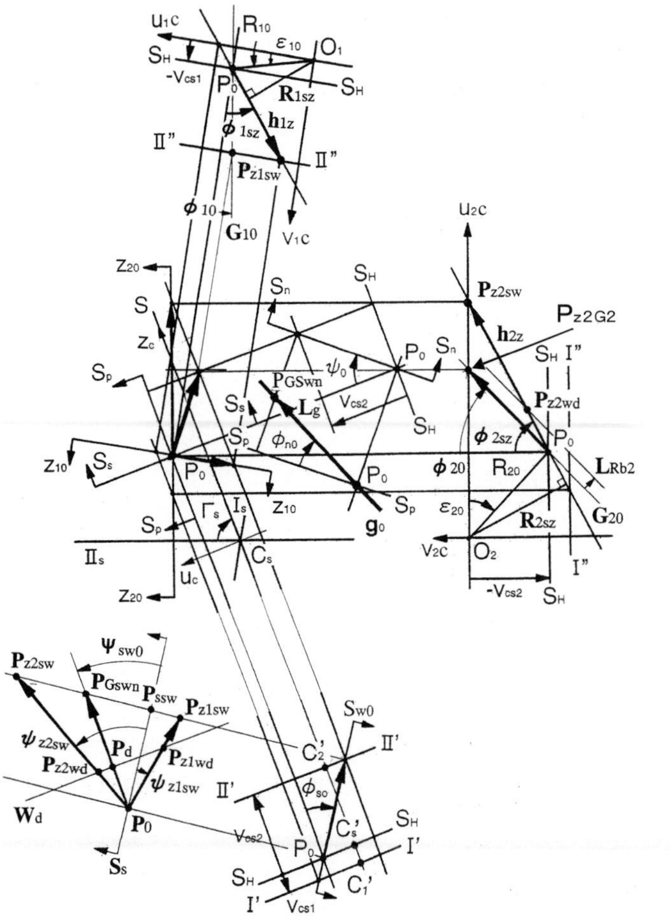

Fig.3.3-2 Relationship among paths of contact $\mathbf{g}_0$, $\mathbf{h}_{1Z}$ and $\mathbf{h}_{2Z}$, when S_n does not include common perpendicular

$L_{uc} = -L_g \sin\phi_{n0}$ (component of L_g in the direction of the axis u_c)

$L_{vc} = L_g \cos\phi_{n0} \cos\psi_0$ (component of L_g in the direction of the axis v_c)

$L_{zc} = L_g \sin\phi_{n0} \tan\Gamma_S$ (component of L_g in the direction of the axis z_c)

Transforming them into the coordinate system C_2, the inclination angle h_{2Z} (ϕ_{2sZ}, 0 ; C_2) of $\mathbf{h}_{2Z}$ is obtained as follows.

$$L_{u2c} = -L_{uc}\cos\Gamma_S + L_{zc}\sin\Gamma_S = L_g(\sin\phi_{n0}\cos\Gamma_S + \sin\phi_{n0}\tan\Gamma_S\sin\Gamma_S)$$
$$L_{v2c} = L_{vc} = L_g\cos\phi_{n0}\cos\psi_0$$
$$L_{z2c} = 0$$
$$\tan\phi_{2SZ} = L_{u2c}/L_{v2c} = \tan\phi_{n0}/(\cos\psi_0\cos\Gamma_S) \qquad (3.3\text{-}1)$$

The limit of action of $\mathbf{h}_{2Z}$ is obtained as a point tangent to a cylinder around the gear axis II as shown in 2.4, so that the radius R_{2SZ} of the cylinder tangent to $\mathbf{h}_{2Z}$ is obtained as follows, where $R_{20} = O_2P_0$ (Fig. 3.3-2).

$$R_{2SZ} = R_{20}\cos(\phi_{2SZ} + \varepsilon_{20})$$

When the design point P_0 ($u_{2p0} \neq 0$, $-v_{cs2}$, z_{2p0}; C_2) is the limit of action, the inclination angle g_{2S} (ψ_0, ϕ_{n2S}; C_S) of the limit path of contact $\mathbf{g}_{2S}$ is obtained as follows because of $R_{2SZ} = R_{20}$.

$$\left.\begin{aligned}\phi_{2SZ} &= -\varepsilon_{20} = -\tan^{-1}(-v_{cs2}/u_{2p0})\\ \tan\phi_{n2S} &= -\tan\varepsilon_{20}\cos\psi_0\cos\Gamma_S\end{aligned}\right\} \qquad (3.3\text{-}2)$$

In the same way, h_{1Z} (ϕ_{1SZ}, 0 ; C_1) and g_{1S} (ψ_0, ϕ_{n1S} ; C_S) of $\mathbf{g}_{1S}$ are obtained as follows.

$$\left.\begin{aligned}\tan\phi_{1SZ} &= -\tan\phi_{n0}/\{\cos\psi_0\cos(\Sigma - \Gamma_S)\}\\ \tan\phi_{n1S} &= \tan\varepsilon_{10}\cos\psi_0\cos(\Sigma - \Gamma_S)\end{aligned}\right\} \qquad (3.3\text{-}3)$$

In bevel gears, substituting $\varepsilon_{10} = \varepsilon_{20} = 0$ for Eqs. (3.3-2) and (3.3-3), $\phi_{n2S} = \phi_{n1S} = 0$ are obtained.

(2) When the plane S_n includes the common perpendicular (crossed helical, worm and cylindrical gears)

Figure 3.3-3 shows the case in which the plane S_n includes the common perpendicular, where only the definitions of the plane S_n and ψ_0 are different from those in Fig. 3.3-2. The components of $\mathbf{h}_{2Z}$ in the direction of each axis of the coordinate system C_S are expressed by a directed segment $P_0P_{GSWn} = L_g$ as follows.

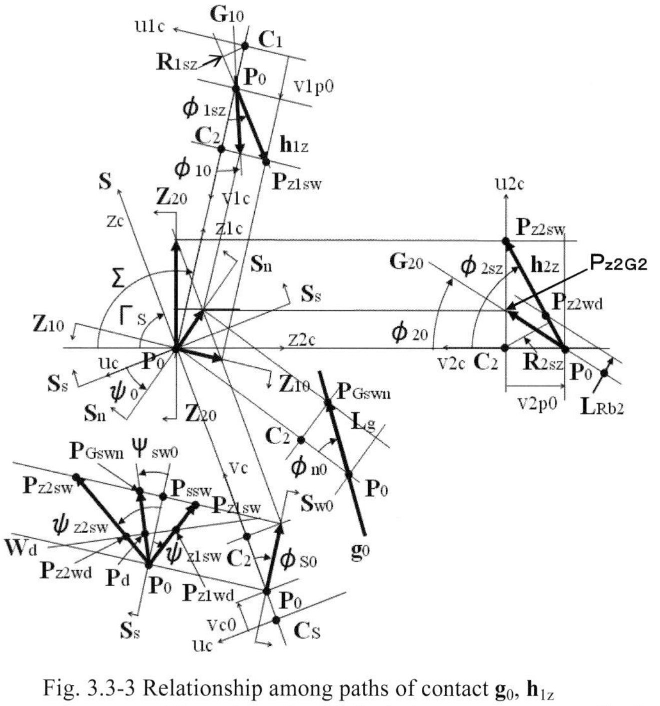

Fig. 3.3-3 Relationship among paths of contact $\mathbf{g}_0$, $\mathbf{h}_{1z}$ and $\mathbf{h}_{2z}$, when plane S_n includes common perpendicular

$L_{uc} = -L_g \sin\phi_{n0} \cos\varphi_0$ (component of $\mathbf{L}_g$ in the direction of the axis u_c)
$L_{vc} = L_g \cos\phi_{n0}$ (component of $\mathbf{L}_g$ in the direction of the axis v_c)
$L_{zc} = L_g \sin\phi_{n0} \cos\varphi_0 \tan\Gamma_s$ (component of $\mathbf{L}_g$ in the direction of the axis z_c)

Transforming them into the coordinate system C_2, the inclination angle h_{2Z} (ϕ_{2SZ}, 0 ; C_2) of $\mathbf{h}_{2Z}$ is obtained as follows.

$$L_{u2c} = -L_{uc}\cos\Gamma_S + L_{zc}\sin\Gamma_S = L_g \sin\phi_{n0}\cos\varphi_0(\cos\Gamma_S + \tan\Gamma_S \sin\Gamma_S)$$
$$L_{v2c} = L_{vc} = L_g\cos\phi_{n0}$$
$$L_{z2c} = 0$$
$$\tan\phi_{2SZ} = L_{u2c}/L_{v2c} = \tan\phi_{n0}\cos\varphi_0/\cos\Gamma_S \qquad (3.3\text{-}4)$$

Because the design point P_0 ($0, v_{2p0} \neq 0, 0$; C_2) is on the common perpendicular, substituting $\phi_{2SZ} = -\varepsilon_{20} = \pi/2$ for Eq. (3.3-4), the limit path of contact $\mathbf{g}_{2S}$ (φ_0, ϕ_{n2S} ; C_S) is obtained as follows.

$$\phi_{n2S} = \pm\pi/2 \quad (+ \text{ for } \Gamma_S \leqq \pi/2,\ - \text{ for } \Gamma_S > \pi/2) \qquad (3.3\text{-}5)$$

The inclination angle h_{1Z} (ϕ_{1SZ}, 0 ; C_1) of $\mathbf{h}_{1Z}$ is obtained in the same way.

$$\tan\phi_{1SZ} = -\tan\phi_{n0}\cos\varphi_0/\cos(\Sigma - \Gamma_S) \qquad (3.3\text{-}6)$$

Because of P_0 (0, $v_{1p0} \neq 0$, 0 ; C_1), substituting $\phi_{1SZ} = -\varepsilon_{10} = \pm\pi/2$ ($-$ for $v_{1p0} > 0$, $+$ for $v_{1p0} < 0$) for Eq. (3.3-6), $\mathbf{g}_{1S}$ (φ_0, ϕ_{n1S} ; C_S) is obtained as follows.

$$\phi_{n1S} = \pm\pi/2 \quad (+ \text{ for } v_{1p0}\cos(\Sigma - \Gamma_S) \geqq 0,\ - \text{ for } v_{1p0}\cos(\Sigma - \Gamma_S) < 0) \qquad (3.3\text{-}7)$$

3.3.3 Limit paths of contact $\mathbf{g}_{1k}$ and $\mathbf{g}_{2k}$

(1) When the plane S_n does not include the common perpendicular (bevel and hypoid gears)

The inclination angle ϕ_{2SZ} of P_0P_{Z2G2} ($= \mathbf{h}_{2Z}$) shown in Fig. 3.3-2 is ϕ_{20} of the plane G_{20}, therefore the relation of the inclination angles of ϕ_{2SZ} and ϕ_{n0} of $\mathbf{g}_0$ on the plane S_n are already given by Eq. (3.2-3) as follows.

$$\tan\phi_{2SZ} = \tan\phi_{20} = \tan\phi_{n0}\cos\Gamma_S/\cos\varphi_0 + \tan\varphi_0\sin\Gamma_S$$

The radius of the cylinder tangent to $\mathbf{h}_{2Z}$ (radius of a limit of action), namely $R_{2SZ} = R_{b2Z} = R_{b20}$ is obtained as follows.

$$R_{2SZ} = R_{b2Z} = R_{b20} = R_{20}\cos(\phi_{2SZ} + \varepsilon_{20})$$

Because the design point P_0 ($u_{2p0} \neq 0$, $-v_{cs2}$, z_{2p0} ; C_2) is the limit of action, the limit path of contact $\mathbf{g}_{2k}$ (φ_0, ϕ_{n2k} ; C_S) is obtained as follows because of $R_{2SZ} = R_{20}$.

$$\left.\begin{aligned}\phi_{2sz} &= -\varepsilon_{20} = -\tan^{-1}(-v_{cs2}/u_{2p0})\\ \tan\phi_{n2k} &= -(\tan\varepsilon_{20}+\tan\phi_0\sin\Gamma_s)\cos\phi_0/\cos\Gamma_s\end{aligned}\right\} \quad (3.3\text{-}8)$$

In the same way, the inclination angle $\mathbf{g}_{1k}$ (ϕ_0, ϕ_{n1k} ; C_s) of $\mathbf{g}_{1k}$ is obtained as follows.

$$\left.\begin{aligned}\tan\phi_{1sz} &= \tan\phi_{10} = -\tan\phi_{n0}\cos(\Sigma-\Gamma_s)/\cos\phi_0+\tan\phi_0\sin(\Sigma-\Gamma_s)\\ \tan\phi_{n1k} &= \{\tan\varepsilon_{10}+\tan\phi_0\sin(\Sigma-\Gamma_s)\}\cos\phi_0/\cos(\Sigma-\Gamma_s)\\ \text{where } \tan\varepsilon_{10} &= -v_{cs1}/u_{1p0} \text{ at } P_0(u_{1p0}\neq 0, -v_{cs1}, z_{1p0}; C_1)\end{aligned}\right\} \quad (3.3\text{-}9)$$

In bevel gears, they are obtained by substituting $\varepsilon_{10}=\varepsilon_{20}=0$ for Eqs. (3.3-8) and (3.3-9).

(2) When the plane S_n includes the common perpendicular (crossed helical, worm and cylindrical gears)

The inclination angle ϕ_{2sz} of P_0P_{z2G2} ($=\mathbf{h}_{2z}$) shown in Fig. 3.3-3 is ϕ_{20} of the plane G_{20}, therefore the relation of the inclination angles of ϕ_{2sz} and ϕ_{n0} of $\mathbf{g}_0$ on the plane S_n are already given by Eq. (3.2-4) as follows. In the same way, ϕ_{1sz} is given by Eq. (3.2-5).

$$\left.\begin{aligned}\tan\phi_{2sz} &= \tan\phi_{20} = \tan\phi_{n0}\cos(\Gamma_s-\phi_0)\\ \tan\phi_{1sz} &= \tan\phi_{10} = -\tan\phi_{n0}\cos(\Sigma-\Gamma_s+\phi_0)\end{aligned}\right\} \quad (3.3\text{-}10)$$

Because the design point P_0 (0, $v_{2p0}\neq 0$, 0 ; C_2) is the limit of action, the limit path of contact $\mathbf{g}_{2k}$ (ϕ_0, ϕ_{n2k} ; C_s) is obtained by substituting $\phi_{2sz}=-\varepsilon_{20}=\pi/2$ for Eq. (3.3-10) as follows.

$$\phi_{n2k}=\pm\pi/2 \quad (+ \text{ for } \Gamma_s-\phi_0\leqq\pi/2,\ - \text{ for } \Gamma_s-\phi_0>\pi/2) \quad (3.3\text{-}11)$$

In the same way, the inclination angle $\mathbf{g}_{1k}$ (ϕ_0, ϕ_{n1k} ; C_s) of $\mathbf{g}_{1k}$ is obtained as follows.

$$\left.\begin{aligned}\phi_{1sz} &= -\varepsilon_{10} = \pm\pi/2 \quad (- \text{ for } v_{1p0}>0,\ + \text{ for } v_{1p0}<0)\\ \phi_{n1k} &= \pm\pi/2 \ (+ \text{ for } v_{1p0}\cos(\Sigma-\Gamma_s+\phi_0)\geqq 0,\ - \text{ for } v_{1p0}\cos(\Sigma-\Gamma_s+\phi_0)<0)\end{aligned}\right\} \quad (3.3\text{-}12)$$

3．3．4 Schematic views of limit paths of contact $\mathbf{g}_{1s}$, $\mathbf{g}_{2s}$, $\mathbf{g}_{1k}$ and $\mathbf{g}_{2k}$

Figures from 3.3-4 to 3.3-8 shows schematically the limit surfaces of action S_{w1s} and S_{w2s}, the limit planes of action G_{1k} and G_{2k} and the limit paths of contact $\mathbf{g}_{1s}$, $\mathbf{g}_{2s}$, $\mathbf{g}_{1k}$ and $\mathbf{g}_{2k}$ with the planes $S_H(v_c=0)$, $S_p(u_c=u_{c0})$, $S_s(z_c=z_{c0})$ and S_n according to the kinds of gears.

(1) Hypoid gears (Fig. 3.3-4)

The design point P_0 is chosen on the plane S_H and the plane S_n is determined by the relative velocity V_{rs0}. The limit surfaces of action S_{W1S} and S_{W2S} (nearly equal in this example) and the limit planes of action G_{1k} and G_{2k} are determined as different planes and their intersections with the plane S_n are the limit paths of contact $\mathbf{g}_{1S}$, $\mathbf{g}_{2S}$, $\mathbf{g}_{1k}$ and $\mathbf{g}_{2k}$ drawn as different directed lines.

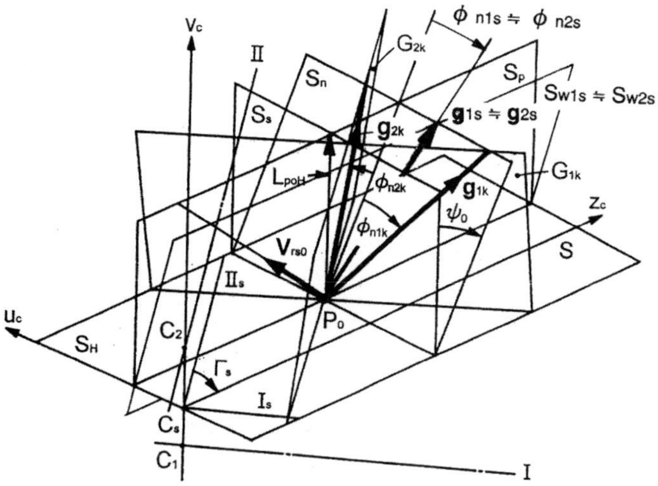

Fig.3.3-4 Limit paths of contact g_{1s}, g_{2s}, g_{1k} and g_{2k} of hypoid gears

Table 3.3-1 Dimensions of hypoid gears

Shaft angle	Σ(°)	90
Ratio of angular velocities	i_0	4.1
Offset (below center)	E(mm)	35
Reference radius of gear	R_{20}(mm)	95

Figure 3.3-5 shows the inclination angles ϕ_{n1S}, ϕ_{n2S}, ϕ_{n1k} and ϕ_{n2k} on the plane S_n of the limit paths of contact g_{1S}, g_{2S}, g_{1k} and g_{2k} of hypoid gears whose dimensions are shown in Table 3.3-1. In this example, the absolute values and the difference of ϕ_{n1S} and ϕ_{n2S} are comparatively small because the shaft angle is 90° and E/R_{20} is also comparatively small. The amount of variation of ϕ_{n1k} is smaller than that of ϕ_{n2k} because $i_0=4.1$ is comparatively large. There is the limit path of contact g_k (19°, 8°; C_S) where $g_{1k} = g_{2k}$ is satisfied, which corresponds to L_{pOB} (0, 0; C_S) in bevel gears as shown in Fig. 3.3-6.

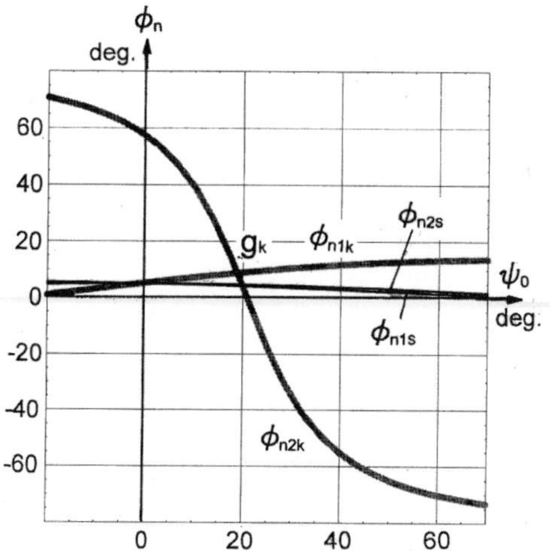

Fig.3.3-5 Example of variation of inclination angles of ϕ_{n1s}, ϕ_{n2s}, ϕ_{n1k} and ϕ_{n2k} of hypoid gears

(2) Bevel gears (Fig. 3.3-6)

Figure 3.3-6 is the special case of Fig. 3.3-4 with the offset $E = 0$, namely $\varepsilon_{10} = \varepsilon_{20} = 0$. The limit surfaces of action S_{w1s} and S_{w2s} coincide with $S_p(u_c = 0)$. The limit planes of action G_{1k} and G_{2k} are perpendicular to the plane S_H and parallel to each gear axis, the intersection of which coincides with the intersection L_{pOB} of the planes S_p and S_S. Because of the relative velocity $V_{rs0} = 0$ at P_0, the plane S_n which is perpendicular to the plane S_p and inclines to the plane S_S is chosen as

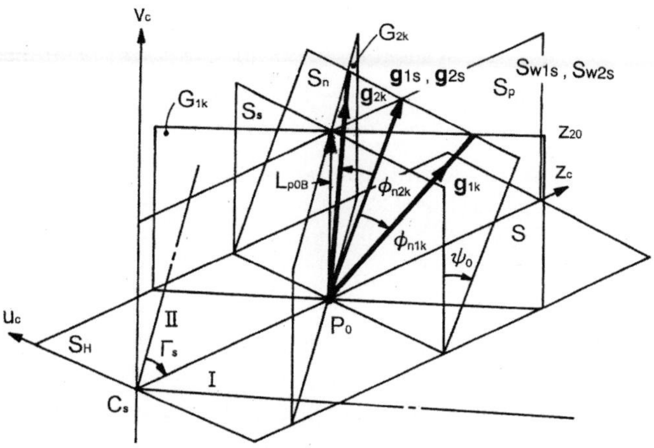

Fig.3.3-6 Limit paths of contact g_{1s}, g_{2s}, g_{1k} and g_{2k} of bevel gears

mentioned in 3.2.2(2). In spiral bevel gears ($\psi_0 \neq 0$), the limit paths of contact $\mathbf{g}_{1S}$ and $\mathbf{g}_{2S}$ become the intersection of the planes S_p and S_n and the limit paths of contact $\mathbf{g}_{1k}$ and $\mathbf{g}_{2k}$ are the intersections of the different planes of action G_{1k} and G_{2k} and the plane S_n. In straight bevel gears ($\psi_0 = 0$), the plane S_n coincides with the plane S_S, the planes G_{1k}, G_{2k}, S_n and S_p intersect along L_{p0B} and the limit paths of contact $\mathbf{g}_{1S} = \mathbf{g}_{2S} = \mathbf{g}_{1k} = \mathbf{g}_{2k} = L_{p0B}$ are obtained.

(3) Crossed helical and worm gears (Fig. 3.3-7)

Because the design point P_0 is on the common perpendicular, $\varepsilon_{10} = \pm\pi/2$ and $\varepsilon_{20} = -\pi/2$ are obtained, so that the limit surfaces of action S_{W1S} and S_{W2S} and the limit planes of action G_{1k} and G_{2k} become the same plane parallel to the plane S_H, which means that the base cylinders of gears I and II make point contact at P_0. The intersection of the planes $S_{W1S} = S_{W2S} = G_{1k} = G_{2k}$ and the plane S_n which is perpendicular to the relative velocity V_{rs0} at P_0 and includes the common perpendicular are the limit paths of contact $\mathbf{g}_{1S} = \mathbf{g}_{2S} = \mathbf{g}_{1k} = \mathbf{g}_{2k}$.

(4) Cylindrical gears (Fig. 3.3-8)

Because the design point P_0 is the intersection C_S of the common perpendicular and the plane S_H, the limit surfaces of action S_{W1S} and S_{W2S} and the limit planes of action G_{1k} and G_{2k} become the same plane S_H, which means that the base cylinders of gears I and II make contact along the instantaneous axis S. Because of the relative velocity $V_{rs0} = 0$ at P_0, as mentioned in 3.2.2(4), there is any plane S_n-helical including the common perpendicular for helical gears or any plane S_n-tapered for tapered gears which is perpendicular to the axial plane at P_0 and inclines to the plane of rotation. The intersection of the planes S_n and S_H ($= S_{W1S} = S_{W2S} = G_{1k} = G_{2k}$) are the limit paths of contact $\mathbf{g}_{1S} = \mathbf{g}_{2S} = \mathbf{g}_{1k} = \mathbf{g}_{2k}$.

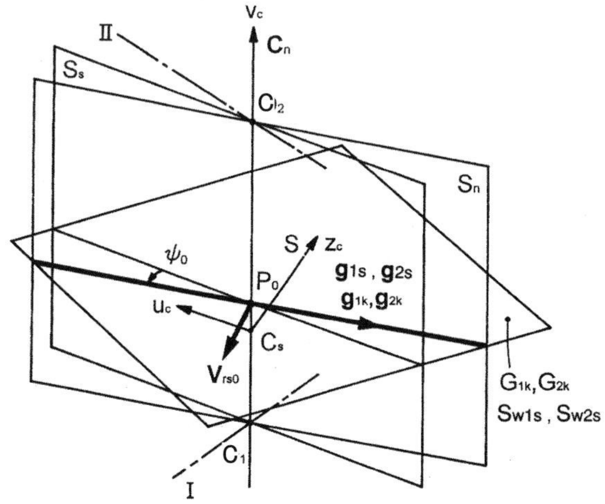

Fig.3.3-7 Limit paths of contact $\mathbf{g}_{1s}$, $\mathbf{g}_{2s}$, $\mathbf{g}_{1k}$ and $\mathbf{g}_{2k}$ of worm and crossed helical gears

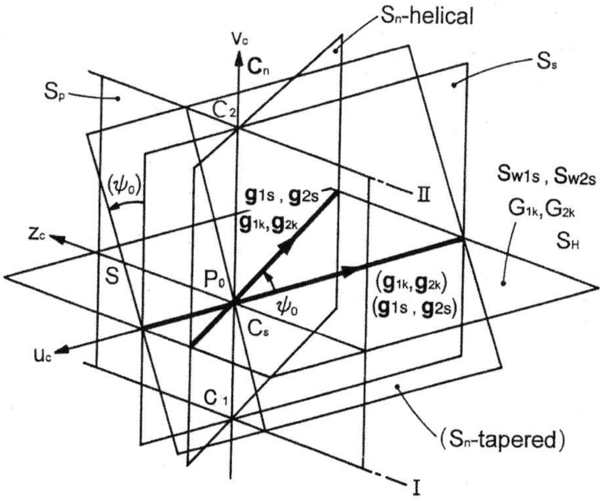

Fig.3.3-8 Limit paths of contact $\mathbf{g}_{1s}$, $\mathbf{g}_{2s}$, $\mathbf{g}_{1k}$ and $\mathbf{g}_{2k}$ of cylindrical gears

3. 3. 5 Selection of inclination angle ϕ_{n0} of path of contact $\mathbf{g}_0$ to have smooth tooth surfaces in the vicinity of design point P_0

(1) When a generating gear axis coincides with the mating gear axis (line contact)

Because the limit path of contact $\mathbf{g}_{1S}$ or $\mathbf{g}_{2S}$ is determined on the plane S_n, it is enough to give a path of contact $\mathbf{g}_0(\psi_0, \phi_{n0}; C_S)$ through P_0 as a directed line which is inclined adequately from $\mathbf{g}_{1S}$ or $\mathbf{g}_{2S}$ on the plane S_n. Therefore, when a tooth surface is given on gear I, $\mathbf{g}_0$ is inclined from $\mathbf{g}_{2S}$ because gear II has its conjugate tooth surface. Conversely, when a tooth surface is given on gear II, $\mathbf{g}_0$ is inclined from $\mathbf{g}_{1S}$ because gear I has its conjugate tooth surface. However, in cylindrical, crossed helical, worm and bevel gears, always $\mathbf{g}_{1S} = \mathbf{g}_{2S}$, while in hypoid gears with the shaft angle $= 90°$, the difference between $\mathbf{g}_{1S}$ and $\mathbf{g}_{2S}$ is not so large. To work out how much $\mathbf{g}_0$ should be inclined from $\mathbf{g}_{1S}$ or $\mathbf{g}_{2S}$, it is necessary to determine the amount individually by considering the limit line of action on the surface of action made by a given tooth surface, because a gear pair must have a sufficient tooth depth and facewidth besides a smooth tooth surface in the vicinity of P_0.

(2) When a generating gear axis is parallel to the generated gear axis (point contact)

Because the limit path of contact $\mathbf{g}_{1k}$ or $\mathbf{g}_{2k}$ is determined on the plane S_n, it is enough to give a path of contact $\mathbf{g}_0(\psi_0, \phi_{n0}; C_S)$ through P_0 as a directed line which is inclined adequately from $\mathbf{g}_{1k}$ or $\mathbf{g}_{2k}$ on the plane S_n. While in cylindrical, crossed helical and worm gears, always $\mathbf{g}_{1k} = \mathbf{g}_{2k}$, however, in bevel and hypoid gears, $\mathbf{g}_{1k} \neq \mathbf{g}_{2k}$ and their variation are large as shown in Fig. 3. 3-5, so that it depends on the individual case whether $\mathbf{g}_0$ should be inclined from $\mathbf{g}_{1k}$ or $\mathbf{g}_{2k}$. In gear pairs ($i_0 \geqq 1$) dealt with in this theory, the limit of action of gear I is nearer to the design point P_0 than that of gear II, therefore $\mathbf{g}_0$ should be inclined from $\mathbf{g}_{1k}$ and the sufficient inclination of $\mathbf{g}_0$ from $\mathbf{g}_{2k}$ should be made sure. How much $\mathbf{g}_0$ should be inclined from $\mathbf{g}_{1k}$ or $\mathbf{g}_{2k}$ depends on how much allowance from the limit of action is given to the tooth profiles.

By selecting ϕ_{n0} as mentioned above, a path of contact $\mathbf{g}_0(\psi_0, \phi_{n0}; C_S)$ which has smooth tooth surfaces around the design point P_0 $(u_{c0}, v_{c0}, z_{c0}; C_S)$ is determined.

3. 3. 6 Limit paths of contact and Wildhaber's limit normal

When a path of contact (Eq. (2. 3-2)) and its surface of action made by a generating gear are given, the necessary and sufficient condition for a point P on the given path of contact to be a limit of action on the generated tooth surface II is that Eq. (2. 4-3) is satisfied on the path of contact $\mathbf{h}_{2Z}$ which is on the plane of rotation through P.

$$q_{2c}(1-d\chi_2/d\theta_2)+dR_{b2Z}/d\theta_2 = 0 \qquad (2.4\text{-}3)$$

In Fig. 3. 3-9, examples of the solution of Eq. (2. 4-3) are shown schematically. Using that Eq. (2. 4-3) is the condition for $\mathbf{h}_{2Z}$ to make contact with a cylinder around gear axis II at P, the limit paths of contact $\mathbf{g}_{2S}$ and $\mathbf{g}_{2k}$ are obtaind. $\mathbf{g}_{2S}$, $\mathbf{g}_{2k}$ and Wildhaber's limit normal are the solutions of Eq. (2. 4-3) for the design point P_0 to be a limit of action on the generated tooth surface II when $\mathbf{h}_{2Z}$ with its common contact normal has special conditions.

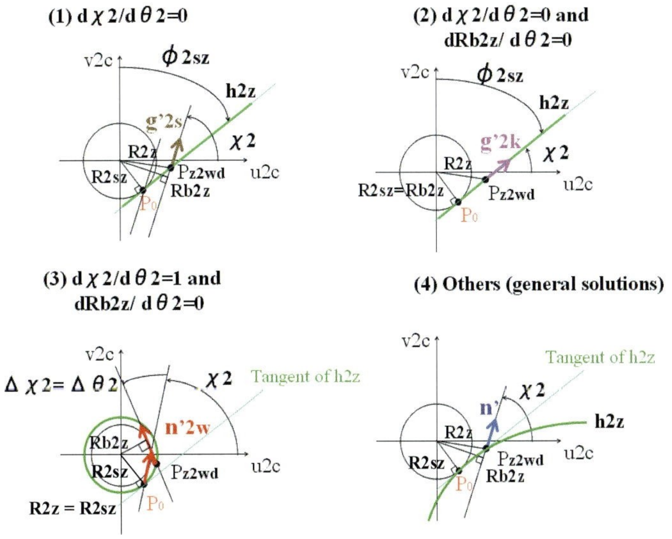

Fig.3.3-9 Examples of solution at P_0 of $q_{2c}(1-d\chi_2/d\theta_2)+dR_{b2z}/d\theta_2=0$

(1) When common contact normal of h_{2Z} satisfies $d\chi_2/d\theta_2=0$ (g_{2S})

When the path of contact g_0 is given, the infinitesimal tooth surface in the vicinity of g_0 is approximated by the tangent plane and a generating gear axis coincides with the mating gear axis, then the surface of action becomes the plane made by moving g_0 in parallel to the instantaneous axis. Therefore, the intersection h_{2Z} with the plane of rotation is a straight line, the common contact normal g_0 satisfies $d\chi_2/d\theta_2=0$ and the path of contact g_0 which has h_{2Z} that satisfies Eq. (2.4-3) ($q_{2c}+dR_{b2Z}/d\theta_2=0$) at P_0 is the limit path of contact g_{2S} (Fig. 3.3-9(1)).

(2) When common contact normal of h_{2Z} satisfies $d\chi_2/d\theta_2=0$ and $dR_{b2Z}/d\theta_2=0$ (g_{2k})

When the path of contact g_0 is given, the infinitesimal tooth surface in the vicinity of g_0 is approximated by the tangent plane and a generating gear axis is parallel to the generated gear axis, then the surface of action becomes the plane of action made by moving g_0 in parallel to the gear axis II. Therefore, the intersection h_{2Z} with the plane of rotation is a straight line, the common contact normal g_0 satisfies the above conditions and the path of contact g_0 which has h_{2Z} that satisfies Eq. (2.4-3) ($q_{2c}=0$) at P_0 is the limit path of contact g_{2k} (Fig. 3.3-9(2)).

(3) When common contact normal of h_{2Z} satisfies $d\chi_2/d\theta_2=1$ and $dR_{b2Z}/d\theta_2=0$ (Wildhaber's limit normal)

Because the same point on tooth surface II is assumed to make contact during an infinitesimal angle of rotation, the path of contact h_{2Z} becomes a circular arc around the axis II, its common contact normal is always tangent to the same cylinder around the axis II ($dR_{b2Z}/d\theta_2=0$) and the amount of variation $\Delta\chi_2$ equals the infinitesimal angle $\Delta\theta_2$ of rotation ($d\chi_2/d\theta_2=1$). Therefore, Eq. (2.4-3) is always valid for any q_{2c} (Fig. 3.3-9(3)). Using the above conditions, the variation $\Delta\chi_2$ of the inclination angle of common contact normal can be obtained easily from $\Delta\theta_2$, so that the common contact normal (limit normal n_{2W}) and its inclination angle (limit pressure angle) are obtained at P_0 from the fundamental requirement for contact [37], [38].

(4) General solutions

Because Eq. (2.4-3) has innumerable solutions, it is easier to obtain numerically a point on h_{2Z} tangent to a cylinder around the axis II (Fig. 3.3-9(4)).

The limit paths of contact in this theory and Wildhaber's limit normal are ones of the solutions of Eq. (2.4-3) in which different conditions are given to $\mathbf{h}_{2Z}$ and the common contact normal at each point. Therefore, Wildhaber's limit normal is not considered to be an inherent value of hypoid gears which indicates the limit of action (the state of pressure angle=zero).

A limit of action of tooth surface I is obtained in the same way.

3. 3. 7 Differential coefficients of radius of base cylinder $dR_{b1Z}/d\theta_1$ of $\mathbf{h}_{1Z}$ and $dR_{b2Z}/d\theta_2$ of $\mathbf{h}_{2Z}$ when a path of contact $\mathbf{g}_0$ is given

When a path of contact $\mathbf{g}_0$ is given, $dR_{b1Z}/d\theta_1$ and $dR_{b2Z}/d\theta_2$ are obtained as follows.

(1) When a generating gear axis coincides with the mating gear axis

(a) When the plane S_n does not include the common perpendicular (bevel and hypoid gears)

Figure 3.3-2 shows a point of contact P_d in the vicinity of P_0 on $\mathbf{g}_0$ and the intersection of the tangent plane W_d and the surface of action S_{W0} through P_d, where the intersection of the plane $v_{2c}=0$ and the plane S_{W0} intersects the plane S_S at P_{SSW}. Figure 3.3-2 is shown in the case of $\phi_{n0} \geqq 0$ ($\phi_{S0} \leqq 0$). The inclination angle g_0 (ϕ_{S0}, ϕ_{SW0} ; C_S) of $\mathbf{g}_0$ on the plane S_{W0} is given by Eq. (3.2-3) as follows.

$$\left.\begin{aligned} \tan\phi_{S0} &= -\tan\phi_{n0}/\cos\phi_0 \\ \sin\phi_{SW0} &= \cos\phi_{n0}\sin\phi_0 \end{aligned}\right\} \qquad (3.3\text{-}13)$$

The inclination angles h_{1Z} (ϕ_{1SZ}, 0 ; C_1) of $\mathbf{h}_{1Z}$ (P_0P_{Z1SW}) and h_{2Z} (ϕ_{2SZ}, 0 ; C_2) of $\mathbf{h}_{2Z}$ (P_0P_{Z2SW}) are given by Eqs. (3.3-3) and (3.3-1) respectively. Therefore, transforming them into the coordinate system C_S through Eqs. (2.1-16) and (2.1-15), the inclination angles h_{1Z} (ϕ_{S0}, ϕ_{Z1SW} ; C_S) and h_{2Z} (ϕ_{S0}, ϕ_{Z2SW} ; C_S) are obtained as follows, where ϕ_{Z1SW} and ϕ_{Z2SW} are angles of $\mathbf{h}_{1Z}$ (P_0P_{Z1SW}) and $\mathbf{h}_{2Z}$ (P_0P_{Z2SW}) to the plane S_S on the plane S_{W0}.

$$\left.\begin{aligned} \sin\phi_{Z1SW} &= \sin\phi_{1SZ}\sin(\Sigma-\Gamma_S) \\ \sin\phi_{Z2SW} &= \sin\phi_{2SZ}\sin\Gamma_S \end{aligned}\right\} \qquad (3.3\text{-}14)$$

When the tangent plane W_d intersects $\mathbf{h}_{1Z}$ and $\mathbf{h}_{2Z}$ at points P_{Z1Wd} and P_{Z2Wd} respectively, P_0P_d, P_dP_{Z1Wd} and P_dP_{Z2Wd} are obtained as follows.

$$\begin{aligned} P_0P_d &= L_{gd} = R_{b10}\cos\phi_{b10}\triangle\theta_1 = R_{b20}\cos\phi_{b20}\triangle\theta_2 \\ P_dP_{Z1Wd} &= L_{Wd1} = L_{gd}\tan(\phi_{Z1SW}-\phi_{SW0}) \\ P_dP_{Z2Wd} &= L_{Wd2} = L_{gd}\tan(\phi_{Z2SW}-\phi_{SW0}) \end{aligned}$$

Because P_0P_d is on the plane G_{20} and has no component in the direction of R_{b2}, the component in the R_{b2} direction of P_0P_{Z2Wd} (= $P_0P_d + P_dP_{Z2Wd}$) is obtained from P_dP_{Z2Wd} as follows. The components of $P_dP_{Z2Wd} = L_{Wd2}$ in the direction of each axis of the coordinate system C_S are expressed as follows.

$$\begin{aligned} L_{uc2} &= -L_{Wd2}\sin\phi_{SW0}\sin\phi_{S0} \\ L_{vc2} &= -L_{Wd2}\sin\phi_{SW0}\cos\phi_{S0} \\ L_{zc2} &= L_{Wd2}\cos\phi_{SW0} \end{aligned}$$

Transforming into the coordinate system C_2,

$$L_{u2c} = -L_{uc2}\cos\Gamma_s + L_{zc2}\sin\Gamma_s = L_{Wd2}(\sin\phi_{sw0}\sin\phi_{s0}\cos\Gamma_s + \cos\phi_{sw0}\sin\Gamma_s)$$
$$L_{v2c} = L_{vc2} = -L_{Wd2}\sin\phi_{sw0}\cos\phi_{s0}$$

And transforming into the coordinate system C_{q2}, the component L_{Rb2} of L_{Wd2} in the direction of R_{b2} and $dR_{b2Z}/d\theta_2$ are obtained as follows.

$$L_{Rb2} = L_{u2c}\sin\chi_{20} - L_{v2c}\cos\chi_{20}$$
$$= L_{Wd2}\{(\sin\phi_{sw0}\sin\phi_{s0}\cos\Gamma_s + \cos\phi_{sw0}\sin\Gamma_s)\sin\chi_{20} + \sin\phi_{sw0}\cos\phi_{s0}\cos\chi_{20}\}$$
$$dR_{b2Z}/d\theta_2 = L_{Rb2}/\triangle\theta_2$$
$$= R_{b20}\cos\phi_{b20}\tan(\phi_{Z2sw} - \phi_{sw0})$$
$$\times\{(\sin\phi_{sw0}\sin\phi_{s0}\cos\Gamma_s + \cos\phi_{sw0}\sin\Gamma_s)\sin\chi_{20} + \sin\phi_{sw0}\cos\phi_{s0}\cos\chi_{20}\} \quad (3.3\text{-}15)$$

When $\mathbf{g}_0$ is given, $R_{b2Z}/d\theta_2$ becomes constant by Eq. (3. 3-15).

$dR_{b1Z}/d\theta_1$ is obtained as follows in the same way.
The components of $P_dP_{Z1Wd} = L_{Wd1}$ in the direction of each axis of the coordinate system C_s are expressed as follows.

$$L_{uc1} = -L_{Wd1}\sin\phi_{sw0}\sin\phi_{s0}$$
$$L_{vc1} = -L_{Wd1}\sin\phi_{sw0}\cos\phi_{s0}$$
$$L_{zc1} = L_{Wd1}\cos\phi_{sw0}$$

Transforming into the coordinate systems C_1 and C_{q1}, the component L_{Rb1} of L_{Wd1} in the direction of R_{b1} and $dR_{b1Z}/d\theta_1$ are obtained as follows.

$$L_{u1c} = L_{uc1}\cos(\Sigma-\Gamma_s) + L_{zc1}\sin(\Sigma-\Gamma_s) = L_{Wd1}(-\sin\phi_{sw0}\sin\phi_{s0}\cos(\Sigma-\Gamma_s) + \cos\phi_{sw0}\sin(\Sigma-\Gamma_s))$$
$$L_{v1c} = L_{vc1} = -L_{Wd1}\sin\phi_{sw0}\cos\phi_{s0}$$
$$L_{Rb1} = L_{u1c}\sin\chi_{10} - L_{v1c}\cos\chi_{10}$$
$$= L_{Wd1}\{(-\sin\phi_{sw0}\sin\phi_{s0}\cos(\Sigma-\Gamma_s) + \cos\phi_{sw0}\sin(\Sigma-\Gamma_s))\sin\chi_{10} + \sin\phi_{sw0}\cos\phi_{s0}\cos\chi_{10}\}$$

$$dR_{b1Z}/d\theta_1 = L_{Rb1}/\triangle\theta_1 = R_{b10}\cos\phi_{b10}\tan(\phi_{Z1sw} - \phi_{sw0})$$
$$\times\{(-\sin\phi_{sw0}\sin\phi_{s0}\cos(\Sigma-\Gamma_s) + \cos\phi_{sw0}\sin(\Sigma-\Gamma_s))\sin\chi_{10} + \sin\phi_{sw0}\cos\phi_{s0}\cos\chi_{10}\} \quad (3.3\text{-}16)$$

(b) When the plane S_n includes the common perpendicular
(crossed helical, worm and cylindrical gears)

Figure 3. 3-3 shows a point of contact P_d in the vicinity of P_0 on $\mathbf{g}_0$ and the intersection of the tangent plane W_d and the surface of action S_{W0} through P_d, where the intersection of the plane $v_{2C} = 0$ and the plane S_{W0} intersects the plane S_S at P_{SSW}. Figure 3. 3-3 is shown in the case of $\phi_{n0} \geqq 0$ ($\phi_{s0} \leqq 0$) and the plane S_n (ϕ_0) is different from that in Fig. 3. 3-2. The inclination angle $\mathbf{g}_0$ (ϕ_{s0}, ϕ_{sw0} ; C_S) of $\mathbf{g}_0$ on the plane S_{W0} is given by Eq. (3. 2-6) as follows.

$$\left.\begin{aligned}\tan\phi_{s0} &= -\tan\phi_{n0}\cos\psi_0 \\ \sin\psi_{sw0} &= \sin\phi_{n0}\sin\psi_0\end{aligned}\right\} \quad (3.3\text{-}17)$$

The inclination angles h_{1z} (ϕ_{1sz}, 0 ; C_1) of $\mathbf{h}_{1z}$ (P_0P_{z1sw}) and h_{2z} (ϕ_{2sz}, 0 ; C_2) of $\mathbf{h}_{2z}$ (P_0P_{z2sw}) are given by Eqs. (3.3-6) and (3.3-4) respectively. Therefore, transforming them into the coordinate system C_s through Eqs. (2.1-16) and (2.1-15), the inclination angles h_{1z} (ϕ_{s0}, ψ_{z1sw} ; C_s) and h_{2z} (ϕ_{s0}, ψ_{z2sw} ; C_s) are obtained as follows, where ψ_{z1sw} and ψ_{z2sw} are angles of $\mathbf{h}_{1z}$ (P_0P_{z1sw}) and $\mathbf{h}_{2z}$ (P_0P_{z2sw}) to the plane S_s on the plane S_{w0}.

$$\left.\begin{aligned}\sin\psi_{z1sw} &= \sin\phi_{1sz}\sin(\Sigma-\Gamma_s) \\ \sin\psi_{z2sw} &= \sin\phi_{2sz}\sin\Gamma_s\end{aligned}\right\} \quad (3.3\text{-}18)$$

Substituting Eqs. (3.3-17) and (3.3-18) for Eqs. (3.3-15) and (3.3-16) respectively, $dR_{b1z}/d\theta_1$ and $R_{b2z}/d\theta_2$ are obtained.

(2) When a generating gear axis is parallel to the generated gear axis

When a generating gear axis is parallel to the generated gear axis, the plane S_{w0} coincides with the plane G_{20} and $\mathbf{g}_0$ and $\mathbf{h}_{2z}$ are straight lines on the plane G_{20}, so that $dR_{b2z}/d\theta_2 = 0$ is obtained, which is also obtained by substituting $\Gamma_s = 0$ for Eq. (3.3-15).

In the same way, $dR_{b1z}/d\theta_1 = 0$ is obtained.

(Note 3.3-1) : In Eqs. (3.3-15) and (3.3-16), the sign of ϕ_{s0} is taken in the opposite direction, therefore the sign of the first term in the right side is opposite to that in reference (33).

3. 4 Equivalent rack (34)

When a pair of paths of contact $\mathbf{g}_{0D}$ (drive side) and $\mathbf{g}_{0C}$ (coast side) which intersect at the design point P_0 (u_{c0}, v_{c0}, z_{c0} ; C_s) is given, an equivalent rack which has $\mathbf{g}_{0D}$ and $\mathbf{g}_{0C}$ is introduced and a tooth shape which has sufficient top and bottom lands is discussed to determine the tip and the root surfaces.

3. 4. 1 Design reference line g_t and design vertical C_n

(1) When the plane S_n does not include the common perpendicular (bevel and hypoid gears)

Figure 3.4-1 shows schematically the relation among the planes S_H, S_p and S_n of hypoid gears, the plane S_t made by the peripheral velocities $\mathbf{V}_{10}$ and $\mathbf{V}_{20}$ at P_0 and the plane W_N perpendicular to the planes S_t and S_n, where S_H is the plane $v_c = 0$, S_p is the plane $u_c = u_{c0}$ in the coordinate system C_s and S_n is the plane perpendicular to the relative velocity $\mathbf{V}_{rs0} = \mathbf{V}_{10} - \mathbf{V}_{20}$ at P_0.

The design point P_0 is selected on the plane S_H, therefore $\mathbf{V}_{rs0}$ is on the plane S_p. Because the plane S_t includes $\mathbf{V}_{rs0}$, the planes S_t and S_p intersect along the relative velocity $\mathbf{V}_{rs0}$, and this intersection on which $\mathbf{V}_{rs0}$ exists is defined as the design reference line V_{rs0}. The planes S_t and S_n intersect at right angles along the velocity component $\mathbf{V}_{gt0}$ of $\mathbf{V}_{10}$ and $\mathbf{V}_{20}$ on the plane S_n, and this intersection on which $\mathbf{V}_{gt0}$ exists is defined as the design reference line $\mathbf{g}_t$ (positive in the direction of V_{gt0}). The plane S_t is obtained by rotating the plane S_p around V_{rs0} by the inclination angle ϕ_{nt} of $\mathbf{g}_t$ on the plane S_n, which corresponds to the pitch plane in the present theory of hypoid gears.

When the gear axes I and II intersect the plane S_n at O_{1n} and O_{2n} respectively, the peripheral velocities V_{10} and V_{20} at P_0 are expressed as follows.

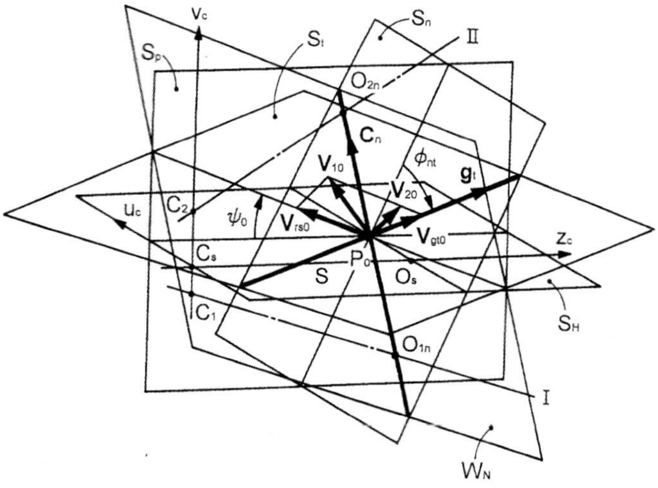

Fig.3.4-1 Relationship among planes S_H, S_p, S_n, S_t and W_N of bevel and hypoid gears

$$V_{10} = \omega_{10} \times O_{1n}P_0 \ , \qquad V_{20} = \omega_{20} \times O_{2n}P_0 \tag{3.4-1}$$

Because $O_{1n}P_0$ is on the plane S_n, it is perpendicular to V_{rs0} and also to V_{10} from Eq. (3.4-1), therefore it is also perpendicular to the plane S_t at P_0. In the same way, $O_{2n}P_0$ is perpendicular to V_{rs0} and V_{20}, therefore it is perpendicular to the plane S_t at P_0. That means that $O_{1n}P_0O_{2n}$ makes a straight line which is defined as the design vertical C_n (positive in the direction from O_{1n} to O_{2n}). C_n, V_{rs0} and $\mathbf{g}_t$ are at right angles at P_0 and the plane made by C_n and V_{rs0} is W_N whose normal is $\mathbf{g}_t$. In hypoid gears, these three perpendiculars, C_n, V_{rs0} and $\mathbf{g}_t$ at P_0, are inclined to the axes of the coordinate system C_S which represents the field of relative velocity, while in crossed helical, worm and cylindrical gears, C_n coincides with the axis v_c and in bevel gears, C_n is parallel to the axis u_c. That is why analyzing tooth contact is more difficult in hypoid gears than in others.

The inclination angle $g_t(\psi_0, \phi_{nt}; C_S)$ of $\mathbf{g}_t$ is obtained as follows.

Figure 3.4-2 shows the design vertical C_n and the design reference line $\mathbf{g}_t$ on the plane S_n. When the points O_{1np} and O_{2np} are the orthographic projections from the points O_{1n} and O_{2n} toward the plane S_p (perpendicular to the plane S_n) respectively, ϕ_{nt} is obtained as follows, where each directed segment is positive in the direction of the axis of the coordinate system C_S.

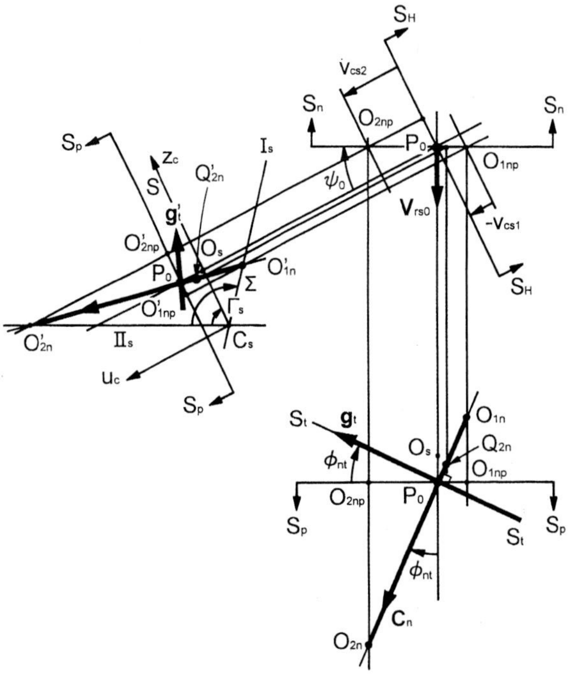

Fig.3.4-2 Design vertical $\mathbf{C}_n$ and design reference line $\mathbf{g}_t$ of bevel and hypoid gears

$$\left.\begin{aligned}\tan\phi_{nt} &= P_0O_{2np}/O_{2np}O_{2n} = O_{1np}P_0/O_{1n}O_{1np}\\ &= v_{cs2}/[\cos\phi_0\{(z_{c0}+v_{cs2}\tan\phi_0)\tan\Gamma_s-u_{c0}\}]\\ &= -v_{cs1}/[\cos\phi_0\{(z_{c0}+v_{cs1}\tan\phi_0)\tan(\Sigma-\Gamma_s)+u_{c0}\}]\end{aligned}\right\} \quad (3.4\text{-}2)$$

(a) Hypoid gears

v_{cs2}, v_{cs1}, u_{c0} and z_{c0} are expressed in the coordinate systems C_1 and C_2 as follows.

$$\begin{aligned}v_{cs2} &= E\tan\Gamma_s/\{\tan\Gamma_s+\tan(\Sigma-\Gamma_s)\}\\ u_{c0} &= E\sin(\Sigma-\Gamma_s)\sin\Gamma_s\tan\phi_0/\sin\Sigma\\ z_{c0} &= (u_{2p0}+u_{c0}\cos\Gamma_s)/\sin\Gamma_s\\ u_{2p0} &= -v_{cs2}/\tan\varepsilon_{20} \quad (\varepsilon_{20}\neq 0)\\ v_{cs1} &= -E\tan(\Sigma-\Gamma_s)/\{\tan\Gamma_s+\tan(\Sigma-\Gamma_s)\}\\ z_{c0} &= \{u_{1p0}-u_{c0}\cos(\Sigma-\Gamma_s)\}/\sin(\Sigma-\Gamma_s)\\ u_{1p0} &= -v_{cs1}/\tan\varepsilon_{10} \quad (\varepsilon_{10}\neq 0)\end{aligned}$$

Using the equations above, v_{cs2}, v_{cs1}, u_{c0} and z_{c0} are eliminated from Eq. (3.4-2) and simplified,

$$\left.\begin{aligned}\tan\phi_{nt} &= \cos\Gamma_s/(-\cos\phi_0/\tan\varepsilon_{20}+\sin\Gamma_s\sin\phi_0)\\ &= \cos(\Sigma-\Gamma_s)/\{\cos\phi_0/\tan\varepsilon_{10}-\sin(\Sigma-\Gamma_s)\sin\phi_0\}\end{aligned}\right\} \quad (3.4\text{-}3)$$

$\mathbf{g}_t$ is inclined by ϕ_{nt} to the plane S_p on the plane S_n, where ϕ_{nt} is positive in the clockwise direction seen from the negative to positive direction of the axis z_c.

(b) Bevel gears

It is obtained by ε_{10} and $\varepsilon_{20}\rightarrow 0$ in Eq. (3.4-3).

$$\phi_{nt} = 0 \quad (3.4\text{-}4)$$

The plane S_t coincides with the plane S_p.

(c) Tapered gears (cylindrical gears)

It is obtained by $\varepsilon_{20}\rightarrow -\pi/2$ and $\Gamma_s\rightarrow 0$ (external gears) or π (internal gears) in Eq. (3.4-3).

$$\phi_{nt} = \pm\pi/2 \text{ (+ for external gears, − for internal gears)} \quad (3.4\text{-}5)$$

(2) When the plane S_n includes the common perpendicular (crossed helical, worm and cylindrical gears)

Figure 3.4-3 shows schematically the design reference line $\mathbf{g}_t$ of crossed helical and worm gears. The plane S_s includes the common perpendicular and is perpendicular to the instantaneous axis S. The plane S_t is perpendicular to the common perpendicular because the point P_0 is chosen on the common perpendicular. The plane S_n includes the common perpendicular, so that $\mathbf{C}_n$ always coincides with the common perpendicular. Therefore, $\mathbf{g}_t$ (the normal of the plane W_N made by $\mathbf{V}_{rs0}$ and $\mathbf{C}_n$) becomes the intersection of the planes

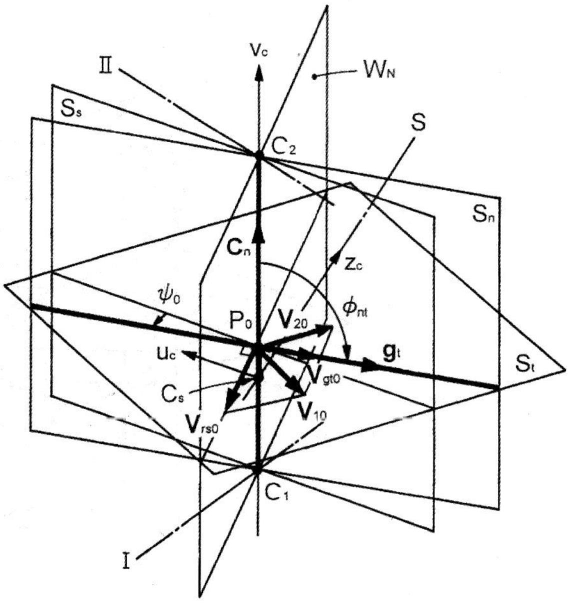

Fig.3.4-3 Relationship among planes S_H, S_p, S_n, S_t and W_N of crossed helical, worm and cylindrical gears

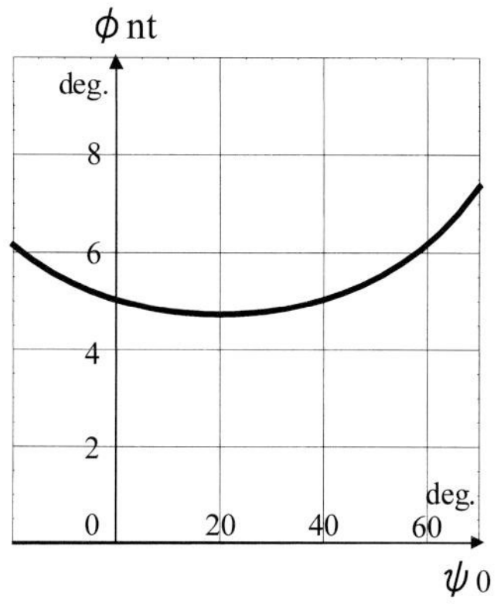

Fig.3.4-4 An example of variation of inclination angle ϕ_{nt} of hypoid gears

S_t and S_n and the inclination angle ϕ_{nt} is obtained as follows.

$$\phi_{nt} = \pm\pi/2 \ (+ \text{ for } \Gamma_S - \psi_0 \leqq \pi/2, \ - \text{ for } \Gamma_S - \psi_0 > \pi/2) \qquad (3.4\text{-}6)$$

In cylindrical gears, where P_0 coincides with C_S ($v_{c0} = 0$) and $\Sigma = 0$ or π, the plane S_t coincides with the plane S_H.

Figure 3.4-4 shows the inclination angle ϕ_{nt} of the design reference line $\mathbf{g}_t$ of hypoid gears which have the dimensions shown in Table 3.3-1. The variation of ϕ_{nt} is not so large in the practical range of ψ_0.

3．4．2　Equivalent rack

(1) Definition of equivalent rack

Figure 3.4-5 shows an equivalent rack on the plane S_n seen from the positive direction of the relative velocity $\mathbf{V}_{rs0}$, where the peripheral velocities $\mathbf{V}_{10}$ and $\mathbf{V}_{20}$ at P_0, the relative velocity $\mathbf{V}_{rs0} = \mathbf{V}_{s10} - \mathbf{V}_{s20}$, the velocity $\mathbf{V}_{gt0}$ in the direction of $\mathbf{g}_t$, the pair of paths of contact $\mathbf{g}_{0D}$ and $\mathbf{g}_{0C}$ on the plane S_n and the tangent planes W_D and W_C of the tooth surfaces D and C which

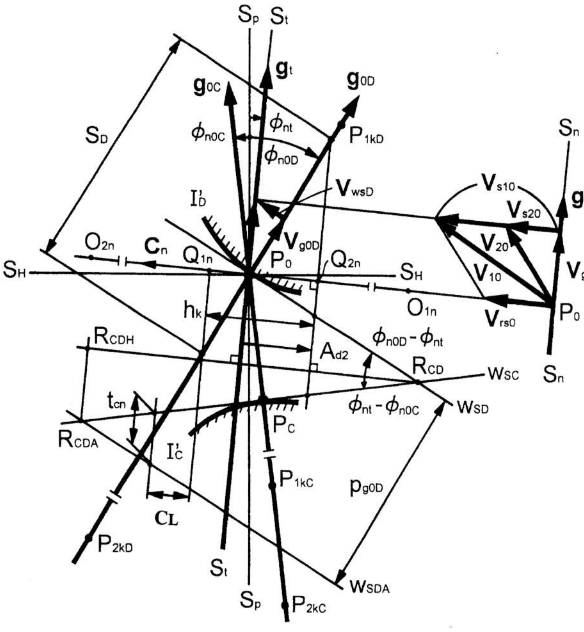

Fig.3.4-5 Equivalent rack on plane S_n

have $\mathbf{g}_{0D}$ and $\mathbf{g}_{0C}$ and are represented by the intersections w_{SD} and w_{SC} with the plane S_n are drawn.

$\mathbf{V}_{10}$ and $\mathbf{V}_{20}$ are expressed by $\mathbf{V}_{gt0}$ and the velocity components $\mathbf{V}_{S10}$ and $\mathbf{V}_{S20}$ perpendicular to the plane S_n as follows.

$$\mathbf{V}_{10} = \mathbf{V}_{gt0} + \mathbf{V}_{S10}$$
$$\mathbf{V}_{20} = \mathbf{V}_{gt0} + \mathbf{V}_{S20}$$

The normal velocities $\mathbf{V}_{g0D}$ and $\mathbf{V}_{g0C}$ of the tangent planes W_D and W_C at P_0 are expressed as follows because they are the components of $\mathbf{V}_{gt0}$ in the directions of $\mathbf{g}_{0D}$ and $\mathbf{g}_{0C}$ respectively.

$$\left.\begin{aligned} V_{g0D} &= V_{gt0}\cos(\phi_{n0D} - \phi_{nt}) \\ V_{g0C} &= V_{gt0}\cos(\phi_{nt} - \phi_{n0C}) \end{aligned}\right\} \qquad (3.4\text{-}7)$$

While V_{gt0}, V_{g0D} and V_{g0C} are also expressed as follows.

$$\left.\begin{aligned} V_{gt0} &= R_{b2gt}(d\theta_2/dt)\cos\phi_{b2gt} \\ V_{g0D} &= R_{b20D}(d\theta_2/dt)\cos\phi_{b20D} \\ V_{g0C} &= R_{b20C}(d\theta_2/dt)\cos\phi_{b20C} \end{aligned}\right\} \qquad (3.4\text{-}8)$$

where, R_{b2gt}, R_{b20D} and R_{b20C} : radii of base cylinders of gear II tangent to $\mathbf{g}_t$, $\mathbf{g}_{0D}$ and $\mathbf{g}_{0C}$

ϕ_{b2gt}, ϕ_{b20D} and ϕ_{b20C} : inclination angles of $\mathbf{g}_t$, $\mathbf{g}_{0D}$ and $\mathbf{g}_{0C}$ on the planes of action of gear II

Because $\mathbf{g}_t$, $\mathbf{g}_{0D}$ and $\mathbf{g}_{0C}$ are fixed straight lines in the static space and $(d\theta_2/dt)$ is constant, the normal velocities V_{gt0}, V_{g0D} and V_{g0C} are always constant from Eq. (3.4-8) at any point on $\mathbf{g}_t$, $\mathbf{g}_{0D}$ and $\mathbf{g}_{0C}$ including P_0, which means that the normal velocities $\mathbf{V}_{g0D}$ and $\mathbf{V}_{g0C}$ are always represented by Eq. (3.4-7) at any point on $\mathbf{g}_{0D}$ and $\mathbf{g}_{0C}$. That leads to the idea of a rack which has the reference line $\mathbf{g}_t$ and the pair of paths of contact $\mathbf{g}_{0D}$ and $\mathbf{g}_{0C}$ (tooth profiles w_{SD} and w_{SC}), and which moves in the direction of $\mathbf{g}_t$ at $\mathbf{V}_{gt0}$ and makes contact on $\mathbf{g}_{0D}$ and $\mathbf{g}_{0C}$ according to the rotation of gears. The equivalent rack is a generalized rack of involute spur gears. The tooth contact in the vicinity of the paths of contact $\mathbf{g}_{0D}$ and $\mathbf{g}_{0C}$ of all kinds of gears from cylindrical to hypoid gears can be analyzed through the equivalent rack.

(2) Dimensions of equivalent rack

Dimensions of the equivalent rack are obtained as follows.

(a) Pressure angles (inclination angles of rack profiles w_{SD} and w_{SC} to $\mathbf{C}_n$)

w_{SD} and w_{SC} are inclined to $\mathbf{C}_n$ by $\phi_{n0D} - \phi_{nt}$ and $\phi_{nt} - \phi_{n0C}$ respectively. $\phi_{n0D} = \phi_{n0C} = \phi_{nt}$ ($\mathbf{g}_{0D} = \mathbf{g}_{0C} = \mathbf{g}_t$) means that the pressure angle of the equivalent rack equals 0. However, in hypoid gears, P_0 does not mean the limit of action even if the pressure angle of the equivalent rack equals 0, because $\mathbf{g}_t$ is not the limit path of contact $\mathbf{g}_{2S}$.

The pressure angles of the equivalent rack $\phi_{n0D} - \phi_{nt}$ and $\phi_{nt} - \phi_{n0C}$ are chosen under the following conditions in this theory.

$$\left.\begin{aligned}\phi_{n0D}-\phi_{n0C} &\fallingdotseq 40^\circ\\ \phi_{n0D}-\phi_{nt} &\geqq 0\end{aligned}\right\}\qquad(3.4\text{-}9)$$

Equation(3.4-9) shows the conditions for the equivalent rack to make a trapezoid inclined adequately to the design vertical $\mathbf{C}_n$.

(b) Reference pitch p_{gt} and normal pitches p_{g0D} and p_{g0C}

The reference pitch p_{gt} on $\mathbf{g}_t$ of the equivalent rack and the normal pitches p_{g0D} and p_{g0C} are expressed as follows, where N_2 is number of teeth of gear II.

$$\left.\begin{aligned}p_{gt} &= 2\pi R_{b2gt}\cos\phi_{b2gt}/N_2 && \text{(in the direction of } \mathbf{g}_t)\\ p_{g0D} &= p_{gt}\cos(\phi_{n0D}-\phi_{nt}) = 2\pi R_{b20D}\cos\phi_{b20D}/N_2 && \text{(in the direction of } \mathbf{g}_{0D})\\ p_{g0C} &= p_{gt}\cos(\phi_{nt}-\phi_{n0C}) = 2\pi R_{b20C}\cos\phi_{b20C}/N_2 && \text{(in the direction of } \mathbf{g}_{0C})\end{aligned}\right\}\qquad(3.4\text{-}10)$$

(c) Working depth h_k and addendums

In Fig. 3.4-5, when w_{SDA} is located 1 pitch (p_{g0D}) apart from w_{SD} and the intersections of w_{SC} and both w_{SD} and w_{SDA} are R_{CD} and R_{CDA} respectively, the maximum tooth depth $h_{cr}(R_{CD}\ R_{CDH})$ which makes the cusp at the tip of the rack is expressed as follows because $R_{CD}\ R_{CDH}$ is perpendicular to $\mathbf{g}_t$.

$$h_{cr} = R_{CD}R_{CDH} = p_{g0D}\cos(\phi_{nt}-\phi_{n0C})/\sin(\phi_{n0D}-\phi_{n0C})$$

Therefore, the working depth h_k (in the direction of $\mathbf{C}_n$) is obtained as follows.

$$h_k = h_{cr} - 2t_{cn}/\{\tan(\phi_{n0D}-\phi_{nt})+\tan(\phi_{nt}-\phi_{n0C})\} - 2c_L \qquad (3.4\text{-}11)$$

where t_{cn} : normal tooth thickness at the tip of the equivalent rack
c_L : clearance

Giving t_{cn} and c_L adequately, the working depth h_k is determined by Eq. (3.4-11), through which top lands of a designed gear pair are assured indirectly in this theory.

In Fig. 3.4-5, putting points Q_{1n} and Q_{2n} on C_n in the neighborhood of P_0 so as to be $h_k = Q_{1n}Q_{2n}$, addendums A_{d2} and A_{d1} are defined as follows.

$$\begin{aligned}P_0Q_{2n} &= A_{d2} && \text{(addendum of gear II)}\\ P_0Q_{1n} &= A_{d1} = h_k - A_{d2} && \text{(addendum of gear I)}\end{aligned}$$

where $A_{d2} \geqq 0$ when Q_{2n} is on O_{1n} side of C_n to P_0. When the working depth determined by choosing A_{d2} is restricted by the limit of action, the addendum A_{d2} or the inclination angles ϕ_{n0D} and ϕ_{n0C} of the paths of contact $\mathbf{g}_{0D}$ and $\mathbf{g}_{0C}$ need to be adapted.

(d) Phase angle of the rack profile w_{SC} to w_{SD} through P_0

When $\mathbf{g}_{0C}$ intersects w_{SC} at P_C, P_0P_C is obtained as follows.

$$P_0P_C = -\{A_{d2}+(h_{cr}-h_k)/2\}\sin(\phi_{n0D}-\phi_{n0C})/\cos(\phi_{n0D}-\phi_{nt})$$

where P_0P_C is positive in the direction of $\mathbf{g}_{0C}$.

When P_0 is in contact at $\theta_2 = 0$, the phase angle θ_{2wsC} of the point P_C is expressed as follows.

$$\theta_{2wsC} = (P_0P_C/p_{g0C})(2\theta_{2p}) \qquad (3.4\text{-}12)$$

where $2\theta_{2p}$: angular pitch of gear II

Equation(3.4-12) means that w_{sC} is located by θ_{2wsC} late to w_{sD} (P_0). The phase angle θ_{2wsC} of w_{sC} to w_{sD} generalizes the tooth thickness of the rack in the present theory, which is determined by giving the working depth and the addendum.

3. 4. 3 Tip and root surfaces of a gear pair

Tip surfaces of a gear pair are considered to be any surface of revolution around each gear axis through the points Q_{1n} and Q_{2n} introduced in 3.4.2, so that they are chosen as follows in this theory.

(1) When the plane S_n does not include the common perpendicular (bevel and hypoid gears)

The points $Q_{2n}(u_{cQ2n}, v_{cQ2n}, z_{cQ2n}\,;\,C_s)$ and $Q_{1n}(u_{cQ1n}, v_{cQ1n}, z_{cQ1n}\,;\,C_s)$ are obtained from Fig.3.4-2 as follows.

$$\left.\begin{aligned} u_{cQ2n} &= u_{c0} - A_{d2}\cos\phi_{nt} \\ v_{cQ2n} &= -A_{d2}\sin\phi_{nt}\cos\psi_0 \\ z_{cQ2n} &= z_{c0} - A_{d2}\sin\phi_{nt}\sin\psi_0 \end{aligned}\right\} \qquad (3.4\text{-}13)$$

$$\left.\begin{aligned} u_{cQ1n} &= u_{c0} + (h_k - A_{d2})\cos\phi_{nt} \\ v_{cQ1n} &= (h_k - A_{d2})\sin\phi_{nt}\cos\psi_0 \\ z_{cQ1n} &= z_{c0} + (h_k - A_{d2})\sin\phi_{nt}\sin\psi_0 \end{aligned}\right\} \qquad (3.4\text{-}14)$$

Transforming them into the coordinate systems C_2 and C_1, the radii R_{2Q2n} and R_{1Q1n} of Q_{2n} and Q_{1n} are obtained as follows.

$$\left.\begin{aligned} &Q_{2n}(u_{2Q2n}, v_{2Q2n}, z_{2Q2n}\,;\,C_2) \\ &Q_{1n}(u_{1Q1n}, v_{1Q1n}, z_{1Q1n}\,;\,C_1) \\ &R_{2Q2n} = \sqrt{(u_{2Q2n}{}^2 + v_{2Q2n}{}^2)} \\ &R_{1Q1n} = \sqrt{(u_{1Q1n}{}^2 + v_{1Q1n}{}^2)} \end{aligned}\right\} \qquad (3.4\text{-}15)$$

In this theory, tip surfaces of gears I and II are assumed to be cones through the points Q_{1n} and Q_{2n} whose cone angles γ_f and Γ_f are chosen adequately so as to be $\Gamma_f = \Gamma_s$ and $\gamma_f = \Sigma - \Gamma_s$ or near to those values.

(2) When the plane S_n includes the common perpendicular (crossed helical, worm and cylindrical gears)

From Fig. 3.4-3, the points Q_{2n} and Q_{1n} are on the common perpendicular, so that cylinders are adopted as tip surfaces, radii R_{2Q2n} and R_{1Q1n} of which are as follows.

$$\left.\begin{aligned} R_{2Q2n} &= |v_{2p0}| + A_{d2} \\ R_{1Q1n} &= |v_{1p0}| + (h_k - A_{d2}) && \text{(external gears)} \\ &= |v_{1p0}| + A_{d2} && \text{(internal gears)} \end{aligned}\right\} \quad (3.4\text{-}16)$$

Concerning root surfaces of a gear pair, putting points Q_{2nR} and Q_{1nR} outside of the working depth h_k on C_n so as to have the clearance c_L apart from the points Q_{2n} and Q_{1n} in both cases (1) and (2), cones or cylinders through the points Q_{2nR} and Q_{1nR} are adopted as root surfaces in this theory.

3．5 Surface of action, conjugate tooth surfaces, design reference bodies of revolution and tooth traces [35]

When the design point P_0 and the paths of contact $\mathbf{g}_{0D}$ (drive side) and $\mathbf{g}_{0C}$ (coast side) are given in the static space, the following two kinds of tooth surfaces are discussed.

(1) Tooth surfaces in point contact

They are obtained by giving a pair of involute helicoids which have $\mathbf{g}_{0D}$ (or $\mathbf{g}_{0C}$) to gears I (pinion) and II (gear) and are realized already as crossed helical and conical gears.

(2) Tooth surfaces in line contact

Those which have $\mathbf{g}_{0D}$ (or $\mathbf{g}_{0C}$) are innumerable, therefore in this theory, the following two cases which are basic and theoretically simple are discussed.

(a) A surface of action which has $\mathbf{g}_{0D}$ (or $\mathbf{g}_{0C}$) and a straight line of contact on a tangent plane is given and the conjugate tooth surfaces are obtained.

(b) An involute helicoid which has $\mathbf{g}_{0D}$ (or $\mathbf{g}_{0C}$) is given and the surface of action and the conjugate tooth surface are obtained.

Item (2a) is discussed in this section, since items (1) and (2b) will be discussed in Chapter 4.

3．5．1 Surface of action and pitch line element

Figure 3.5-1 shows surfaces of action of hypoid gears on the left-hand side and crossed helical and worm gears on the right-hand side. In hypoid gears, according to the definition of 3.2, the design point P_{0H} and the path of contact $\mathbf{g}_{0D}$ are given. Because P_{0H} is on the plane S_H, the plane S_{nH} which includes $\mathbf{g}_{0D}$ and is parallel to the axis u_c is perpendicular to the relative velocity $\mathbf{V}_{rs0H}$ at P_{0H} and becomes the plane S_n.

A point of contact P is chosen at will on the path of contact $\mathbf{g}_{0D}$ and the tangent plane at P is shown as W. A point of contact Q is chosen at will on the tangent plane W and the common contact normal through Q is $\mathbf{g}_D$, which intersects the plane S_H at Q_H. Because $\mathbf{g}_D$ is parallel to $\mathbf{g}_{0D}$, $P_{0H}Q_H$ becomes parallel to the instantaneous axis z_c from Eq. (2.4-1). $P_{0H}Q_H$ is a line element of a cylinder with radius r_{sc} which is called pitch line element L_{p0H} on which all

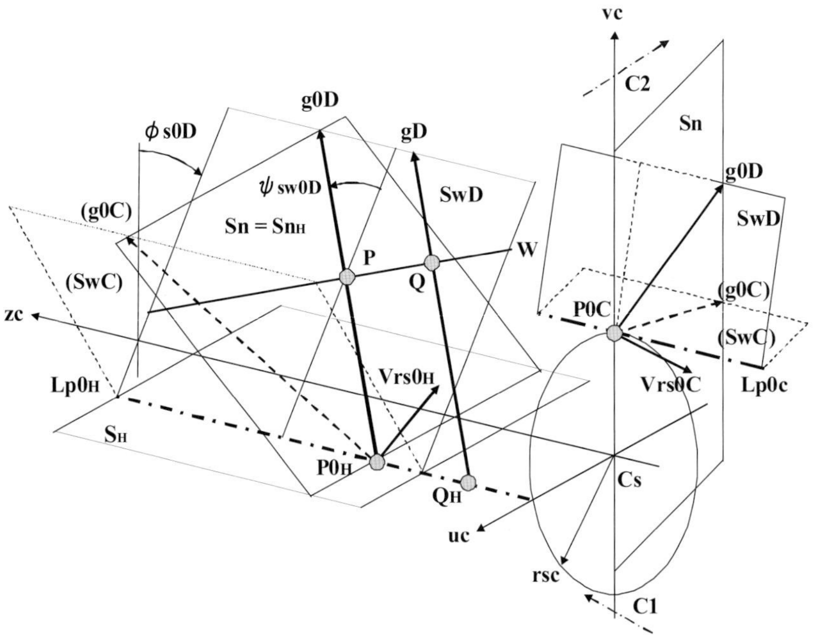

Fig.3.5-1 Relationship among L_{p0H}, S_n, g_{0D} and S_{wD} at P_{0H}

the relative velocities V_{rs}s are the same from Eq. (2.2-3). Conversely, when a point Q_H is chosen at will on the pitch line element L_{p0H} and g_D through Q_H is parallel to g_{0D}, the relative velocity at Q_H is equal to that at P_{0H}. Therefore, g_D becomes the common contact normal from Eqs. (2.2-7) and (2.2-8) and the intersection Q of g_D and the tangent plane W is a point of contact, which means that the plane made by the path of contact g_{0D} and the pitch line element L_{p0H} is the surface of action S_{WD} and the intersection of S_{WD} and the tangent plane W is the line of contact w.

In the same way, when another path of contact g_{0C} (broken line) through P_{0H} different from g_{0D} is given on the plane S_n, the surface of action S_{WC} (broken line) in the coast side is obtained, where the pitch line element L_{p0H} becomes the intersection of the surfaces of action S_{WD} and S_{WC}.

In crossed helical and worm gears, according to the definition of 3.2, when the design point P_{0C} on the common perpendicular, the plane S_n perpendicular to the relative velocity V_{rs0C} at P_{0C} and the paths of contact g_{0D} and g_{0C} are given, the surfaces of action S_{WD} and S_{WC} which intersect along the pitch line element L_{p0C} are obtained.

In bevel and cylindrical gears, as mentioned in the basic theory 2.2.3, the intersection Q_H of the common contact normal g_D at the point of contact Q and the plane S_H is on the instantaneous axis z_c so that the pitch line element becomes the instantaneous axis z_c and the surfaces of action S_{WD} and S_{WC} are the planes which intersect along the instantaneous axis z_c.

The surfaces of action S_{WD} and S_{WC} mentioned above, are common to all kinds of gears and are the easiest ones theoretically, which become the planes of action of involute helical gears in cylindrical gears. In other gears, from the point of view of no variation of bearing loads, it is practical to realize those surfaces of action approximately by a tool and the generating roll of a machine. The present generating method of bevel and hypoid gears which uses cone types of cutter can be considered to be such an approximate example.

There are innumerable surfaces of action besides those mentioned above. For example, any g_D which passes through Q_H on the plane S_n can be a path of contact, so that a surface of action where the inclination angle of $\mathbf{g}_D$ varies on the plane S_n with the movement of $\mathbf{g}_D$ along the pitch line element can be considered, though this is not discussed in this theory.

3．5．2 Equations of surface of action and limit lines of action

Figure 3.5-2 shows a surface of action S_W and the limit lines of action in the coordinate systems C_S and C_{qS}, where S_W and $\mathbf{g}_0$ represent S_{WC} and $\mathbf{g}_{0C}$ in Fig. 3.5-1 respectively. It shows the case where the design point P_0 represents P_{0H} in hypoid gears, but it is also effective in other cases where P_0 represents P_{0C} on the common perpendicular in worm and crossed helical gears, P_{0B} on the instantaneous axis in bevel gears and C_S in cylindrical gears.

The path of contact $\mathbf{g}_0$ is given by Eq. (3.2-9) and the design point P_0 at $\theta_2 = 0$ is given as follows.

$$P_0(u_{c0}=u_c(0), v_{c0}=v_c(0), z_{c0}=z_c(0); C_S), \quad P_0(q_{c0}=q_c(0), -R_{bc0}, z_{c0}=z_c(0); C_{qS})$$

(1) Equations of surface of action

The intersection of the tangent plane W_0 at P_0 and the surface of action S_W is shown by the line of contact w_0. When gear II rotates by θ_2, the point of contact P_0 moves to P and w_0 to w. A common contact normal $\mathbf{g}$ which passes any point Q on w and is perpendicular to w on the surface (plane) of action S_W intersects w_0 at Q_0.

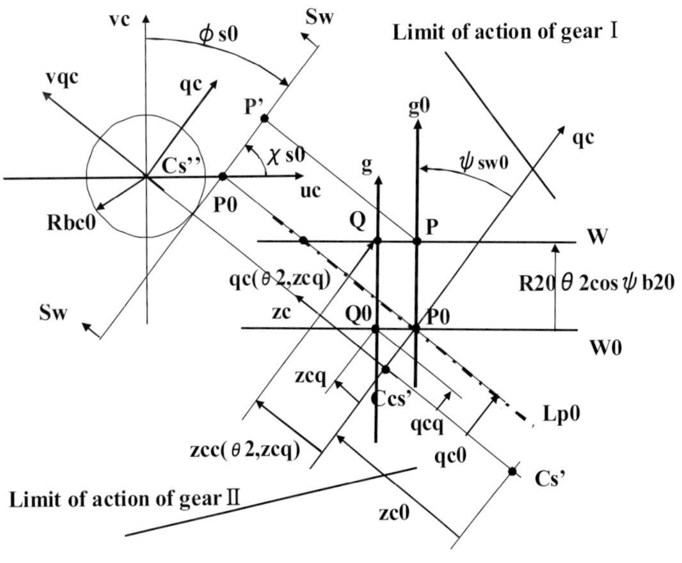

Fig.3.5-2 Surface of action S_W in the coordinate systems C_S and C_{qS}

When the axis z_c intersects the plane $z_c = z_{c0}$ at C_{CS}, the coordinate systems C_{CS} (u_c, v_c, z_{cc}) and C_{CqS} $(q_c, -R_{bc0}, z_{cc})$ are defined by transferring the origin of the coordinate systems C_S and C_{qS} to the point C_{CS} along the z_c axis, therefore

$$z_{cc} = z_c - z_{c0}$$

The point Q_0 $(q_{Cq}, -R_{bc0}, z_{Cq}; C_{CqS})$ is expressed in the coordinate system C_{CqS} as follows.

$$q_{Cq} = q_{C0} - z_{Cq}\tan\psi_{sw0}$$

Because a normal displacement P_0P on $\mathbf{g}_0$ is expressed by $R_{b20}\theta_2\cos\psi_{b20}$ $(=R_{b10}\theta_1\cos\psi_{b10})$, any point Q on the surface of action S_W is expressed in the coordinate systems C_S and C_{qS} as follows.

$$
\left.\begin{aligned}
q_c(\theta_2, z_{Cq}) &= R_{b20}\,\theta_2\cos\phi_{b20}\cos\phi_{sw0} + q_{c0} - z_{Cq}\tan\phi_{sw0} \\
R_{bc}(\theta_2, z_{Cq}) &= R_{bc0} \\
z_{cc}(\theta_2, z_{Cq}) &= R_{b20}\,\theta_2\cos\phi_{b20}\sin\phi_{sw0} + z_{Cq} \\
u_c(\theta_2, z_{Cq}) &= q_c(\theta_2, z_{Cq})\cos\chi_{s0} + R_{bc0}\sin\chi_{s0} \\
v_c(\theta_2, z_{Cq}) &= q_c(\theta_2, z_{Cq})\sin\chi_{s0} - R_{bc0}\cos\chi_{s0} \\
z_c(\theta_2, z_{Cq}) &= z_{cc}(\theta_2, z_{Cq}) + z_{c0} \\
&\mathbf{g}_0(\phi_{s0} = \pi/2 - \chi_{s0},\ \phi_{sw0};\ C_s)
\end{aligned}\right\} \qquad (3.5\text{-}1)
$$

Equation (3.5-1) is the surface of action S_W expressed by θ_2 and z_{Cq} in the coordinate systems C_S and C_{qS}, which represents a line of contact w when θ_2 = constant, a common contact normal **g** when z_{Cq} = constant and the path of contact $\mathbf{g}_0$ when $z_{Cq} = 0$.

Using Eqs. (2.1-11), (2.1-2) and (2.1-18), Eq. (3.5-1) is transformed into the coordinate systems C_2 and C_{q2} as follows.

$$
\left.\begin{aligned}
u_{2c}(\theta_2, z_{Cq}) &= -u_c(\theta_2, z_{Cq})\cos\Gamma_s + z_c(\theta_2, z_{Cq})\sin\Gamma_s \\
v_{2c}(\theta_2, z_{Cq}) &= v_c(\theta_2, z_{Cq}) - v_{cs2} \\
z_{2c}(\theta_2, z_{Cq}) &= -u_c(\theta_2, z_{Cq})\sin\Gamma_s - z_c(\theta_2, z_{Cq})\cos\Gamma_s \\
q_{2c}(\theta_2, z_{Cq}) &= u_{2c}(\theta_2, z_{Cq})\cos\chi_{20} + v_{2c}(\theta_2, z_{Cq})\sin\chi_{20} \\
R_{b2}(\theta_2, z_{Cq}) &= u_{2c}(\theta_2, z_{Cq})\sin\chi_{20} - v_{2c}(\theta_2, z_{Cq})\cos\chi_{20} \\
z_{2c}(\theta_2, z_{Cq}) &= z_{2c}(\theta_2, z_{Cq}) \\
u_{2c}(\theta_2, z_{Cq}) &= q_{2c}(\theta_2, z_{Cq})\cos\chi_{20} + R_{b2}(\theta_2, z_{Cq})\sin\chi_{20} \\
v_{2c}(\theta_2, z_{Cq}) &= q_{2c}(\theta_2, z_{Cq})\sin\chi_{20} - R_{b2}(\theta_2, z_{Cq})\cos\chi_{20} \\
\mathbf{g}_0(\phi_{20} = \pi/2 - \chi_{20},\ \phi_{b20};\ C_2) &= \mathbf{g}_0(\phi_{s0} = \pi/2 - \chi_{s0},\ \phi_{sw0};\ C_s)
\end{aligned}\right\} \qquad (3.5\text{-}2)
$$

Equation (3.5-2) represents the surface of action S_W in the coordinate systems C_2 and C_{q2}.

In the same way, Eq. (3.5-1) is expressed by θ_1 and z_{Cq} as follows.

$$
\left.\begin{aligned}
\theta_1 &= i_0\,\theta_2 \quad (\text{when } \theta_2 = 0,\ \ \theta_1 = 0.) \\
q_c(\theta_1, z_{Cq}) &= R_{b10}\,\theta_1\cos\phi_{b10}\cos\phi_{sw0} + q_{c0} - z_{Cq}\tan\phi_{sw0} \\
R_{bc}(\theta_1, z_{Cq}) &= R_{bc0} \\
z_{cc}(\theta_1, z_{Cq}) &= R_{b10}\,\theta_1\cos\phi_{b10}\sin\phi_{sw0} + z_{Cq} \\
u_c(\theta_1, z_{Cq}) &= q_c(\theta_1, z_{Cq})\cos\chi_{s0} + R_{bc0}\sin\chi_{s0} \\
v_c(\theta_1, z_{Cq}) &= q_c(\theta_1, z_{Cq})\sin\chi_{s0} - R_{bc0}\cos\chi_{s0} \\
z_c(\theta_1, z_{Cq}) &= z_{cc}(\theta_1, z_{Cq}) + z_{c0} \\
&\mathbf{g}_0(\phi_{s0} = \pi/2 - \chi_{s0},\ \phi_{sw0};\ C_s)
\end{aligned}\right\} \qquad (3.5\text{-}3)
$$

Equation (3.5-3) is transformed into the coordinate systems C_1 and C_{q1} using Eqs. (2.1-12), (2.1-3) and (2.1-19) as follows.

$$\left.\begin{aligned}
u_{1c}(\theta_1, z_{Cq}) &= u_c(\theta_1, z_{Cq})\cos(\Sigma-\Gamma_s)+ z_c(\theta_1, z_{Cq})\sin(\Sigma-\Gamma_s)\\
v_{1c}(\theta_1, z_{Cq}) &= v_c(\theta_1, z_{Cq})- v_{cs2}\\
z_{1c}(\theta_1, z_{Cq}) &= -u_c(\theta_1, z_{Cq})\sin(\Sigma-\Gamma_s) + z_c(\theta_1, z_{Cq})\cos(\Sigma-\Gamma_s)\\
q_{1c}(\theta_1, z_{Cq}) &= u_{1c}(\theta_1, z_{Cq})\cos\chi_{10}+ v_{1c}(\theta_1, z_{Cq})\sin\chi_{10}\\
R_{b1}(\theta_1, z_{Cq}) &= u_{1c}(\theta_1, z_{Cq})\sin\chi_{10}- v_{1c}(\theta_1, z_{Cq})\cos\chi_{10}\\
z_{1c}(\theta_1, z_{Cq}) &= z_{1c}(\theta_1, z_{Cq})\\
u_{1c}(\theta_1, z_{Cq}) &= q_{1c}(\theta_1, z_{Cq})\cos\chi_{10}+ R_{b1}(\theta_1, z_{Cq})\sin\chi_{10}\\
v_{1c}(\theta_1, z_{Cq}) &= q_{1c}(\theta_1, z_{Cq})\sin\chi_{10}- R_{b1}(\theta_1, z_{Cq})\cos\chi_{10}\\
g_0(\phi_{10}&= \pi/2 -\chi_{10}, \phi_{b10}; C_1) = g_0(\phi_{s0}= \pi/2 -\chi_{s0}, \phi_{sw0}; C_s)
\end{aligned}\right\}\quad(3.5\text{-}4)$$

Equation (3.5-4) represents the surface of action S_W in the coordinate systems C_1 and C_{q1}.

(2) Limit lines of action

Obtaining the intersection of the surface of action given by Eq. (3.5-2) and the plane of rotation of gear II and applying Eq. (2.4-4) to it, the limit of action of gear II is obtained, which draws the limit line of action by moving the plane of rotation. In the same way, from the surface of action given by Eq. (3.5-4), the limit line of action of gear I is obtained.

Figure 3.5-2 shows schematically the limit lines of action of gears I and II transformed into the coordinate system C_s.

3．5．3 Equations of conjugate tooth surfaces and modified tooth surfaces

(1) Equations of conjugate tooth surfaces

Transforming Eq. (3.5-2) into the coordinate system C_{r2} through Eq. (2.3-6), the conjugate tooth surface II and the surface normal **n** at each point are obtained as follows.

$$\left.\begin{aligned}
\chi_{r2} &= \chi_{20}-\theta_2 = \pi/2 -\phi_{20}-\theta_2 = \pi/2 -\phi_{r2}\\
u_{r2c} &= q_{2c}(\theta_2, z_{Cq})\cos\chi_{r2}+ R_{b2}(\theta_2, z_{Cq})\sin\chi_{r2}\\
v_{r2c} &= q_{2c}(\theta_2, z_{Cq})\sin\chi_{r2}- R_{b2}(\theta_2, z_{Cq})\cos\chi_{r2}\\
z_{r2c} &= z_{2c}(\theta_2, z_{Cq})\\
\mathbf{n}(\phi_{r2}&= \phi_{20}+\theta_2, \phi_{b20}; C_{r2})
\end{aligned}\right\}\quad(3.5\text{-}5)$$

Transforming Eq. (3.5-4) into the coordinate system C_{r1} through Eq. (2.3-7), the conjugate tooth surface I and the surface normal **n** at each point are obtained as follows.

$$\left.\begin{aligned}
\chi_{r1} &= \chi_{10}-\theta_1 = \pi/2 -\phi_{10}-\theta_1\\
u_{r1c} &= q_{1c}(\theta_1, z_{Cq})\cos\chi_{r1}+ R_{b1}(\theta_1, z_{Cq})\sin\chi_{r1}\\
v_{r1c} &= q_{1c}(\theta_1, z_{Cq})\sin\chi_{r1}- R_{b1}(\theta_1, z_{Cq})\cos\chi_{r1}\\
z_{r1c} &= z_{1c}(\theta_1, z_{Cq})\\
\mathbf{n}(\phi_{r1}&= \phi_{10}+\theta_1, \phi_{b10}; C_{r1})
\end{aligned}\right\}\quad(3.5\text{-}6)$$

(2) Equations of modified tooth surfaces

Figure 3.5-3 shows schematically the relation among a point of contact Q on the conjugate tooth surface II represented by a plane of θ_2 and z_{cq}, a point Q_{2f} on the modified tooth surface II, the normal $\mathbf{n}$ (= $\mathbf{g}$), the surface of action and the line of contact. The point Q_{2f} (u_{r2cf}, v_{r2cf}, z_{r2cf} ; C_{r2}) on the modified tooth surface II has an amount of modification $\Delta p_{2n}(\theta_2, z_{cq})$ in the direction of $\mathbf{n}$ corresponding to the point Q (u_{r2c}, v_{r2c}, z_{r2c} ; C_{r2}) on the conjugate tooth surface II, so that it is expressed in the coordinate system C_{r2} as follows.

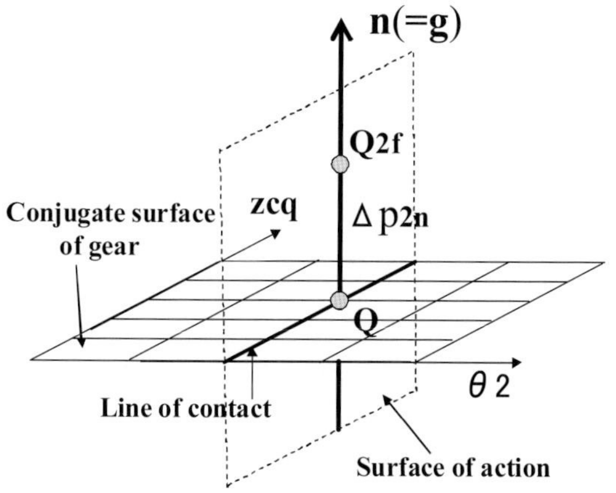

Fig.3.5-3 Definition of modified surface Q_{2f} of gear

$$\left.\begin{aligned} u_{r2cf} &= u_{r2c} + \Delta p_{2n}(\theta_2, z_{cq})\cos\phi_{b20}\cos\chi_{r2} \\ v_{r2cf} &= v_{r2c} + \Delta p_{2n}(\theta_2, z_{cq})\cos\phi_{b20}\sin\chi_{r2} \\ z_{r2cf} &= z_{r2c} + \Delta p_{2n}(\theta_2, z_{cq})\sin\phi_{b20} \\ \mathbf{n}&(\phi_{20}+\theta_2,\ \phi_{b20};\ C_{r2}),\ \text{where } \chi_{r2} = \pi/2 - \phi_{20} - \theta_2 \end{aligned}\right\} \quad (3.5\text{-}7)$$

In the same way, the point Q_{1f} (u_{r1cf}, v_{r1cf}, z_{r1cf} ; C_{r1}) on the modified tooth surface I corresponding to Q (u_{r1c}, v_{r1c}, z_{r1c} ; C_{r1}) on the conjugate tooth surface I is expressed in the coordinate system C_{r1} as follows.

$$\left.\begin{aligned} u_{r1cf} &= u_{r1c} + \Delta p_{1n}(\theta_1, z_{cq})\cos\phi_{b10}\cos\chi_{r1} \\ v_{r1cf} &= v_{r1c} + \Delta p_{1n}(\theta_1, z_{cq})\cos\phi_{b10}\sin\chi_{r1} \\ z_{r1cf} &= z_{r1c} + \Delta p_{1n}(\theta_1, z_{cq})\sin\phi_{b10} \\ \mathbf{n}&(\phi_{10}+\theta_1,\ \phi_{b10};\ C_{r1}),\ \text{where } \chi_{r1} = \pi/2 - \phi_{10} - \theta_1 \end{aligned}\right\} \quad (3.5\text{-}8)$$

The amount of modification $\Delta p_{1n}(\theta_1, z_{cq})$ or $\Delta p_{2n}(\theta_2, z_{cq})$ can be supposed to be any function, but from a practical point of view, it is a function of z_{cq} only for a tooth surface to be modified along the line of contact.

3.5.4 Design reference bodies of revolution (pitch hyperboloids)

As shown in Fig. 3.5-1, when the design point P_0 (P_{0H} or P_{0C}) is given, the pitch line element L_{p0} (L_{p0H} or L_{p0C}) is determined and when the paths of contact $\mathbf{g}_{0D}$ and $\mathbf{g}_{0C}$ through P_0 are given, the surfaces of action S_{WD} and S_{WC} which intersect along L_{p0} are necessarily determined. A pair of hyperboloids obtained by rotating the intersection L_{p0} of the surfaces of action around each gear axis makes line contact along L_{p0}. According to the gear rotation, the point of contact moves on the same L_{p0} in both the drive and the coast sides. Therefore, the pair of hyperboloids is suitable for the design reference bodies of revolution (pitch surfaces) to determine the outside shapes of gears, which are called pitch hyperboloids and are common to all kinds of gears because they have a common definition that they are formed by the rotation of the intersection of the surfaces of action.

The pitch hyperboloids vary according to the location of the pitch line element as follows (see Fig. 3. 2-1).

(1) L_{p0H} through P_{0H} : pitch hyperboloids of hypoid gears.
(2) The instantaneous axis S through P_{0B} when $E = 0$: pitch cones of bevel gears.
(3) L_{p0C} through P_{0C} : pitch hyperboloids of crossed helical and worm gears.
(4) The instantaneous axis S through C_S when $\Sigma = 0$ or π : pitch cylinders of cylindrical gears.

The pitch line element L_{p0} through the design point P_0 is expressed in the coordinate system C_S as follows, where z_{Cq} is a parameter.

$$u_c = u_{c0}, \quad v_c = v_{c0}, \quad z_c = z_{Cq} + z_{c0} \qquad (3.5\text{-}9)$$

Transforming Eq. (3. 5-9) into the coordinate systems C_2 and C_{r2}, the pitch line element and the pitch hyperboloid of gear II are obtained as functions of θ_2 and z_{Cq} as follows.

$$\left.\begin{aligned} u_{2c} &= -u_{c0}\cos\Gamma_S + (z_{Cq} + z_{c0})\sin\Gamma_S \\ v_{2c} &= v_{c0} - v_{cs2} \\ z_{2c} &= -u_{c0}\sin\Gamma_S - (z_{Cq} + z_{c0})\cos\Gamma_S \\ u_{r2c} &= u_{2c}\cos\theta_2 + v_{2c}\sin\theta_2 \\ v_{r2c} &= -u_{2c}\sin\theta_2 + v_{2c}\cos\theta_2 \\ z_{r2c} &= z_{2c} \end{aligned}\right\} \qquad (3.5\text{-}10)$$

In the same way, transforming Eq. (3. 5-9) into the coordinate systems C_1 and C_{r1}, the pitch line element and the pitch hyperboloid of gear I are obtained as functions of θ_1 and z_{Cq}.

3．5．5 Tooth trace

In this new theory, the tooth trace is defined as follows. Because the pitch line element L_{p0} defined by Eq. (3. 5-9) is the intersection of the surfaces of action S_{WD} and S_{WC}, L_{p0} is defined as a tooth trace on the surface of action and the intersection of the pitch hyperboloid drawn by L_{p0} and a tooth surface is defined as a tooth trace on the tooth surface. The present tooth trace (ISO) is defined as an intersection of a reference pitch surface and a tooth surface, so that the tooth trace in this theory coincides with those in cylindrical and bevel gears, but is different from those in other gears. Especially in present hypoid gears, the selected pitch cones have no relation with the L_{p0}, therefore the tooth traces on the surface of action and on the tooth surface are different from those in this theory.

Equations of the tooth trace L_{p0} on the surface of action are obtained by combining Eqs. (3. 5-9) and (3. 5-1) as follows.

$$\left.\begin{aligned} v_c(\theta_2, z_{Cq}) &= 0 \quad (\text{for } P_0 = P_{0H} \text{ or } P_{0B} \text{ or } C_S) \\ &= v_{c0} \quad (\text{for } P_0 = P_{0C}) \end{aligned}\right\} \qquad (3.5\text{-}11)$$

Substituting the solution $(\theta_2, z_{cq}(\theta_2))$ of Eq. (3. 5-11) for Eqs. (3. 5-1) and (3. 5-2), the tooth trace L_{p0} on the surface of action is obtained and substituting it for Eq. (3. 5-5), the tooth trace L_{p0} on the conjugate tooth surface II is obtained.

In the same way, using Eqs. (3.5-3), (3.5-4) and (3.5-6), the tooth trace on the conjugate tooth surface I is obtained.

The surface of action, the conjugate tooth surfaces, the design reference bodies of revolution and the tooth trace which are defined here are an idea common to all kinds of gears which have the path of contact $\mathbf{g}_0$.

3.5.6 Radii of curvature of conjugate tooth surfaces and angles between curves on conjugate tooth surfaces (36)

When any point Q on the conjugate tooth surfaces is given by Eqs. (3.5-5) and (3.5-6), the conjugate tooth surfaces are expressed by position vectors $\mathbf{r}_2$ and $\mathbf{r}_1$ which have the origins C_2 or C_1 respectively as follows.

$$\left.\begin{aligned} C_2Q &= \mathbf{r}_2(u_{r2c}(\theta_2, z_{Cq}), v_{r2c}(\theta_2, z_{Cq}), z_{r2c}(\theta_2, z_{Cq}) ; C_{r2}) \\ C_1Q &= \mathbf{r}_1(u_{r1c}(\theta_1, z_{Cq}), v_{r1c}(\theta_1, z_{Cq}), z_{r1c}(\theta_1, z_{Cq}) ; C_{r1}) \end{aligned}\right\} \quad (3.5\text{-}12)$$

When a curve on the conjugate tooth surface II is given by $z_{Cq} = z_{Cq}(\theta_2)$ (or $\theta_2 = \theta_2(z_{Cq})$), the radius of curvature ρ_{n2} of the curve in the direction of principal normal $\mathbf{n}$ is obtained as follows.

$$\rho_{n2} = |d\mathbf{r}_2/d\theta_2|^3 / |d\mathbf{r}_2/d\theta_2 \times d^2\mathbf{r}_2/d\theta_2^2| \quad (3.5\text{-}13)$$

In the same way, the radius of curvature ρ_{n1} of a curve on the conjugate tooth surface I in the direction of principal normal $\mathbf{n}$ is obtained as follows.

$$\rho_{n1} = |d\mathbf{r}_1/d\theta_1|^3 / |d\mathbf{r}_1/d\theta_1 \times d^2\mathbf{r}_1/d\theta_1^2| \quad (3.5\text{-}14)$$

The angle between curves on the conjugate tooth surface is obtained from the tangent vector $d\mathbf{r}_2/d\theta_2$ (or $d\mathbf{r}_1/d\theta_1$). For example, the angle between a tangent vector $d\mathbf{r}_2/d\theta_2$ and a line of contact vector $d\mathbf{r}_2/dz_{Cq}$ at a point is obtained from the triangle which has these two vectors as the sides through the cosine theorem.

3.6 Contact ratios (35)

3.6.1 Effective surface of action

Figure 3.6-1 shows the effective surface of action which is the surface of action S_W shown in Fig. 3.5-2 enclosed by tip lines of gears I and II and a facewidth, where the tip points Q_{1n} and Q_{2n} of gears I and II on the design

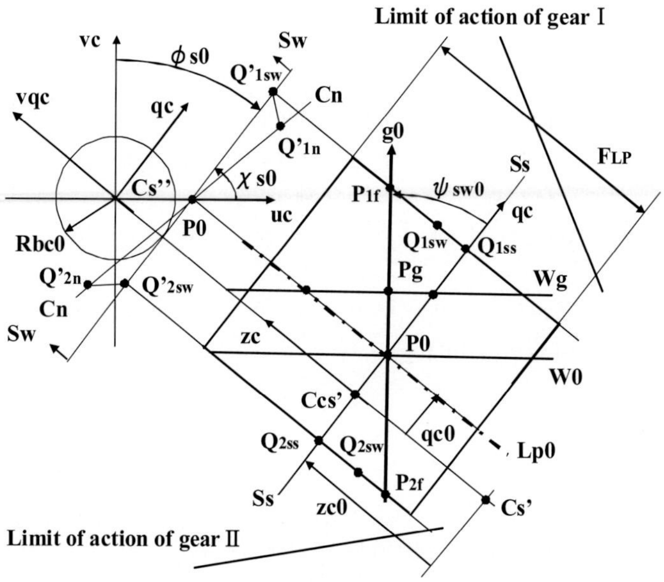

Fig.3.6-1 Effective surface of action S_w in the coordinate systems C_S and Cqs

vertical C_n move to points Q_{1SW} and Q_{2SW} on the surface of action S_W respectively according to the rotation of gears. In this theory, lines $Q_{1SW}Q_{1SS}$ and $Q_{2SW}Q_{2SS}$ which pass through Q_{1SW} and Q_{2SW} on the surface of action S_W and are parallel to the pitch line element L_{p0} are defined as tip lines of gears I and II respectively, where points Q_{1SS} and Q_{2SS} are the intersections of the tip lines of gears I and II and the plane S_S which passes the design point P_0 and is perpendicular to L_{p0}. Therefore, tip hyperboloids are obtained by rotating the tip lines $Q_{1SW}Q_{1SS}$ and $Q_{2SW}Q_{2SS}$ around each gear axis. The effective surface of action is a region of a rectangle which is enclosed by the tip lines $Q_{1SW}Q_{1SS}$ and $Q_{2SW}Q_{2SS}$ and both ends of the facewidth F_{LP} perpendicular to L_{p0}, where it is supposed that $\mathbf{g}_0$ ($\mathbf{g}_{0D}$ or $\mathbf{g}_{0C}$) is chosen for the limit lines of action not to enter in the rectangular region, the top lands have no cusp at the points Q_{1n} and Q_{2n} and the gear pair have a sufficient facewidth.

When the gear pair rotates by one pitch, P_0 moves to P_g and the tangent plane W_0 to W_g, so that the distance P_0P_g is obtained as follows.

$$P_0P_g = p_{0g} = R_{b20}(2\theta_{2p})\cos\phi_{b20} \tag{3.6-1}$$

where p_{0g} : 1 pitch on $\mathbf{g}_0$
$2\theta_{2p}$: angular pitch of gear II

3. 6. 2 Definition of contact ratios

The following four contact ratios are defined based on Fig. 3. 6-1 in this theory.

(1) Transverse contact ratio ms : the ratio between the effective length $Q_{1SS}Q_{2SS}$ on the plane S_S and the pitch in the direction of $Q_{1SS}Q_{2SS}$.
(2) Lengthwise contact ratio mf : the ratio between the facewidth F_{LP} and the pitch in the direction of L_{p0}.
(3) Contact ratio on the path of contact $\mathbf{g}_0$: $m0 = P_{1f}P_{2f} / P_0P_g$.
(4) Total contact ratio : ms + mf

3. 6. 3 Equations of contact ratios

(1) Tip points Q_{1SW} and Q_{2SW} on the surface of action

Tip points $Q_{2n}(u_{cQ2n}, v_{cQ2n}, z_{cQ2n}; C_S)$ of gear II and $Q_{1n}(u_{cQ1n}, v_{cQ1n}, z_{cQ1n}; C_S)$ of gear I on the design vertical C_n are given by Eqs. (3. 4-13) and (3. 4-14). Transforming them into the coordinate systems C_2 and C_1, $Q_{2n}(u_{2Q2n}, v_{2Q2n}, z_{2Q2n}; C_2)$, $Q_{1n}(u_{1Q1n}, v_{1Q1n}, z_{1Q1n}; C_1)$ and the radii R_{2Q2n} and R_{1Q1n} are given by Eq. (3. 4-15). Figure 3. 6-1 shows the case of bevel and hypoid gears, but in the case of crossed helical, worm and cylindrical gears, points Q_{2n} and Q_{1n} are on the common perpendicular, so that the tip surfaces become cylinders and the radii R_{2Q2n} and R_{1Q1n} are given by Eq. (3. 4-16).

Using Eq. (3. 5-2) of the surface of action S_W, the intersection $Q_{2SW}(u_{2SW}, v_{2SW}, z_{2SW}; C_2)$ of the tip circle through Q_{2n} and the surface of action S_W is obtained as the solution of the following equation (3. 6-2).

$$\left.\begin{aligned} u_{2c}(\theta_2, z_{Cq})^2 + v_{2c}(\theta_2, z_{Cq})^2 &= R_{2Q2n}^2 \\ z_{2c}(\theta_2, z_{Cq}) &= z_{2Q2n} \end{aligned}\right\} \tag{3.6-2}$$

Q_{1SW} (u_{1SW}, v_{1SW}, z_{1SW} ; C_1) is obtained in the same way.

(2) Transverse contact ratio ms

Transforming Q_{2SW} and Q_{1SW} into the coordinate system C_S, Q_{2SW} (u_{c2SW}, v_{c2SW}, z_{c2SW} ; C_S) and Q_{1SW} (u_{c1SW}, v_{c1SW}, z_{c1SW} ; C_S) are obtained, so that the intersection $Q_{1SS}Q_{2SS}$ of the plane S_S and the effective surface of action is obtained as follows.

$$Q_{1SS}Q_{2SS} = |v_{c1SW} - v_{c2SW}|/\cos\phi_{SO}$$

Therefore, the transverse contact ratio ms is obtained as follows.

$$\begin{aligned} ms &= Q_{1SS}Q_{2SS}/(P_0P_g/\cos\phi_{SWO}) \\ &= (|v_{c1SW} - v_{c2SW}|/\cos\phi_{SO})/(2R_{b20}\,\theta_{2p}\cos\phi_{b20}/\cos\phi_{SWO}) \end{aligned} \tag{3.6-3}$$

(3) Lengthwise contact ratio mf

The effective length F_{LP} of the tooth trace is the length of the pitch line element cut by the outside and inside cylinders of gear II, therefore,

$$F_{LP} = \left(\sqrt{(R_{2h}^2 - v_{2p0}^2)} - \sqrt{(R_{2t}^2 - v_{2p0}^2)}\right)/\sin\Gamma_S \tag{3.6-4}$$

where R_{2h} and R_{2t} : outside and inside radii of gear II.

Therefore, the lengthwise contact ratio mf is obtained as follows.

$$\begin{aligned} mf &= F_{LP}/(P_0P_g/\sin\phi_{SWO}) \\ &= F_{LP}/(2R_{b20}\,\theta_{2p}\cos\phi_{b20}/\sin\phi_{SWO}) \end{aligned} \tag{3.6-5}$$

(4) Contact ratio m0 on the path of contact $\mathbf{g}_0$

It is obtained as follows.

$$\begin{aligned} m0 &= (Q_{1SS}Q_{2SS}/\cos\phi_{SWO})/P_0P_g \\ &= ms/\cos^2\phi_{SWO} \end{aligned} \tag{3.6-6}$$

The contact ratios shown here are common to all kinds of gears which have $\mathbf{g}_0$ ($\mathbf{g}_{0D}$ or $\mathbf{g}_{0C}$) and the plane surface of action S_W (S_{WD} or S_{WC}).

3.6.4 Correction of design point P_0

Practically, it is the easiest method to put the design point P_0 on the plane S_H according to 3.2 and give a pair of paths of contact $\mathbf{g}_{0D}$ and $\mathbf{g}_{0C}$ intersecting at P_0. However, in this method, the lengthwise pitches of the drive and the coast sides are slightly different. In order to obtain the same lengthwise contact ratios in the drive and the coast sides, P_0 must be corrected as follows.

According to 3.2, P_0 and the path of contact $\mathbf{g}_{0D}$ in the drive side are given as a function of θ_2 (or θ_1). When any point P_w (u_{cpw}, v_{cpw}, z_{cpw} ; C_S) on $\mathbf{g}_{0D}$ and the path of contact $\mathbf{g}_{0C}$ (ϕ_{ShC}, ϕ_{n0C} ; C_S) through P_w in the coast side are given, $\mathbf{g}_{0C}$ intersects the plane S_H at P_{0ShC} (u_{cShC}, 0, z_{cShC} ; C_S). The geometric relation of P_w and P_{0ShC} leads the following equation.

$$u_{cShC} = u_{cpw} + (v_{cpw}/\cos\phi_{ShC})\tan\phi_{n0C} \quad (3.6\text{-}7)$$

For $\mathbf{g}_{0C}$ to be the common contact normal at P_{0ShC}, the fundamental requirement for contact Eq. (2.2-8) must be effective, therefore,

$$u_{cShC} = E\tan\phi_{ShC}\sin(\Sigma-\Gamma_S)\sin\Gamma_S/\sin\Sigma \quad (3.6\text{-}8)$$

In Eq. (3.6-5) for the lengthwise contact ratio, in order to have the same contact ratios in the drive and coast sides (mfD = mfC), the following equation (3.6-9) must be effective, where F_{LP} is equal and R_{b2wC}, ϕ_{b2wC} and ϕ_{SWwC} mean the radius of base cylinder and the inclination angles of g_{0C} through P_w in the coordinate systems C_{q2}, C_2 and C_S respectively.

$$R_{b2wC}\cos\phi_{b2wC}/\sin\phi_{SWwC} = R_{b20D}\cos\phi_{b20D}/\sin\phi_{SW0D} \quad (3.6\text{-}9)$$

In Eq. (3.6-9), the right-hand item is already known and the left-hand one is expressed as a function of θ_2 and ϕ_{ShC} by giving ϕ_{n0C} of $\mathbf{g}_{0C}$, therefore eliminating u_{cShC} from Eqs. (3.6-7) and (3.6-8) and obtaining the solutions θ_2 and ϕ_{ShC} satisfying Eq. (3.6-9), the corrected design point P_w and the path of contact $\mathbf{g}_{0C}$ through P_w are determined.

The paths of contact $\mathbf{g}_{0D}$ and $\mathbf{g}_{0C}$ intersect at P_w and the surfaces of action S_{WD} and S_{WC} intersect along the pitch line element L_{Pw} through P_w, so that the bodies of revolution obtained by rotating L_{Pw} become the new pitch hyperboloids, the new design reference bodies of revolution.

When P_0 is given on the plane S_H, the lengthwise contact ratios are practically equal, so that it is almost unnecessary to correct P_0 to obtain the same lengthwise contact ratios in the drive and the coast sides.

3. 7 Design method for tooth surfaces of hypoid gears (35), (37), (38)

In the following discussion in Sections 3.7 and 3.8, gears I and II are called pinion and gear respectively.

3. 7. 1 Flow of designing tooth surfaces of hypoid gears

Figure 3.7-1 shows a flow of designing tooth surfaces of hypoid

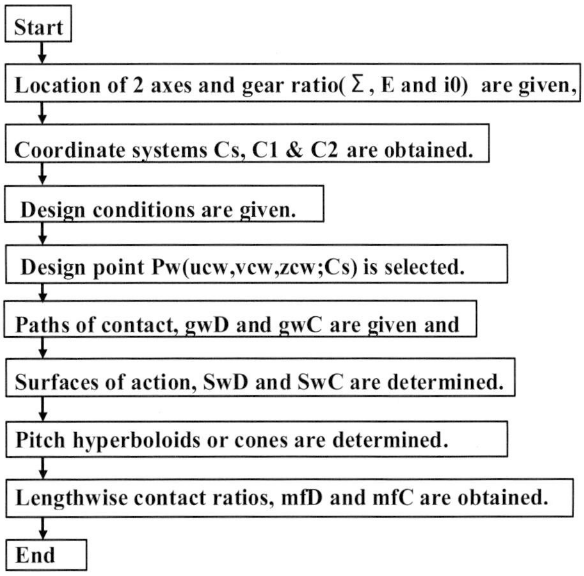

Fig.3.7-1 Flow of designing tooth surfaces of hypoid gears

gears, which is applicable to designing almost all the tooth surfaces in line contact. When the shaft angle and the offset of the two axes and the gear ratio are given, the coordinate systems are determined. According to the design conditions, when a design point P_w and paths of contact $\mathbf{g}_{wD}$ and $\mathbf{g}_{wC}$ through P_w are chosen, the pitch line element, the surfaces of action, the pitch hyperboloids, the tooth surfaces and the contact ratios are designed by the results of the sections 3.5 and 3.6. Therefore, the selection methods of the design point P_w and the paths of contact $\mathbf{g}_{wD}$ and $\mathbf{g}_{wC}$ through P_w are discussed in this section.

3. 7. 2 Selection of design point P_w of hypoid gears

The design point P_w is chosen by the following two methods in this theory.

(1) P_w is directly given on the plane S_H

As shown in sections 3.2 to 3.6, when the design point P_0 on the plane S_H is chosen instead of P_w and the paths of contact $\mathbf{g}_{0D}$ and $\mathbf{g}_{0C}$ intersecting at P_0 are given instead of $\mathbf{g}_{wD}$ and $\mathbf{g}_{wC}$, the pitch line element L_{p0}, the pitch hyperboloids (or approximate pitch cones through P_0), the surfaces of action and the contact ratios are calculated according to 3.3 to 3.6 and the tooth surfaces of hypoid gears are designed more easily like bevel gears. If the pitch line element L_{p0} is replaced by the instantaneous axis, this design method becomes almost the same as in bevel gears. It is not necessary to make the approximate cones contact at P_0 in this method, because the inclination angles of common contact normals $\mathbf{g}_{0D}$ and $\mathbf{g}_{0C}$ are given directly in the coordinate systems and the plane tangent to a pair of cones is not used as the pitch plane. In this method, the troublesome work of choosing a pair of cones circumscribed by each other in space is not necessary, so that hypoid gears can be designed as easily as bevel gears. However, this approach has just been developed, therefore actual results must be accumulated from now on.

(2) P_w is chosen as a point of contact of circumscribed cones

It is most generally used as Wildhaber (Gleason)'s method. Here, based on this new theory, the design method for circumscribed cones (the pitch cones) at the design point P_w is discussed and another method different from Wildhaber's one is clarified.

3. 7. 3 Design method of circumscribed cones (pitch cones) at design point P_w

Figure 3.7-2(a) shows the basic dimensions of Wildhaber's method based on this new theory. The shaft angle Σ and the offset E of the two axes and the gear ratio i_0 (constant) are given and the coordinate systems C_1, C_2 and C_s are determined. When the design point P_w is chosen at will, the pitch line element L_{pw}, the plane S_{tw} (pitch plane) made by the peripheral veloc-

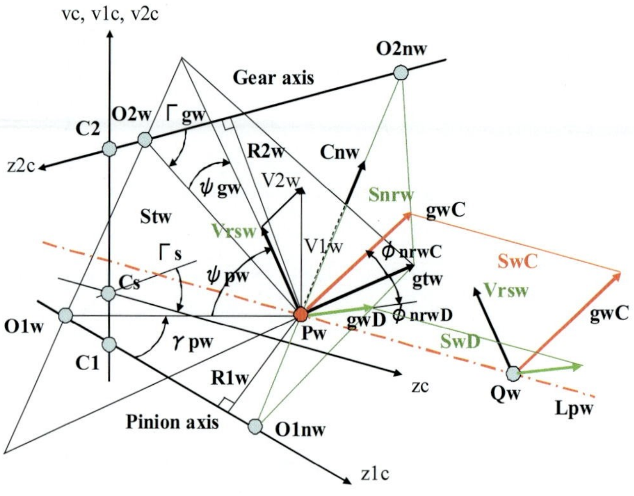

Fig.3.7-2(a) Pitch line element L_{pw}, planes S_{tw} and S_{nrw} and surfaces of action S_{wC} and S_{wD} at P_w in the coordinate systems C_1, C_2 and C_S

ities V_{1w} and V_{2w}, the relative velocity V_{rsw} and the plane S_{nrw} perpendicular to V_{rsw} are determined, where the paths of contact $\mathbf{g}_{wC}$ and $\mathbf{g}_{wD}$ on the plane S_{nrw} through P_w and the surfaces of action S_{wC} and S_{wD} made by $\mathbf{g}_{wC}$ and L_{Pw} or $\mathbf{g}_{wD}$ and L_{Pw} are drawn.

Figure 3.7-2(b) shows Fig. 3.7-2(a) seen from the positive direction of the axis z_c of the coordinate system C_s. The relative velocity V_{rsw} at P_w is on the plane S_{Prw} tangent to the cylinder which has the axis z_c and the radius r_{sCw} and passes P_w, on which V_{rsw} inclines by ψ_{rw} to L_{Pw}.

The relation between r_{sCw} and ψ_{rw} is already given by Eq. (2.2-3).

The design point P_w is given in the coordinate systems C_1, C_2 and C_s as follows.

$P_w(u_{Cw},\ v_{Cw},\ z_{Cw};\ C_s)$
$P_w(u_{1cw},\ v_{1cw},\ z_{1cw};\ C_1)$
$P_w(u_{2Cw},\ v_{2Cw},\ z_{2Cw};\ C_2)$

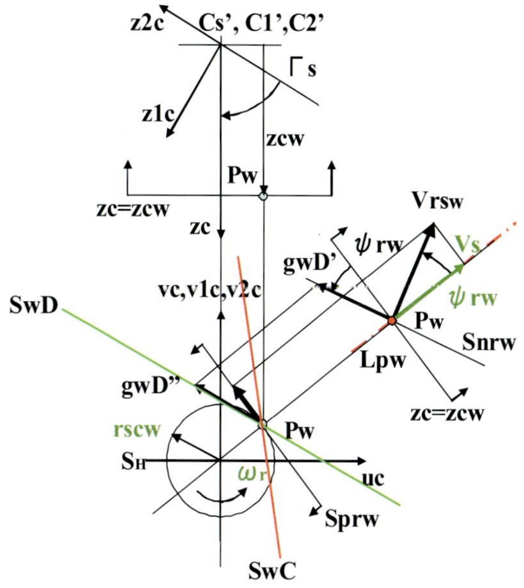

Fig.3.7-2(b) Relationship between r_{scw} and ψ_{rw} of $\mathbf{V}_{rsw}$ at P_w

(1) Reference radii R_{1w} and R_{2w} of the pinion and gear

The reference radii R_{1w} and R_{2w} of the pinion and gear are obtained as follows.

$$\left.\begin{aligned} R_{1w}^2 &= u_{1cw}^2 + v_{1cw}^2 \\ R_{2w}^2 &= u_{2Cw}^2 + v_{2Cw}^2 \end{aligned}\right\} \tag{3.7-1}$$

(Note 3.7-1) : Giving $R_{2w} = R_{20}$ and $v_{cw} = 0$ in Eq. (3.7-1) and using Eq. (2.2-3) or (3.2-1), the design point P_0 and the pair of hyperboloids (or approximate cones whose pitch angles are Γ_s and $\Sigma-\Gamma_s$) are obtained, which is the design method based on 3.2 to 3.6 in this theory. The approximate cones obtained here do not make contact at P_0. It is not the cones but the tooth surfaces that must make contact at the design point P_0.

(2) Pitch angles γ_{pw} and Γ_{gw}

Figure 3.7-3 shows the intersections O_{1nw} and O_{2nw} of each axis and the plane S_{nrw} shown in Fig. 3.7-2(a) and seen from the positive directions of the axes z_{1c}

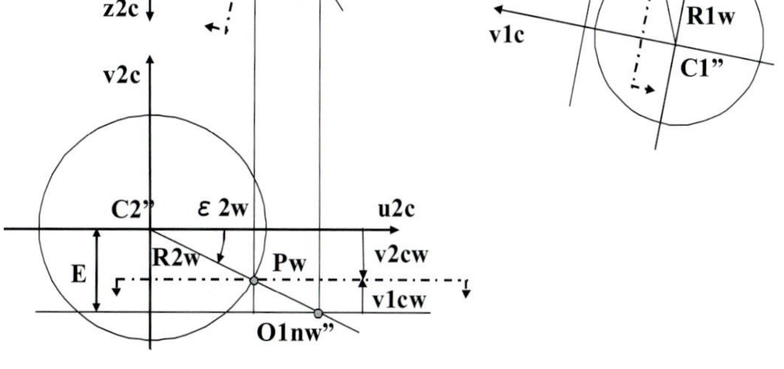

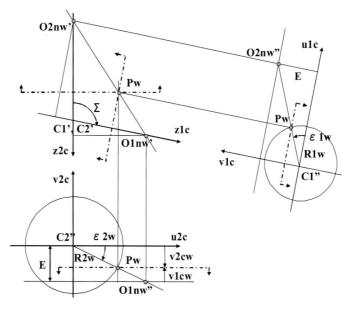

Fig.3.7-3 Relationship among P_w, O_{1nw} and O_{2nw}

and z_{2c}. O_{1nw} and O_{2nw} are expressed as follows.

$$O_{1nw}(0,\ 0,\ -E/(\tan\varepsilon_{2w}\sin\Sigma);\ C_1)$$
$$O_{2nw}(0,\ 0,\ -E/(\tan\varepsilon_{1w}\sin\Sigma);\ C_2)$$

where $\sin\varepsilon_{1w} = v_{1cw}/R_{1w}$ and $\sin\varepsilon_{2w} = v_{2CW}/R_{2W}$.

Therefore, $O_{1nw}P_w$ and $O_{2nw}P_w$ are obtained as follows.

$$O_{1nw}P_w = \sqrt{(R_{1w}{}^2 + (-E/(\tan\varepsilon_{2w}\sin\Sigma) - z_{1cw})^2)}$$
$$O_{2nw}P_w = \sqrt{(R_{2w}{}^2 + (-E/(\tan\varepsilon_{1w}\sin\Sigma) - z_{2Cw})^2)}$$

Pitch angles γ_{pw} and Γ_{gw} of the pinion and gear are obtained as follows, because $O_{1nw}P_w$ and $O_{2nw}P_w$ are the back cone elements.

$$\left.\begin{aligned} \cos\gamma_{pw} &= R_{1w}/O_{1nw}P_w \\ \cos\Gamma_{gw} &= R_{2w}/O_{2nw}P_w \end{aligned}\right\} \qquad (3.7\text{-}2)$$

From Eqs. (3.7-1) and (3.7-2), a pair of cones are obtained, which have the tangent plane S_{tw} in common, apexes O_{1w} and O_{2w} which are the intersections of the plane S_{tw} and each gear axis, the cone elements P_wO_{1w} and P_wO_{2w} and the radii R_{1w} and R_{2w} at P_w.

(3) Spiral angles ψ_{pw} and ψ_{gw} of the pinion and gear

For a pair of tooth surfaces to make contact at P_w, the plane tangent to the tooth surfaces at P_w must include the relative velocity $\mathbf{V}_{rsw}$, so that the intersection of the plane tangent to the tooth surfaces and the plane S_{tw} must coincide with $\mathbf{V}_{rsw}$.

The relative velocity $\mathbf{V}_{rsw}$ and the peripheral ones $\mathbf{V}_{1w}$ and $\mathbf{V}_{2w}$ are expressed as follows.

$$V_{rsw}/\omega_2 = \sqrt{((E\sin\Gamma_S)^2 + (r_{SCw}\sin\Sigma/\sin(\Sigma-\Gamma_S))^2)}$$
$$V_{1w}/\omega_2 = i_0 R_{1w}$$
$$V_{2w}/\omega_2 = R_{2w}$$
$$r_{SCw} = \sqrt{(u_{cw}{}^2 + v_{cw}{}^2)}$$

The angle $\psi v12w$ between V_{1w} and V_{2w} on the plane S_{tw} and the angle $\psi vrs1w$ between V_{rsw} and V_{1w} are obtained as follows.

$$\cos(\psi v12w) = (V_{1w}{}^2 + V_{2w}{}^2 - V_{rsw}{}^2)/(2V_{1w}V_{2w})$$
$$\cos(\psi vrs1w) = (V_{rsw}{}^2 + V_{1w}{}^2 - V_{2w}{}^2)/(2V_{1w}V_{rsw})$$

The spiral angles ψ_{pw} and ψ_{gw} of the pinion and gear are obtained as the inclination angles of V_{rsw} to the cone elements P_wO_{1w} and P_wO_{2w} on the plane S_{tw} as follows.

$$\left.\begin{aligned} \psi_{pw} &= \pi/2 - \psi vrs1w \\ \psi_{gw} &= \pi/2 - \psi v12w - \psi vrs1w \end{aligned}\right\} \qquad (3.7\text{-}3)$$

When the design point P_w $(u_{Cw}, v_{Cw}, z_{Cw}\ ;\ C_S)$ is given, the dimensions of the pair of cones which make contact and have the spiral angle ψ_{pw} or ψ_{gw} under the fundamental requirement for contact at P_w are obtained through Eqs. (3.7-1), (3.7-2) and (3.7-3).

(4) Selection of a pair of pitch cones (design point)

The simultaneous equations to select a pair of pitch cones are obtained from Eqs. (3.7-1), (3.7-2) and (3.7-3) as follows.

$$\left.\begin{aligned} R_{2w}{}^2 &= u_{2Cw}{}^2 + v_{2Cw}{}^2 \\ \cos\Gamma_{gw} &= R_{2w}/O_{2nw}P_w \\ \psi_{gw} &= \pi/2 - \psi\, v12w - \psi\, vrs1w \end{aligned}\right\} \qquad (3.7\text{-}4)$$

Equation (3.7-4) consists of three equations with six unknown quantities, therefore when three independent variables, for examples R_{2w}, ψ_{gw} and Γ_{gw} are given, a pair of cones (or a design point P_w $(u_{Cw}, v_{Cw}, z_{Cw}\ ;\ C_S)$) is obtained.

(a) Selection method based on this theory

Giving $\Gamma_{gw} = \Gamma_S$ besides R_{2w} and ψ_{gw}, $P_w(u_{Cw}, v_{Cw}, z_{Cw}; C_S)$ is determined. As mentioned later, the intersection of the pitch cones and the surfaces of action almost coincides with the pitch line element L_{pw} which passes P_w and is parallel to the instantaneous axis, so that the tooth traces in the drive and the coast sides are almost the same. In this theory, the reference bodies of revolution are the pitch hyperboloids formed by rotating the pitch line element L_{pw}, but the selected pitch cones here and the pitch hyperboloids have no difference practically. Based on a particular need, for example, to realize a bigger lengthwise contact ratio in the drive side than in the coast side, Γ_{gw} is chosen to be bigger than Γ_S.

(b) Wildhaber's method

Besides R_{2w} and ψ_{gw} (actually ψ_{pw} is given), giving the condition that the radius of curvature of the tooth trace at P_w must coincides with that of the cutter used in production instead of Γ_{gw}, the design point P_w $(u_{Cw}, v_{Cw}, z_{Cw}\ ;\ C_S)$ and the pitch angle Γ_{gw} are determined. The radius of curvature of the tooth trace at P_w is introduced under the special assumption that the path of contact through P_w is a circular arc around the gear axis as mentioned in 3.3.6(3), which is not inherent in hypoid gears. Therefore, it is not suitable for the design condition of constraint for pitch cones. Any radius of curvature can be given to the tooth trace as far as smooth tooth surfaces exist. Essentially there are no problems based on practical experience, if a cutter radius different from Wildhaber's one is chosen for the same pitch cones.

As mentioned later, Wildhaber's pair of pitch cones has no relation with the pitch line element L_{pw} parallel to the instantaneous axis, so that the intersections of the pitch cones and the surfaces of action S_{wC} and S_{wD} become different lines (tooth traces) from L_{pw} in the drive and the coast sides.

3. 7. 4 Selection methods of inclination angle of $\mathbf{g}_w$ ($\mathbf{g}_{wC}$ or $\mathbf{g}_{wD}$) at P_w

(1) Expression of inclination angle of $\mathbf{g}_w$ by planes S_{tw} and S_{nrw} (Wildhaber's method)

As shown in Fig. 3.7-2(a), the relative velocity $\mathbf{V}_{rsw}$ is on the plane S_{tw} tangent to the pitch cones and $\mathbf{g}_w$ is on the plane S_{nrw} perpendicular to $\mathbf{V}_{rsw}$, therefore the inclination angle of $\mathbf{g}_w$ can be expressed by a combination of that (ψ_{pw} or ψ_{gw}) of $\mathbf{V}_{rsw}$ on the plane S_{tw} and that (ϕ_{nrw}) of $\mathbf{g}_w$ on the plane S_{nrw}. That is the reason why the circumscribed cones at P_w are adopted as the pitch cones, but the plane S_{tw} which is the reference one regarding angles varies with the location of P_w, so that the inclination angles are hard to be caught visually and as a result the theory becomes difficult to understand.

Figure 3.7-2(a) shows the common contact normal $\mathbf{g}_w$ ($\mathbf{g}_{wC}$ or $\mathbf{g}_{wD}$) through the design point P_w, using the planes S_{tw} and S_{nrw}. When the intersection of the planes S_{nrw} and S_{tw} is $\mathbf{g}_{tw}$ and the inclination angle of $\mathbf{g}_w$ from $\mathbf{g}_{tw}$ on the plane S_{nrw} is ϕ_{nrw}, the inclination angle of $\mathbf{g}_w$ is expressed by a combination of the angle ψ_{gw} of $\mathbf{V}_{rsw}$ from P_wO_{2w} on the plane S_{tw} and the angle ϕ_{nrw} as $g_w(\psi_{gw}, \phi_{nrw}; S_{tw}, S_{nrw})$.

(2) Expression of inclination angles of $\mathbf{g}_w$ in the coordinate systems C_2 and C_s

In order to give the interchangeability between Wildhaber's method and that in this theory, the inclination angles of $\mathbf{g}_w$ are expressed in the coordinate systems C_2 and C_s as follows.

(a) Transformation from $g_w(\psi_{gw}, \phi_{nrw}; S_{tw}, S_{nrw})$ to $g_w(\phi_{2w}, \psi_{b2w}; C_2)$

Figure 3.7-4 shows the relation between $g_w(\psi_{gw}, \phi_{nrw}; S_{tw}, S_{nrw})$ and $g_w(\phi_{2w}, \psi_{b2w}; C_2)$, where $\mathbf{g}_w$ is given by a directed segment P_wA, the points corresponding to A are indicated by B, C, D and E in order and the sign (' or ") means the orthographic projection from each point toward each plane. The length of each directed segment is obtained as follows, where $P_wA = Lg$.

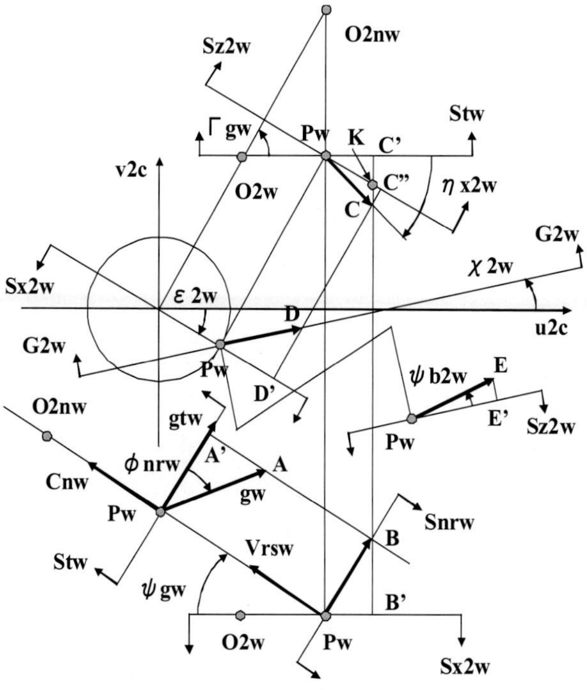

Fig.3.7-4 P_w and g_w shown on planes S_{tw}, S_{nrw} and G_{2w}

$A'A = Lg\sin\phi_{nrw}$
$B'B = Lg\cos\phi_{nrw}\cos\psi_{gw}$
$C'C = A'A$
$P_wC' = P_wB' = Lg\cos\phi_{nrw}\sin\psi_{gw}$
$C'K = P_wC'/\tan\Gamma_{gw}$
$C''C = (C'C - C'K)\sin\Gamma_{gw}$
$\quad = Lg(\sin\phi_{nrw} - \cos\phi_{nrw}\sin\psi_{gw}/\tan\Gamma_{gw})\sin\Gamma_{gw}$
$D'D = B'B$
$P_wE = P_wA$
$E'E = C''C$
$\sin\psi_{b2w} = E'E/P_wE = C''C/Lg$

$$= (\sin\phi_{nrw} - \cos\phi_{nrw}\sin\psi_{gw}/\tan\Gamma_{gw})\sin\Gamma_{gw} \quad (3.7\text{-}5)$$

$$\left.\begin{aligned}
&\tan\eta_{x2w} = C'C/P_wC' = \tan\phi_{nrw}/\sin\psi_{gw}\\
&P_wC = \sqrt{(P_wC'^2 + C'C^2)}\\
&\quad = Lg\sqrt{\{(\cos\phi_{nrw}\sin\psi_{gw})^2 + (\sin\phi_{nrw})^2\}}\\
&P_wC'' = P_wC\cos(\eta_{x2w} - (\pi/2 - \Gamma_{gw})) = P_wC\sin(\eta_{x2w} + \Gamma_{gw})\\
&\tan(\chi_{2w} - \varepsilon_{2w}) = D'D/P_wC'' = \cos\phi_{nrw}\cos\psi_{gw}\\
&\quad /[\sqrt{\{(\cos\phi_{nrw}\sin\psi_{gw})^2 + (\sin\phi_{nrw})^2\}}\sin(\eta_{x2w} + \Gamma_{gw})]\\
&\phi_{2w} = \pi/2 - \chi_{2w}
\end{aligned}\right\} \quad (3.7\text{-}6)$$

(b) Transformation from $g_w(\phi_{2w}, \psi_{b2w}; C_2)$ to $g_w(\psi_w, \phi_{nw}; C_s)$ and $g_w(\phi_{Sw}, \psi_{Sww}; C_S)$
Using Eq. (2.1-20), $g_w(\psi_w, \phi_{nw}; C_S)$ is obtained as follows.

$$\left.\begin{aligned}
&\sin\phi_{nw} = \cos\psi_{b2w}\sin\phi_{2w}\cos\Gamma_S + \sin\psi_{b2w}\sin\Gamma_S\\
&\tan\psi_w = \tan\phi_{2w}\sin\Gamma_S - \tan\psi_{b2w}\cos\Gamma_S/\cos\phi_{2w}
\end{aligned}\right\} \quad (3.7\text{-}7)$$

Using Eq. (2.1-24), $g_w(\phi_{Sw}, \psi_{Sww}; C_S)$ is obtained as follows.

$$\left.\begin{aligned}
&\tan\phi_{Sw} = -\tan\phi_{nw}/\cos\psi_w\\
&\sin\psi_{Sww} = \cos\phi_{nw}\sin\psi_w
\end{aligned}\right\} \quad (3.7\text{-}8)$$

(3) Selection of inclination angle of g_w

(a) Method in this theory

It is clarified in 3.3.5 where P_w is chosen as P_0 on the plane S_H. When P_w is given as a point of contact of the pitch cones, it is practical that it is chosen from the actual design results based on Wildhaber's method and transformed into the coordinate systems as mentioned in item (2).

(b) Wildhaber's method

g_{wD} and g_{wC} are inclined equally (for example, by 19°) from the limit normal n_{2w} which has the limit of action at P_w as mentioned in 3.3.6(3) and inclines by the limit pressure angle from g_{tw} on the plane S_{nrw}. However, the path of contact is not a circular arc around the gear axis generally, therefore this idea is not used now and they are chosen from design experiences.

3. 7. 5 Surfaces of action and contact ratios of Wildhaber's hypoid gears

In Wildhaber's theory and the design methods based on it, when the design point P_w and the common contact normal $\mathbf{g}_w$ are given, the tooth surface is obtained directly by giving the location of a cutter to realize these P_w and $\mathbf{g}_w$ in the machine and the generating roll of the machine. Therefore, there are few theoretical discussions regarding the surface of action and the contact ratios in Wildhaber's theory as far as the author knows. In the present method to estimate contact ratios of hypoid gears, the contact ratios are calculated under the assumption that they are imaginary bevel gears with a spiral angle $(\psi_{pw} + \psi_{gw})/2$ (Fig. 3.7-2(a)), which has no theoretical basis, so that it has no sufficient accuracy in practical use. Therefore, practically, they are estimated from the amount of movement of the line of contact on the tooth surface obtained numerically.

In hypoid gears designed by Wildhaber's method which have the design point P_w and the common contact normal $\mathbf{g}_w$, because the line of contact in the neighborhood of P_w is supposed to be a straight line on the plane tangent to the tooth surface, the plane made by the pitch line element L_{pw} and $\mathbf{g}_w$ through P_w can be assumed to be the surface of action in the neighborhood of P_w. Under the assumption that the surface of action of Wildhaber's hypoid gears can be replaced by the plane made by L_{pw} and $\mathbf{g}_w$ through P_w, the contact ratios are calculated as follows.

Figure 3.7-5 shows the path of contact $\mathbf{g}_{wD}$ and the surface of action S_{wD} based on Fig. 3.5-1, where $\mathbf{g}_{0D}$ is changed to $\mathbf{g}_w = \mathbf{g}_{wD}$, P to P_w and P_{0H} to P_{0D}. The pitch line element through P_w is L_{pw} and the surface of action S_{wC} intersects S_{wD} along L_{pw}. Because L_{pw} is parallel to L_{p0D} $(=L_{p0H})$, the contact ratios ms, mf and m0 can be calculated by the method mentioned in 3.6. Here, the tooth trace is supposed to be the intersection of the pitch cone of the gear and the surface of action mentioned above and the lengthwise contact ratio mfcone, the contact ratio mscone in the direction perpendicular to

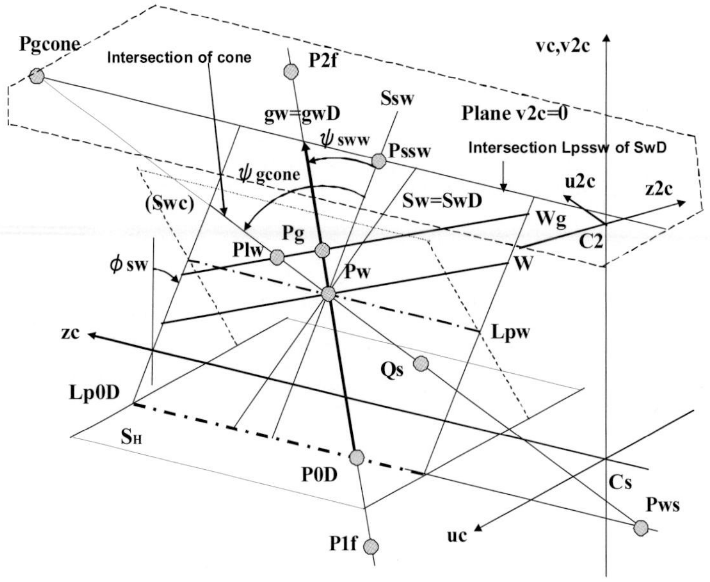

Fig.3.7-5 Relationship among L_{pw}, $\mathbf{g}_w$ and S_w at P_w

the tooth trace and the contact ratio m0cone along $\mathbf{g}_w$ are calculated. The intersection of S_{wD} and the plane $v_{2c} = 0$ is L_{pSSw}, the intersection of L_{pSSw} and the plane S_{Sw} (perpendicular to L_{pw} through P_w) is P_{SSw} and the intersection of L_{pSSw} and the pitch cone is Pgcone. P_wPgcone is the intersection of S_{wD} and the pitch cone and assumed to be a straight line, though it is not strictly straight because it is a conic section.

Figure 3.7-6 shows the relation among S_{wD}, P_wPgcone and the pitch angle Γ_{gw} on the plane $v_{2c} = 0$ seen from the positive direction of the axis v_{2c} in Fig. 3.7-5.

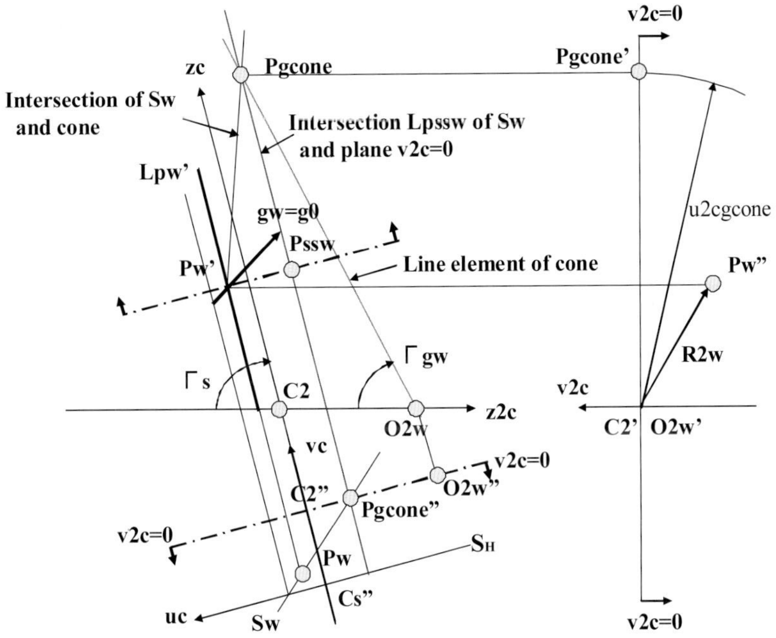

Fig.3.7-6 Intersection P_wPgcone of S_w and a cone through P_w

(1) Equation of path of contact $\mathbf{g}_w$ and one pitch on $\mathbf{g}_w$

When the inclination angle $\mathbf{g}_w$ (ϕ_{2w}, ϕ_{b2w} ; C_2) of $\mathbf{g}_w$ is given, P_w (u_{2Cw}, v_{2Cw}, z_{2Cw} ; C_2) is transformed into P_w (q_{2Cw}, $-R_{b2w}$, z_{2Cw} ; C_{q2}) by Eq. (2.1-2) as follows.

$$q_{2Cw} = u_{2Cw}\cos\chi_{2w} + v_{2Cw}\sin\chi_{2w}$$
$$R_{b2w} = u_{2Cw}\sin\chi_{2w} - v_{2Cw}\cos\chi_{2w}$$
$$\chi_{2w} = \pi/2 - \phi_{2w}$$

Therefore, the path of contact of $\mathbf{g}_w$ is obtained by Eq. (3.1-3) in the coordinate system C_{q2} as follows.

$$\left.\begin{aligned} q_{2c}(\theta_2) &= R_{b2w}\theta_2\cos^2\phi_{b2w} + q_{2cw} \\ R_{b2w}(\theta_2) &= R_{b2w} \\ z_{2c}(\theta_2) &= R_{b2w}\theta_2\cos\phi_{b2w}\sin\phi_{b2w} + z_{2cw} \\ g_w(\phi_{2w} &= \pi/2 - \chi_{2w}, \phi_{b2w} ; C_2) \end{aligned}\right\} \quad (3.7\text{-}9)$$

In the same way, the path of contact $\mathbf{g}_w$ is obtained by Eq. (3.1-5) in the coordinate system C_{q1} as follows.

$$\left.\begin{aligned} q_{1c}(\theta_1) &= R_{b1w}\theta_1\cos^2\phi_{b1w} + q_{1cw} \\ R_{b1w}(\theta_1) &= R_{b1w} \\ z_{1c}(\theta_1) &= R_{b1w}\theta_1\cos\phi_{b1w}\sin\phi_{b1w} + z_{1cw} \\ g_w(\phi_{1w} &= \pi/2 - \chi_{1w},\ \phi_{b1w}\ ;\ C_1) \end{aligned}\right\} \tag{3.7-10}$$

When P_w moves to P_g and the tangent plane W to Wg in parallel by one pitch of rotation of gears, the distance P_wP_g is obtained as follows (Fig. 3.7-5).

$$P_wP_g = p_{wg} = R_{b2w}(2\theta_{2p})\cos\phi_{b2w} = R_{b1w}(2\theta_{1p})\cos\phi_{b1w} \tag{3.7-11}$$

where p_{wg} : 1 pitch on $\mathbf{g}_w$,
$2\theta_{1p}$ and $2\theta_{2p}$: angular pitches of the pinion and gear respectively.

(2) Contact ratio on $\mathbf{g}_w$ m0cone

When $P_{2f}(q_{2cf}, -R_{b2w}, z_{2cf}\ ;\ C_{q2})$ is the intersection of $\mathbf{g}_w$ and the tip cone through Q_{2n} $(u_{2Q2n}, v_{2Q2n}, z_{2Q2n}\ ;\ C_2)$, using that Q_{2n} and P_{2f} are on the tip cone, the angle of rotation of the gear θ_{2f} at P_{2f} is obtained by Eq. (3.7-9) as follows, where the tip angles of the pinion and gear are supposed to be the same as the pitch angles γ_{pw} and Γ_{gw}.

$$R_{2f} = R_{2Q2n} + (z_{2Q2n} - z_{2cf})\tan\Gamma_{gw} = \sqrt{(q_{2cf}^2 + R_{b2w}^2)}$$

where $q_{2cf} = R_{b2w}\theta_{2f}\cos^2\phi_{b2w} + q_{2cw}$
$z_{2cf} = R_{b2w}\theta_{2f}\cos\phi_{b2w}\sin\phi_{b2w} + z_{2cw}$
when $\Gamma_{gw} = \pi/2$, $z_{2cf} = z_{2Q2n}$.

In the same way, the intersection P_{1f} $(q_{1cf}, -R_{b1w}, z_{1cf}\ ;\ C_{q1})$ of $\mathbf{g}_w$ and the tip cone of the pinion through Q_{1n} $(u_{1Q1n}, v_{1Q1n}, z_{1Q1n}\ ;\ C_1)$ and the angle of rotation of the pinion θ_{1f} at P_{1f} are obtained by Eq. (3.7-10) as follows, where the apex is on the negative side of the axis z_{1c}.

$$R_{1f} = R_{1Q1n} - (z_{1Q1n} - z_{1cf})\tan\gamma_{pw} = \sqrt{(q_{1cf}^2 + R_{b1w}^2)}$$

where $q_{1cf} = R_{b1w}\theta_{1f}\cos^2\phi_{b1w} + q_{1cw}$
$z_{1cf} = R_{b1w}\theta_{1f}\cos\phi_{b1w}\sin\phi_{b1w} + z_{1cw}$
when $\gamma_{pw} = \pi/2$, $z_{1cf} = z_{1Q1n}$.

Because points P_{2f} and P_{1f} on $\mathbf{g}_w$ are determined, the contact ratio m0cone is obtained as follows.

$$\begin{aligned} \text{m0cone} &= P_{1f}P_{2f}/p_{wg} \\ &= |\theta_{2f} - \theta_{1f}/i_0| / (2\theta_{2p}) \end{aligned} \tag{3.7-12}$$

(3) Contact ratio mfcone along the intersection of the gear pitch cone and S_W

According to ISO, the tooth trace is defined as the intersection of the tooth surface and the gear pitch cone, so the contact ratio mfcone along the intersection of the gear pitch cone and the surface of action S_w of hypoid gears is calculated here.

When the intersection L_{pSSw} of S_w and the plane $v_{2c} = 0$ intersects the gear pitch cone at Pgcone, Pgcone is expressed as follows.

$$\text{Pgcone}\ (\text{ucgcone},\ v_{cs2},\ \text{zcgcone};\ C_S)$$
$$\text{Pgcone}\ (\text{u2cgcone},\ 0,\ \text{z2cgcone};\ C_2)$$

where

$$\text{ucgcone} = u_{cw} + (v_{cs2} - v_{cw})\tan\phi_{Sw}$$
$$\text{zcgcone} = ((v_{cs2} - v_{cw})/\cos\phi_{Sw})\tan\phi\,\text{gcone} + z_{cw}$$
$$\text{u2cgcone} = -\text{ucgcone}\cos\Gamma s + \text{zcgcone}\sin\Gamma s$$
$$\text{z2cgcone} = -\text{ucgcone}\sin\Gamma s - \text{zcgcone}\cos\Gamma s$$

ϕ gcone : inclination angle of P_wPgcone from P_wP_{SSw} on S_W.

Because Pgcone is on the gear pitch cone with the pitch angle Γ_{gw} through P_w (Fig. 3.7-6), the following equation is obtained.

$$\text{u2cgcone} - R_{2w} = (z_{2cw} - \text{z2cgcone})\tan\Gamma_{gw} \tag{3.7-13}$$

When the pitch angle Γ_{gw} is given, ϕ gcone is obtained by Eq. (3.7-13). Therefore, one pitch pcone along P_wPgcone is obtained as follows.

$$\text{pcone} = p_{wg}/\cos(\phi\,\text{gcone} - \phi_{Sww}) \tag{3.7-14}$$

The length Flwpcone enclosed by the facewidth along P_wPgcone is obtained as follows. When the intersection of P_wPgcone and L_{pOD} in Fig. 3.7-5 is P_{wS} (u_{cwS}, 0, z_{cwS} ; C_S), u_{cwS} and z_{cwS} are obtained as follows.

$$u_{cwS} = u_{cw} - v_{cw}\tan\phi_{Sw}$$
$$z_{cwS} = z_{cw} - (v_{cw}/\cos\phi_{Sw})\tan\phi\,\text{gcone}$$

When a point Qs(ucqs, vcqs, zcqs ; C_S) is chosen at will on P_wPgcone (Fig. 3.7-5), ucqs and vcqs are expressed as functions of zcqs as follows.

$$\text{vcqs} = ((\text{zcqs} - z_{cwS})/\tan\phi\,\text{gcone})\cos\phi_{Sw}$$
$$\text{ucqs} = u_{cwS} + \text{vcqs}\tan\phi_{Sw}$$

Transforming the point Qs into the coordinate system C_2, the radius R2cqs of Qs(u2cqs, v2cqs, z2cqs ; C_2) is obtained as follows.

$$\text{u2cqs} = -\text{ucqs}\cos\Gamma s + \text{zcqs}\sin\Gamma s$$
$$\text{v2cqs} = \text{vcqs} - v_{cs2}$$
$$\text{R2cqs} = \sqrt{(\text{u2cqs}^2 + \text{v2cqs}^2)}$$

Substituting R_{2h} (outside radius of the gear) and R_{2t} (inside radius of the gear) for R2cqs, when the solutions of zcqs are zcqsh and zcqst respectively, the length Flwpcone is obtained as follows.

$$\text{Flwpcone} = (\text{zcqsh} - \text{zcqst})/\sin\phi\text{gcone} \quad (3.7\text{-}15)$$

Therefore, the contact ratio mfcone along P_wPgcone is obtained as follows.

$$\text{mfcone} = \text{Flwpcone}/\text{pcone} \quad (3.7\text{-}16)$$

Figure 3.7-6 shows the relation between Pgcone and the pitch angle Γ_{gw}. When $\Gamma_{gw} \to \Gamma_s$, the intersection L_{pSSw} and the pitch cone element Pgcone O_{2w} on the plane $v_{2c} = 0$ become parallel and the intersection Pgcone goes infinitely far, so that ϕgcone $\to \pi/2$ and P_wPgcone coincides with the pitch line element L_{pw}. Therefore, when $\Gamma_{gw} \to \Gamma_s$, mfcone (Eq. (3.7-16)) coincides with the lengthwise contact ratio mf(Eq. (3.6-5)) and the contact ratios mfcones of the drive and the coast sides become almost equal. In addition, mfcone depends on the pitch angle Γ_{gw}, so that adequate mfcone can be obtained by choosing Γ_{gw}.

(4) Contact ratio mscone in the direction perpendicular to the intersection of S_W and the gear pitch cone

One pitch pscone, the length Fscone and the contact ratio mscone in the direction perpendicular to the intersection of S_W and the gear pitch cone are obtained as follows.

$$\begin{aligned}\text{pscone} &= p_{wg}/\cos(\phi_{Sww} + \pi/2 - \phi\text{gcone})\\ \text{Fscone} &= P_{1f}P_{2f}\cos(\phi_{Sww} + \pi/2 - \phi\text{gcone})\\ \text{mscone} &= \text{Fscone}/\text{pscone} = \text{m0cone}\cos^2(\phi\text{sww} + \pi/2 - \phi\text{gcone})\end{aligned} \quad (3.7\text{-}17)$$

When $\Gamma_{gw} \to \Gamma$s, then ϕgcone $\to \pi/2$ and $\phi_{Sww} \to \phi_{Sw0}$, therefore mscone coincides with ms.

3.8 Design examples of tooth surfaces of hypoid gears (35)

In this section, hypoid gears are designed by both Wildhaber (Gleason)'s method and the new one to compare the dimensions, especially the contact ratios. Table 3.8-1 shows the dimensions of both hypoid gears. A pair of circumscribed cones are adopted as the pitch cones and the following five input data are common.

(1) Shaft angle $\Sigma = 90°$, offset E= 28 mm and gear ratio $i_0 = 47/19$

From Eqs. (2.1-8) and (2.1-10), the intersection C_S of the instantaneous axis and the common perpendicular, the inclination angle Γ_S of the instantaneous axis and the coordinate system C_S are determined to the coordinate systems C_1 and C_2 as follows.

$$C_S(0,\ -24.067,\ 0\ ;\ C_2),\qquad C_S(0,\ 3.933,\ 0\ ;\ C_1),\qquad \Gamma_S = 67.989°$$

(2) Pitch radius of gear $R_{2w} = 89.255$ mm and spiral angle of pinion $\phi_{pw} = 46.988°$.

3.8.1 Wildhaber's method

The left-hand side of Table 3.8-1 shows the dimensions of hypoid gears by Wildhaber's method, where the pitch cones are chosen so as to satisfy that the lengthwise radius of curvature of the gear tooth surface coincides with the radius of the cutter ($rc = 3.75''$).

Based on the new theory, the design point P_w and the paths of contact $\mathbf{g}_{wD}$ and $\mathbf{g}_{wC}$ are obtained as follows.

(1) Design point P_w

When the gear pitch radius $R_{2w} = 89.255$mm, the pinion spiral angle $\psi_{pw} = 46.988°$ and the gear pitch angle $\Gamma_{gw} = 62.784°$ are already given, the design point P_w is obtained by the simultaneous equations (3.7-1), (3.7-2) and (3.7-3) as follows.

$$P_w(9.358,\ 1.836,\ 97.021\ ;\ C_S)$$

Therefore, the pitch line element L_{pw} is determined as a straight line through P_w parallel to the instantaneous axis ($\Gamma_s = 67.989°$) in the coordinate system C_s.

(2) Common contact normal $\mathbf{g}_{wD}$ through P_w and contact ratios of tooth surface D in the drive side

From Table 3.8-1, when the inclination angle g_{wD} ($\psi_{gw} = 30.859°$, $\phi_{nrwD} = 15°$) of $\mathbf{g}_{wD}$ is given, using Eqs. (3.7-5), (3.7-6), (3.7-7) and (3.7-8), it is transformed into the coordinate systems C_S and C_2 as follows.

$$g_{wD}(\phi_{2wD} = 48.41°,\ \phi_{b2wD} = 0.20°\ ;\ C_2),\quad g_{wD}(\phi_{wD} = 46.19°,\ \phi_{nwD} = 16.48°\ ;\ C_S)$$
$$g_{wD}(\phi_{SwD} = -23.13°,\ \phi_{SwwD} = 43.79°\ ;\ C_S)$$

Table 3.8-1 Dimensions of design examples

Dimensions designed	by Gleason's and Pinion	Gear	by new method pinion	gear
Shaft angle Σ	90°			
Offset E	28 (below center)			
Number of teeth N1, N2	19	47		
Angle of instantaneous axis Γs	67.989°			
Radius of cutter Rc (radius of lengthwise curvature of gear)	3.75″			
Reference point Pw	Pw(9.358, 1.836, 97.021: Cs),		Pw(9.607, 0.825, 96.835: Cs)	
Pitch radius R1w, R2w	45.406	89.255	45.449	89.255
Pitch angle of cone γpw, Γgw	26.291°	62.784°	21.214 °	67.989°
Spiral angle on pitch plane ψpw, ψgw	46.988°	30.859°	46.988°	30.768°
Inside and outside radius of gear (facewidth) R2t, R2h (Fg)	73.9, 105, (35)			
Addendum, dedendum and working depth of gear	1.22 6.83 7.15			
Contact ratios	Drive side	Coast side	Drive side	Coast side
Pressure angle φnrw	15°	-27.5°	15°	-27.5°
Transverse(mscone)	1.13 (1.71)	0.78 (0.81)	1.21	0.96
Lengthwise(mfcone)	**2.45** (1.70)	**2.45** (2.92)	2.37	2.40
Total(mscone+mfcone)	3.58 (3.41)	3.23 (3.73)	3.58	3.36

The pitch line element L_{pw} and the common contact normal $\mathbf{g}_{wD}$ determine the plane surface of action S_{wD} in the coordinate system C_s.

According to the assumption that the surface of action of the hypoid gears designed by Wildhaber's method can be replaced by the plane surface of action S_{wD}, the contact ratios mfconeD and msconeD are obtained as follows.

One pitch p_{wgD} on $\mathbf{g}_{wD}$ is obtained by Eq. (3.7-11) as follows.

$$p_{wgD} = 9.894 \text{ mm}$$

From Table 3.8-1, when $\Gamma_{gw} = 62.784°$ and $\gamma_{pw} = 26.291°$ are given,

(a) from Eq. (3.7-12), m0coneD = 2.232,
(b) using Eqs. (3.7-13) to (3.7-16),
ϕgconeD = −74.98°, pconeD = 20.56 mm, FlwpconeD = 34.98 mm, mfconeD = 1.701,
(c) from Eq. (3.7-17), msconeD = 1.715.

(3) Common contact normal $\mathbf{g}_{wC}$ through P_w and contact ratios of tooth surface C in the coast side

In the same way as the tooth surface D, when $\mathbf{g}_{wC}$ ($\phi_{gw} = 30.859°$, $\phi_{nrwC} = -27.5°$) of $\mathbf{g}_{wC}$ is given,

$$g_{wC}(\phi_{2wC} = 28.68°,\ \phi_{b2wC} = -38.22°;\ C_2),\quad g_{wC}(\phi_{wC} = 40.15°,\ \phi_{nwC} = -25.61°;\ C_S)$$
$$g_{wC}(\phi_{SwC} = 32.10°,\ \phi_{SwwC} = 35.55°;\ C_S)$$

One pitch p_{wgC} on g_{wC} and the contact ratios are obtained as follows.

$p_{wgC} = 9.086$ mm, ϕgconeC = 81.08°, pconeC = 12.971 mm, FlwpconeC = 37.86 mm,
m0coneC = 1.591, mfconeC = 2.919, msconeC = 0.810

In hypoid gears designed by Wildhaber's method, the contact ratios along the intersection of the gear pitch cone and the plane surface of action are mfconeD = 1.70 and mfconeC = 2.92 because of $\Gamma_{gw} = 62.784°$, which means the drive side is more disadvantageous to the coast one in contact ratio.

The lengthwise contact ratios by Wildhaber's method in Table 3.8-1 are 2.45 for both the drive and the coast sides because they are calculated as imaginary bevel gears with the spiral angle $(\phi_{pw} + \phi_{gw})/2$, but they have no theoretical basis and insufficient accuracy for practical use.

(4) Generation of tooth surfaces and accuracy of calculated contact ratios

The surface of action and the corresponding tooth surface are obtained by setting up a cutter so as to realize the tooth normal $\mathbf{g}_w$ ($\mathbf{g}_{wC}$ or $\mathbf{g}_{wD}$) at P_w and giving a generating roll in the machine. Strictly speaking, only the point of contact P_w and the tooth normal $\mathbf{g}_w$ of the tooth surface generated above are on the assumed plane surface of action, so that the generated tooth surface is approximated only in the neighborhood of P_w. Therefore, the accuracy of the calculated contact ratios mentioned above needs to be verified.

3. 8. 2 New design method (Γ_{gw} is given directly)

The right-hand side of Table 3.8-1 shows the dimensions of hypoid gears designed by the new method, where the following dimensions are chosen the same as those by Wildhaber's method to compare one to the other.

Inside and outside radii of gear R_{2t} =73.87 mm and R_{2h} = 105 mm
Addendum, dedendum and working depth 1.22 mm, 6.83 mm and 7.15 mm
Pressure angles ϕ_{nrwD} = 15° and ϕ_{nrwC} = -27.5°

(1) Design point P_w and pitch cones

When the pitch radius of gear R_{2w} = 89.255 mm, the pinion spiral angle ψ_{pw} = 46.988° and the pitch angle of gear Γ_{gw} = 67.989° (= Γ_S) are given, the design point P_w and other dimensions are obtained as follows by Eqs. (3.7-1), (3.7-2) and (3.7-3), where the following items are different from Wildhaber's dimensions in Table 3.8-1.

Design point	P_w(9.607, 0.825, 96.835; C_S)
Pitch radius of pinion	R_{1w} = 45.449 mm
Pitch angle of gear	Γ_{gw} = 67.989°
Pitch angle of pinion	γ_{pw} = 21.214°
Spiral angle of gear	ψ_{gw} = 30.768°

(2) Paths of contact $\mathbf{g}_{wD}$ and $\mathbf{g}_{wC}$ and contact ratios of tooth surfaces D and C

Transforming the pressure angles ϕ_{nrwD} and ϕ_{nrwC} which are the same as Wildhaber's dimensions into the coordinate systems C_2 and C_S, the inclination angles of $\mathbf{g}_{wD}$ and $\mathbf{g}_{wC}$ are obtained as follows, using Eqs. (3.7-5) to (3.7-8).

g_{wD}(ψ_{gw} = 30.768°, ϕ_{nrwD} = 15°), g_{wC}(ψ_{gw} = 30.768°, ϕ_{nrwC} = -27.5°)
g_{wD}(ϕ_{2wD} = 48.87°, ψ_{b2wD} = 3.14°; C_2), g_{wC}(ϕ_{2wC} = 33.09°, ψ_{b2wC} = -36.74°; C_2)
g_{wD}(ψ_{wD} = 45.86°, ϕ_{nwD} = 19.43°; C_S), g_{wC}(ψ_{wC} = 43.17°, ϕ_{nwC} = -22.99°; C_S)
g_{wD}(ϕ_{SwD} = -26.86°, ψ_{SwwD} = 42.59°; C_S), g_{wC}(ϕ_{SwC} = 30.19°, ψ_{SwwC} = 39.04°; C_S)

One pitches on $\mathbf{g}_{wD}$ and $\mathbf{g}_{wC}$ are obtained by Eq. (3.7-11) as follows.

p_{wgD} = 9.903 mm, p_{wgC} = 9.094 mm

The contact ratios are obtained by Eqs. (3.7-12), (3.7-16) and (3.7-17) as follows.

Drive side : msconeD = 1.215, mfconeD = 2.371 and m0coneD = 2.241
Coast side : msconeC = 0.960, mfconeC = 2.403 and m0coneC = 1.592

Compared with Wildhaber's dimensions, the lengthwise contact ratios mfconeD ≒ mfconeC are realized. This is because ψ gconeD = -89.99° is under Γ_{gw} = Γ_S = 67.989° and the intersection L_{pw} of the plane surfaces of action nearly coincides with the intersections of the gear pitch cone and the plane surfaces of action S_{WD} and S_{WC} (Figs. 3.7-5, 6).

(3) Pitch hyperboloids, plane surfaces of action and tooth surfaces

Figures 3.8-1 to 3.8-8 show the pitch hyperboloid, the plane surfaces of action and the tooth surfaces calculated by the new method.

Figure 3.8-1 shows the pitch line element L_{pw} and the pitch hyperboloid of the gear calculated by Eqs. (3.5-9) and (3.5-10).

Figures 3.8-2 and 3.8-3 show the effective plane surfaces of action in the drive and the coast sides by Eq. (3.5-1) in the coordinate system C_S which corresponds to Fig. 3.7-5.

Figures 3.8-4 and 3.8-5 show the tooth surfaces of the pinion in the drive and the coast sides by Eq. (3.5-6).

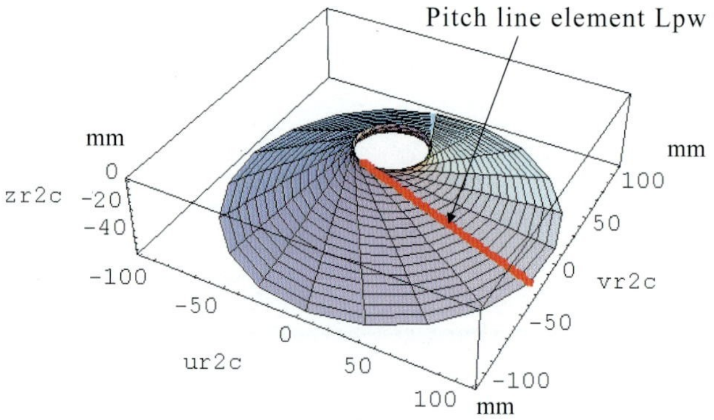

Fig.3.8-1 Pitch line element and hyperboloid of gear

Figures 3.8-6 and 3.8-7 show the tooth surfaces of the gear in the drive and the coast sides by Eq. (3.5-5).

Figure 3.8-8 shows the intersections of the gear tip cone and the tooth surfaces of the drive and coast sides which are nearly the tooth traces because P_w is on L_{pw} and are seen from the positive direction of the gear axis. It shows that they are almost the same curves, where the phase angle of the coast side equals 0 to compare with each other.

(a) Angles between curves on tooth surfaces of the gear at P_w

The results calculated by the method mentioned in 3.5.6 are entered into Figs. 3.8-6 and 3.8-7. Those by Eq. (3.2-13) are almost the same. On the tooth surface C, lines of contact of the gear are nearly parallel to the planes of rotation.

(b) Radii of curvature of curves on tooth surfaces of the gear at P_w

The results calculated by Eq. (3.5-13) are as follows. Those by Eq. (3.2-14) are shown in parentheses for reference.

	Tooth surface D	Tooth surface C
Curves in the plane of rotation	ρ_{n20ZD} = 50.2 mm (55.0),	ρ_{n20ZC} = 8107 mm (20403)
Principle radii of curvature	ρ_{n20D} = 15.8 mm (17.3),	ρ_{n20C} = 79.5 mm (0.20)
(Curves in the direction perpendicular to the line of contact)		
Paths of contact on tooth surface	ρ_{n20GD} = 49.9 mm,	ρ_{n20GC} = 129.3 mm
Tooth traces (Fig. 3.8-8)	$\rho_{n20LpwD}$ = 81.8 mm,	$\rho_{n20LpwC}$ = 81.5 mm

Equation (3.2-14) is effective on the tooth surface D, but it can give no approximate values on the tooth surface C, because it has no reliability in the case of $\eta_{Z2wC} = -89.2°$.

From each radius of curvature mentioned above, the radius R_{n20} of the cylinder which has the line of contact as a cylinder element is calculated by Euler's equation as follows, which means that the tooth surface is not approximated by the cylindrical surface which has the principal radius of curvature R_{n20} and the line of contact (a straight line).

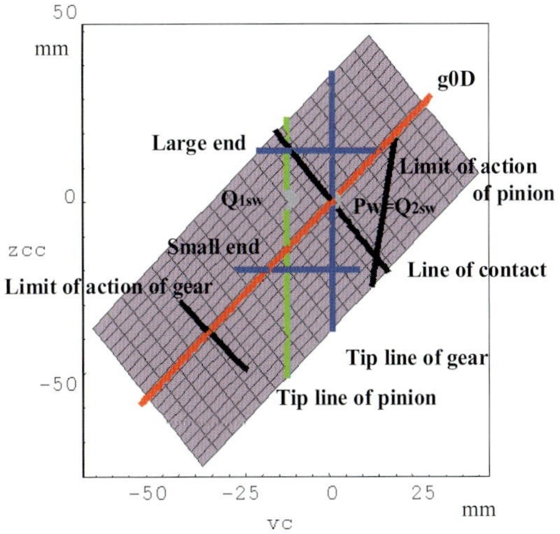

Fig.3.8-2 Effective surface of action SwD

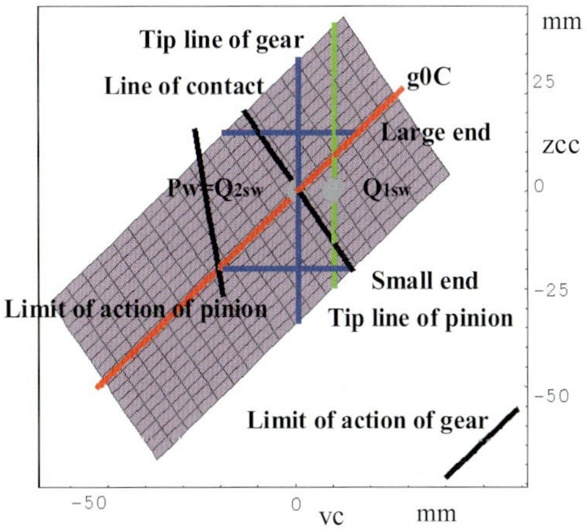

Fig.3.8-3 Effective surface of action SwC

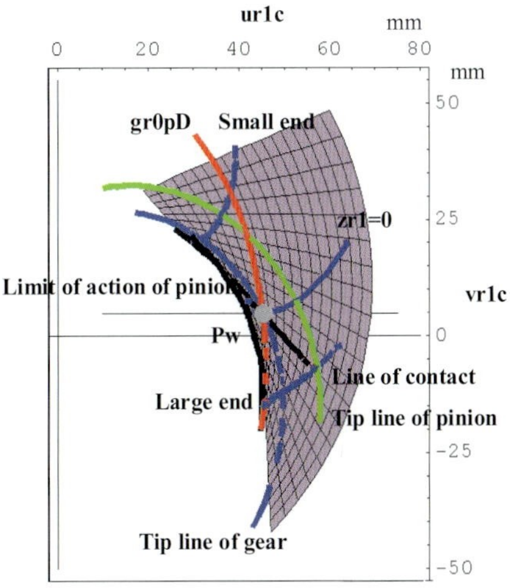

Fig.3.8-4 Tooth surface D of pinion

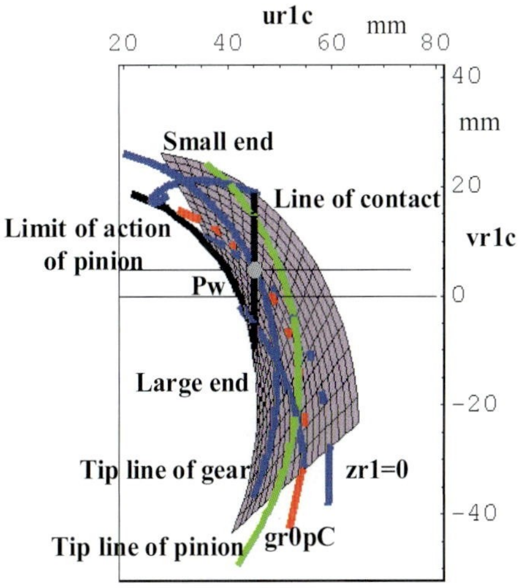

Fig.3.8-5 Tooth surface C of pinion

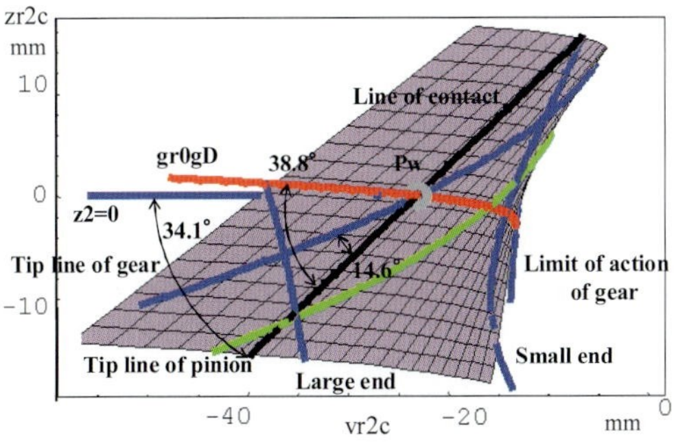

Fig.3.8-6 Tooth surface D of gear

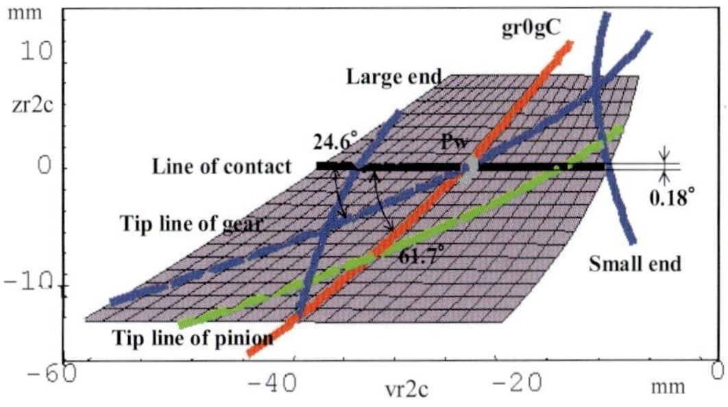

Fig.3.8-7 Tooth surface C of gear

From the curves in the plane of rotation

R_{n20D} = 15.8 mm ($50.2\cos^2 55.9°$)

R_{n20C} = 0.10 mm ($8107\cos^2 89.8°$)

From the paths of contact on tooth surfaces

R_{n20D} = 19.6 mm ($49.9\cos^2 51.2°$)

R_{n20C} = 100 mm ($129.3\cos^2 28.3°$)

From the tooth traces

R_{n20D} = 5.20 mm ($81.8\cos^2 75.4°$)

R_{n20C} = 14.1 mm ($81.5\cos^2 65.4°$)

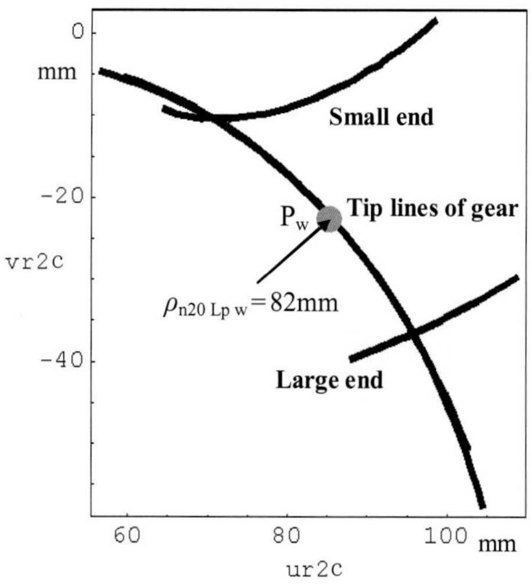

Fig.3.8-8 Tip lines of drive and coast sides of gear

Figures 3.8-2 to 7 show plane surfaces of action and corresponding tooth surfaces common to all kinds of gears, therefore, when the shaft angle equals 0, Figs. 3.8-2 and 3.8-3 show the planes of action of involute helical gears, Figs. 3.8-4 and 3.8-5 show the involute helicoids of the pinion and Figs. 3.8-6 and 3.8-7 show those of the gear.

(4) Correction of design point

Correcting the design point P_w according to 3.6.4, P_w, g_{wC} and mfD = mfC are obtained as follows.

$P_w(9.540,\ 0.957,\ 96.971;\ C_S)$

$g_{wC}(\phi_{wC} = 42.73°,\ \phi_{nwC} = -22.99°;\ C_S)$

mfD = mfC = 2.366

In this case, the difference between mfD and mfC was small at first, so that P_w changes little.

(5) Realization of tooth surfaces corresponding to plane surface of action

Following two methods can be considered.

(a) Cutting by NC machine tool

A ball end mill controlled numerically based on the results shown in Figs. 3.8-4 and 5 and Figs. 3.8-6 and 7 can manufacture the tooth surfaces with the plane surface of action. It has the advantage that one NC machine tool can cut any kind of gears if the quantity is small.

(b) Approximate cutting by Gleason's Formate method

As shown in Fig. 3.8-8, the tip curves (tooth traces) of the gear in both drive and coast sides can be approximated by the same circles, which means that they can be replaced by the cone type cutter, so that the tooth surfaces can be cut by Gleason's Formate method. However, the tooth surfaces are approximated only in the neighborhood of P_w.

(6) Trial results by Gleason's Formate method

Figure 3.8-9 shows the trial results produced by Gleason's Formate method. The left-hand side shows the results related to Wildhaber's hypoid gears and the right-hand side shows the results related to the new one. This figure draws the following conclusions.

(a) The cutter radius in the new method adopts 3.75″ common to Wildhaber's hypoid gears, but the tooth bearings are normal and have no problems caused by the cutter.

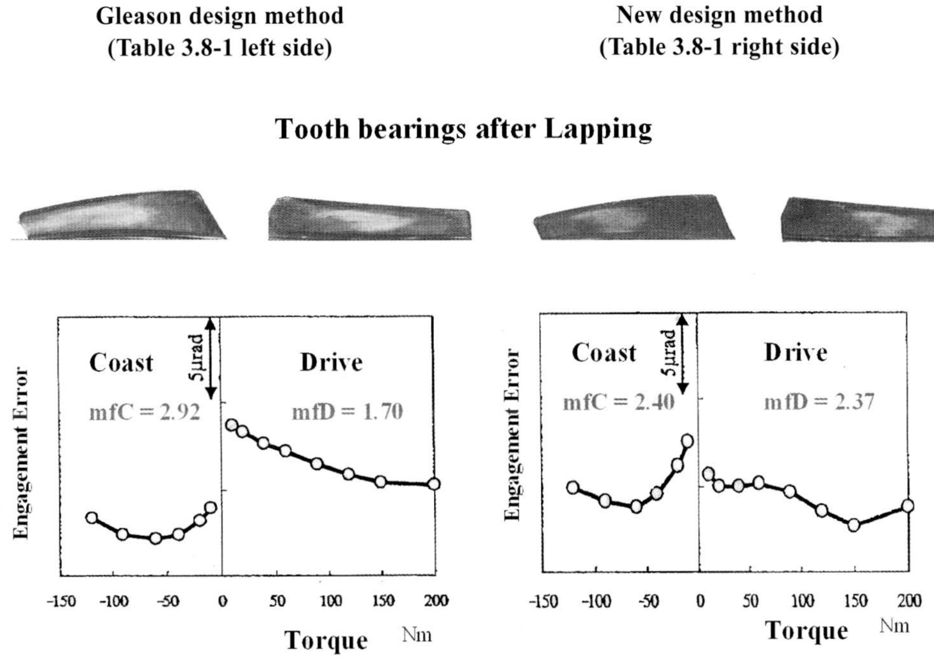

Fig.3.8-9 Tooth surfaces produced by Gleason's Formate method

(b) The engagement errors under load become smaller on the drive side of the new hypoid gears according to the increase of the lengthwise contact ratio. Conversely, on the coast side they become bigger according to the decrease of the lengthwise contact ratio. In practice, the noise in the vehicle is much reduced on the drive side. Each point means the average of engagement errors of five products.

(c) If the cutter radius r_c is chosen to realize the radius of curvature of the tooth trace $\rho_{n20LpwD} \fallingdotseq \rho_{n20LpwC}$ (Fig. 3.8-8), it might be possible for the surface of action to be nearer to the plane one. For example,

$$r_C = \rho_{n20LpwD} \cos \gamma_C = 81.8 \cos 20^\circ = 76.9 \text{ mm } (3.03'')$$

where $\gamma_c = 20^\circ$: half of cone angle made by cutter blades around the cutter axis.

If the cutter radius $r_c = 3.75''$ is changed to $3.0''$, as mentioned in Chapter 5, the surface of action might be nearer to the plane one and the tooth bearings are expected to become more robust and less prone to assembly errors.

3. 9 Summary and references

The ideas necessary to design the tooth surfaces for quieter gear pairs for power transmission are clarified as follows.

(1) A gear pair with a constant gear ratio realizes no variation of bearing loads when the path of contact coincides with the common contact normal which is a straight line fixed in static space.

(2) The equations of the path of contact for no variation of bearing loads are obtained.

(3) The selection methods for the inclination angle of the path of contact to have smooth tooth surfaces around the design point are discussed, and the limit paths of contact $\mathbf{g}_{1S}$ and $\mathbf{g}_{2S}$ in line contact and the ones $\mathbf{g}_{1k}$ and $\mathbf{g}_{2k}$ in point contact are clarified. The path of contact is necessary to incline adequately from these limit paths of contact. These limit paths of contact and Wildhaber's limit normal are ones of the innumerable solutions of Eq. (2. 4-3) which give limits of action to a path of contact.

(4) The design reference line $\mathbf{g}_t$, the design vertical $\mathbf{C}_n$ and the equivalent rack are introduced to determine the dimensions of the tooth shape (tooth depth, tooth thickness, tip and root circles and so on).

(5) The pitch line element which passes the design point and is parallel to the instantaneous axis is introduced, which coincides with the instantaneous axis in bevel and cylindrical gears. The rotating bodies of the pitch line element around each axis are the reference bodies of revolution (pitch hyperboloids) common to all kinds of gears.

(6) It is clarified that the given path of contact $\mathbf{g}_0$ and the pitch line element make the plane surface of action S_W common to all kinds of gears and the pitch line element is the intersection of the plane surfaces of action of the drive and the coast sides.

(7) Equations of the plane surface of action S_W, the corresponding conjugate tooth surfaces and the modified tooth surfaces are obtained.

(8) The tooth trace is defined as the intersection of the plane surface of action and the pitch hyperboloid or the pitch cone and the calculation methods of the contact ratios are clarified.

(9) Wildhaber's design method for hypoid gears is examined to develop new methods based on the new theory, for examples, the selection method of the pitch cones and the calculation method of the contact ratios. According to the new method, it is clarified that the pitch angles chosen so as to coincide with the inclination angle of the instantaneous axis can realize almost the same lengthwise contact ratios for the drive and the coast sides.

(10) When the same inputs are given, the design examples by Wildhaber's method and the new one are compared with each other. The lengthwise contact ratios of Wildhaber's hypoid gears calculated under the assumption that they have the plane surfaces of action are smaller on the drive side than on the coast side, while those of the new hypoid gears are almost the same on both drive and coast sides. In addition, the plane surfaces of action and the corresponding tooth surfaces designed by the new method are shown graphically.

(11) Both the hypoid gears are produced by Gleason's Formate method and it is verified that the cutter radius has no relation to the pitch angles and the engagement errors under load have a correlation with the lengthwise contact ratios calculated by the new method.

References

(31) 本多捷，歯面の接触と動荷重の基礎理論（第２報），機論，62-600，C(1996.8)，3269-3274.
(32) 本多捷，動力伝達用歯車の設計理論（第２報），機論，70-690，C(2004.2)，567-574.
(33) 本多捷，動力伝達用歯車の設計理論（第３報），機論，70-692，C(2004.4)，1190-1198.
(34) 本多捷，動力伝達用歯車の設計理論（第４報），機論，70-696，C(2004.8)，2508-2514.
(35) Honda,S.，“Design of hypoid gears having equal lengthwise contact ratios of drive and coast sides”, Proceedings of JSME Int. Conf. MPT-Sendai，2009，70-75.
(36) 岩堀長慶，数学選書２　ベクトル解析，(1995)，p.118，(株)裳華房.
(37) Wildhaber, E.，“Basic Relationship of Hypoid Gears II”, American Machinist, Feb.28, 1946, 131-134.
(38) Baxter, M. L.，“Basic Geometry and Tooth Contact of Hypoid Gears”, Industrial Mathematics, Vol.11，1961，19-42.

4. Involute helicoid and its conjugate tooth surface

In this chapter, a pair of tooth surfaces are discussed which have the path of contact $\mathbf{g}_0$ ($\mathbf{g}_{0D}$ or $\mathbf{g}_{0C}$) mentioned in items (1) and (2b) of 3.5, namely,

4.2 Equations of an involute helicoid and a pair of involute gears in point contact (mentioned in item 3.5(1)),

4.3 A surface of action and its conjugate tooth surface of an involute helicoid and

4.4 to 4.5 Gear pairs which have involute helicoids and their conjugate tooth surfaces (mentioned in item 3.5(2b)).

In the following discussion from Chapters 4 to 6, new coordinate systems are introduced, because they are more convenient to treat problems in the neighborhood of the design point P_0.

4. 1 Definition of coordinate systems O_2, O_{q2}, O_1 and O_{q1} (40)

Figure 4.1-1 shows the path of contact $\mathbf{g}_0$ in the coordinate systems O_2, O_{q2}, O_1 and O_{q1}. The coordinate systems $O_2(u_2, v_2, z_2)$ and O_{q2} (q_2, v_{q2}, z_2) have the origin O_2 in common which is the intersection of the axis II and the plane of rotation Z_{20} through the design point P_0 around the axis II and are made by transferring by C_2O_2 the coordinate systems C_2 and C_{q2} in the direction of the axis z_{2c}. In the same way, the coordinate systems O_1 (u_1, v_1, z_1) and O_{q1} (q_1, v_{q1}, z_1) have the origin O_1 in common which is the intersection of the axis I and the plane of rotation Z_{10} through the design point P_0 around the axis I and are made by transferring by C_1O_1 the coordinate systems C_1 and C_{q1} in the direction of the axis z_{1c}.

The relations among the coordinate systems C_2 and O_2, C_{q2} and O_{q2} and O_2 and O_{q2} are obtained as follows.

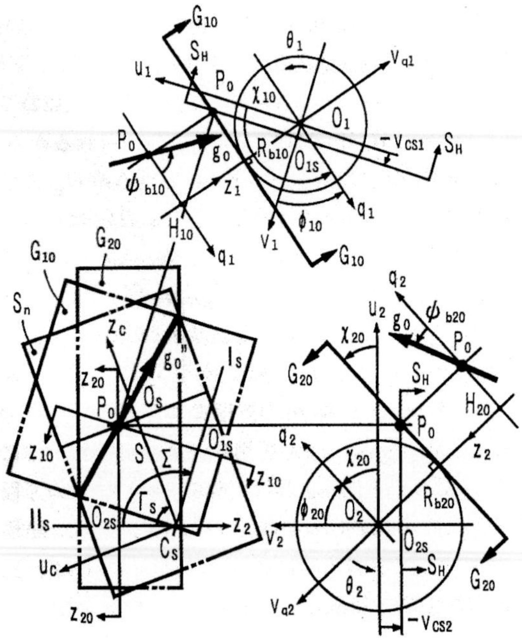

Fig.4.1-1 P_0 and $\mathbf{g}_0$ shown by coordinate systems O_2, O_{q2}, O_1 and O_{q1}

(1) C_2 and O_2

$$u_2 = u_{2c}$$
$$v_2 = v_{2c}$$
$$z_2 = z_{2c} - z_{2c0}$$

(2) C_{q2} and O_{q2}

$$q_2 = q_{2c}$$
$$v_{q2} = v_{q2c} = -R_{b2}$$
$$z_2 = z_{2c} - z_{2c0}$$

(3) O_2 and O_{q2} (z_2 is common)

$$u_2 = q_2 \cos\chi_2 + R_{b2}\sin\chi_2$$
$$v_2 = q_2 \sin\chi_2 - R_{b2}\cos\chi_2$$
$$\chi_2 = \pi/2 - \phi_2$$

where $z_{2c0} = C_S O_{2S} = -(u_{c0}\sin\Gamma_S + z_{c0}\cos\Gamma_S)$.

In the same way, relations among the coordinate systems C_1 and O_1, C_{q1} and O_{q1}, O_1 and O_{q1} and O_1 and O_2 are obtained as follows.

(4) C_1 and O_1

$$u_1 = u_{1c}$$
$$v_1 = v_{1c}$$
$$z_1 = z_{1c} - z_{1c0}$$

(5) C_{q1} and O_{q1}

$$q_1 = q_{1c}$$
$$v_{q1} = v_{q1c} = -R_{b1}$$
$$z_1 = z_{1c} - z_{1c0}$$

(6) O_1 and O_{q1} (z_1 is common)

$$u_1 = q_1 \cos\chi_1 + R_{b1}\sin\chi_1$$
$$v_1 = q_1 \sin\chi_1 - R_{b1}\cos\chi_1$$
$$\chi_1 = \pi/2 - \phi_1$$

where $z_{1c0} = C_S O_{1S} = -u_{c0}\sin(\Sigma - \Gamma_S) + z_{c0}\cos(\Sigma - \Gamma_S)$.

(7) O_1 and O_2

$$u_1 = -u_2\cos\Sigma - (z_2 + z_{2c0})\sin\Sigma$$
$$v_1 = v_2 + E$$
$$z_1 = u_2\sin\Sigma - (z_2 + z_{2c0})\cos\Sigma - z_{1c0}$$

4．2 Equations of involute helicoid [41] and a pair of involute gears in point contact [42], [43]

4．2．1 Equations of path of contact $\mathbf{g}_0$

When the shaft angle Σ and the offset E of the two axes and the gear ratio i_0 (constant) are given, equations of the path of contact $\mathbf{g}_0$ which passes the design point P_0 and realizes no variation of bearing loads are obtained by substituting Eq. (3.2-7) for Eq. (3.1-3) and by transforming them into the coordinate systems O_2 and O_{q2} as follows.

$$\left.\begin{aligned}
q_2(\theta_2) &= R_{b20}\theta_2\cos^2\phi_{b20} + q_{2p0}\\
u_2(\theta_2) &= q_2(\theta_2)\cos\chi_{20} + R_{b20}\sin\chi_{20}\\
v_2(\theta_2) &= q_2(\theta_2)\sin\chi_{20} - R_{b20}\cos\chi_{20}\\
z_2(\theta_2) &= R_{b20}\theta_2\cos\phi_{b20}\sin\phi_{b20}\\
&\mathbf{g}_0(\phi_{20} = \pi/2 - \chi_{20},\ \phi_{b20};\ O_2)
\end{aligned}\right\} \tag{4.2-1}$$

where θ_2 : angle of rotation of gear II (gear)

q_{2p0} : $q_2(0)$ at the design point P_0 ($\theta_2 = 0$)

χ_{20} : inclination angle of the plane of action G_{20} of gear II

ϕ_{b20} : inclination angle of $\mathbf{g}_0$ on the plane of action G_{20}

R_{b20} : radius of cylinder tangent to the plane of action G_{20}

In order to express $\mathbf{g}_0$ by an angle of rotation θ_1 in the coordinate systems O_1 and O_{q1}, it is enough to change the subscript 2 to 1 and the tooth curves corresponding to the path of contact $\mathbf{g}_0$ are called tooth profiles I and II respectively.

4.2.2 Equations of plane of action G_{20}

Figure 4.2-1 shows the plane of action G_{20} with the inclination angle $\chi_{20}=\pi/2-\phi_{20}$ which includes the path of contact $\mathbf{g}_0$ through the design point P_0 given by Eq. (4.2-1) in the coordinate systems O_2 and O_{q2}, where the sign (' or ") means the orthographic projection of a point, a line or a vector toward planes, which is the same in the following figures.

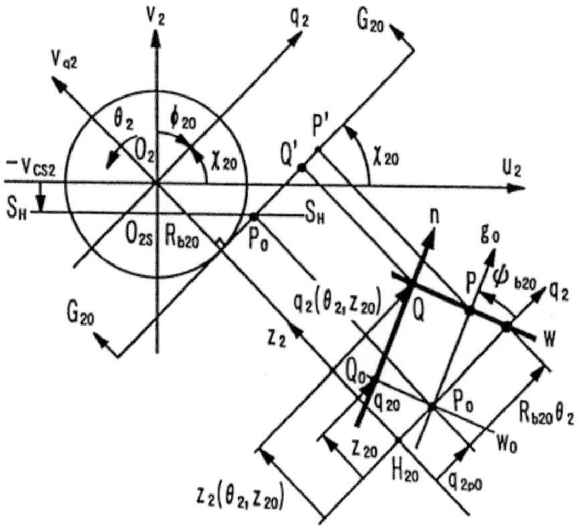

Fig.4.2-1 Plane of action G_{20}

The intersection of the tangent plane W_0 and G_{20} at P_0 is drawn by the line w_0 and when gear II rotates by θ_2, P_0 is supposed to move to P and w_0 to w. When a point Q is chosen at will on w and the directed line $\mathbf{n}$ (positive in the same direction of $\mathbf{g}_0$) which passes Q and is perpendicular to w on G_{20} intersects w_0 at Q_0, $Q_0\,(q_{20}, -R_{b20}, z_{20}\,;\,0_{q2})$ is obtained in the coordinate system O_{q2} as follows.

$$q_{20} = q_{2p0} - z_{20}\tan\phi_{b20}$$

Therefore, expressing Q in the coordinate system O_{q2}, $Q\,(q_2, -R_{b20}, z_2\,;\,0_{q2})$ is obtained as follows.

$$\begin{aligned} q_2(\theta_2, z_{20}) &= R_{b20}\theta_2\cos^2\phi_{b20} + q_{20} \\ &= R_{b20}\theta_2\cos^2\phi_{b20} + q_{2p0} - z_{20}\tan\phi_{b20} \\ z_2(\theta_2, z_{20}) &= R_{b20}\theta_2\cos\phi_{b20}\sin\phi_{b20} + z_{20} \end{aligned}$$

Transforming Q into the coordinate system O_2, $Q\,(u_2, v_2, z_2\,;\,0_2)$ is obtained as follows.

$$\left.\begin{aligned} q_2(\theta_2, z_{20}) &= R_{b20}\theta_2\cos^2\phi_{b20} + q_{2p0} - z_{20}\tan\phi_{b20} \\ u_2(\theta_2, z_{20}) &= q_2(\theta_2, z_{20})\cos\chi_{20} + R_{b20}\sin\chi_{20} \\ v_2(\theta_2, z_{20}) &= q_2(\theta_2, z_{20})\sin\chi_{20} - R_{b20}\cos\chi_{20} \\ z_2(\theta_2, z_{20}) &= R_{b20}\theta_2\cos\phi_{b20}\sin\phi_{b20} + z_{20} \end{aligned}\right\} \qquad (4.2\text{-}2)$$

Equation (4.2-2) expresses the plane of action G_{20} with parameters θ_2 and z_{20}. It also expresses the intersection w of the tangent plane W and the plane of action G_{20} when θ_2 is fixed and the directed line $\mathbf{n}$ when z_{20} is fixed. When $z_{20} = 0$, Eq. (4.2-1) is obtained because the path of contact $\mathbf{g}_0$ is the directed line $\mathbf{n}$ through P_0.

4.2.3 Equations of an involute helicoid

Transforming Eq. (4.2-2) into the coordinate system O_{r2} which rotates with gear II, the line w draws an involute helicoid which has the radius of base cylinder R_{b20} and the helix angle ϕ_{b20} and is expressed by Eq. (4.2-3). It is an involute helicoid of gear II and the directed line **n** at Q becomes the tooth surface normal **n**.

$$\left.\begin{aligned}
&\chi_{r2} = \chi_{20} - \theta_2 = \pi/2 - \phi_{20} - \theta_2 \\
&q_2(\theta_2, z_{20}) = R_{b20}\theta_2\cos^2\phi_{b20} + q_{2p0} - z_{20}\tan\phi_{b20} \\
&u_{r2}(\theta_2, z_{20}) = q_2(\theta_2, z_{20})\cos\chi_{r2} + R_{b20}\sin\chi_{r2} \\
&v_{r2}(\theta_2, z_{20}) = q_2(\theta_2, z_{20})\sin\chi_{r2} - R_{b20}\cos\chi_{r2} \\
&z_{r2}(\theta_2, z_{20}) = z_2(\theta_2, z_{20}) \\
&n(\phi_{20} + \theta_2,\ \phi_{b20};\ O_{r2})
\end{aligned}\right\} \qquad (4.2\text{-}3)$$

4.2.4 Involute helicoid of gear I which has path of contact $\mathbf{g}_0$

(1) Equations of the path of contact $\mathbf{g}_0$ in the coordinate systems O_1 and O_{q1}

The path of contact $\mathbf{g}_0$ given by Eq. (4.2-1) is expressed by an angle of rotation θ_1 of gear I in the coordinate systems O_1 and O_{q1} as follows.

$$\left.\begin{aligned}
&\theta_1 = i_0\theta_2 \quad (\text{when } \theta_2 = 0,\ \ \theta_1 = 0.) \\
&q_1(\theta_1) = R_{b10}\theta_1\cos^2\phi_{b10} + q_{1p0} \\
&u_1(\theta_1) = q_1(\theta_1)\cos\chi_{10} + R_{b10}\sin\chi_{10} \\
&v_1(\theta_1) = q_1(\theta_1)\sin\chi_{10} - R_{b10}\cos\chi_{10} \\
&z_1(\theta_1) = R_{b10}\theta_1\cos\phi_{b10}\sin\phi_{b10} \\
&g_0(\phi_{10} = \pi/2 - \chi_{10},\ \phi_{b10};\ O_1)
\end{aligned}\right\} \qquad (4.2\text{-}4)$$

where
- θ_1 : angle of rotation of gear I (pinion)
- i_0 : gear ratio between gears I and II
- q_{1p0} : $q_1(0)$ of the design point P_0 ($\theta_1 = 0$)
- χ_{10} : inclination angle of plane of action G_{10} of gear I
- ϕ_{b10} : inclination angle of $\mathbf{g}_0$ on plane of action G_{10}
- R_{b10} : radius of base cylinder tangent to plane of action G_{10}

The plane of action G_{10} and the involute helicoid of gear I which has the path of contact $\mathbf{g}_0$ given by Eq. (4.2-4) are obtained in the coordinate systems O_1 and O_{q1} as follows.

(2) Equations of plane of action G_{10}

The plane of action G_{10} which has the path of contact $\mathbf{g}_0$ is expressed as follows in the coordinate systems O_1 and O_{q1}, making reference to Eq. (4.2-2).

$$\left.\begin{aligned}
&q_1(\theta_1, z_{10}) = R_{b10}\theta_1\cos^2\phi_{b10} + q_{1p0} - z_{10}\tan\phi_{b10} \\
&u_1(\theta_1, z_{10}) = q_1(\theta_1, z_{10})\cos\chi_{10} + R_{b10}\sin\chi_{10} \\
&v_1(\theta_1, z_{10}) = q_1(\theta_1, z_{10})\sin\chi_{10} - R_{b10}\cos\chi_{10} \\
&z_1(\theta_1, z_{10}) = R_{b10}\theta_1\cos\phi_{b10}\sin\phi_{b10} + z_{10}
\end{aligned}\right\} \qquad (4.2\text{-}5)$$

z_{10} is a parameter corresponding to z_{20}. Equation (4.2-5) expresses a line parallel to $\mathbf{g}_0$ on G_{10} when z_{10} is fixed and θ_1 varies and a line perpendicular to $\mathbf{g}_0$ on G_{10} when θ_1 is fixed and z_{10} varies. When $z_{10}=0$, it is $\mathbf{g}_0$.

(3) Equations of an involute helicoid of gear I

Transforming Eq. (4.2-5) into the coordinate system O_{r1} which rotates with gear I, the equations of an involute helicoid and its normal $\mathbf{n}$ of gear I are obtained as follows.

$$\left.\begin{aligned}
\chi_{r1} &= \chi_{10}-\theta_1=\pi/2-\phi_{10}-\theta_1\\
q_1(\theta_1,z_{10}) &= R_{b10}\theta_1\cos^2\phi_{b10}+q_{1p0}-z_{10}\tan\phi_{b10}\\
u_{r1}(\theta_1,z_{10}) &= q_1(\theta_1,z_{10})\cos\chi_{r1}+R_{b10}\sin\chi_{r1}\\
v_{r1}(\theta_1,z_{10}) &= q_1(\theta_1,z_{10})\sin\chi_{r1}-R_{b10}\cos\chi_{r1}\\
z_{r1}(\theta_1,z_{10}) &= z_1(\theta_1,z_{10})\\
&n(\phi_{10}+\theta_1,\ \phi_{b10};\ O_{r1})
\end{aligned}\right\}\qquad(4.2\text{-}6)$$

4.2.5 A gear pair in point contact which have path of contact $\mathbf{g}_0$ [(42), (43)]

When an involute helicoid given by Eq. (4.2-3) for tooth surface II and another given by Eq. (4.2-6) for tooth surface I are given, a gear pair in point contact which have the path of contact $\mathbf{g}_0$ is obtained. It has been already in practical use as crossed helical and conical gears.

4.3 Surface of action and conjugate tooth surface of an involute helicoid and a family of involute gear pairs [(25), (41)]

4.3.1 Equations of a group of involute helicoids with the same dimensions

Figure 4.3-1 shows a point P_m chosen at will on an involute helicoid (tooth surface

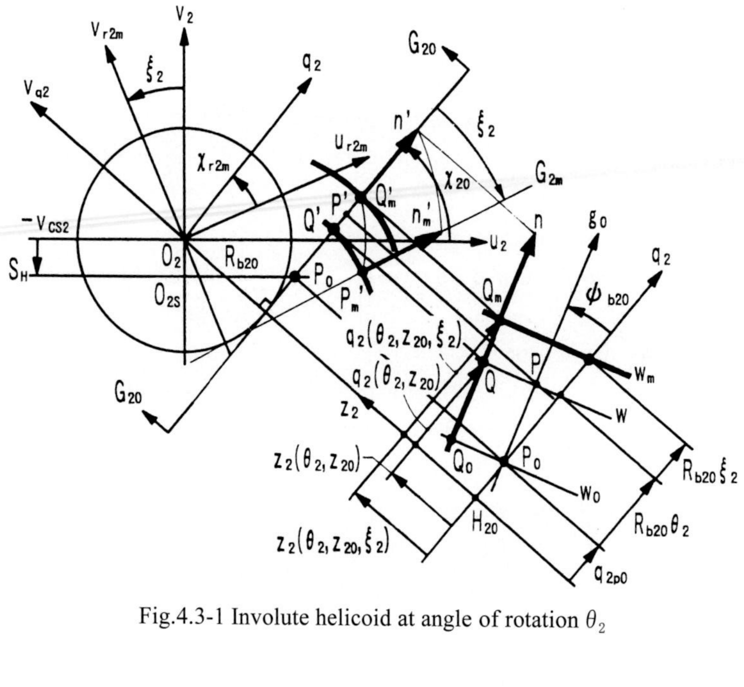

Fig.4.3-1 Involute helicoid at angle of rotation θ_2

II) of gear II which intersects the plane of action G_{20} along a line w at an angle of rotation θ_2 in the coordinate system O_2. It is discussed how to express P_m in the coordinate system O_2.

When gear II rotates additionally by ξ_2 and w moves to w_m, an intersection Q_m $(q_{2m}, -R_{b20}, z_{2m} ; O_{q2})$ of w_m and a normal **n** chosen at will is expressed as follows.

$$
\begin{aligned}
q_{2m}(\theta_2, z_{20}, \xi_2) &= q_2(\theta_2, z_{20}) + R_{b20}\xi_2\cos^2\phi_{b20} \\
z_{2m}(\theta_2, z_{20}, \xi_2) &= z_2(\theta_2, z_{20}) + R_{b20}\xi_2\cos\phi_{b20}\sin\phi_{b20}
\end{aligned}
$$

In the coordinate system O_{r2m} which coincides with the coordinate system O_2 at an angle of rotation θ_2 and rotates by ξ_2 around the axis z_2, $Q_m(u_{r2m}, v_{r2m}, z_{r2m} ; O_{r2m})$ and the inclination angle of its tooth surface normal **n** are expressed as follows.

$$
\begin{aligned}
\chi_{r2m} &= \chi_{20} - \xi_2 = \pi/2 - \phi_{20} - \xi_2 \\
u_{r2m}(\theta_2, z_{20}, \xi_2) &= q_{2m}(\theta_2, z_{20}, \xi_2)\cos\chi_{r2m} + R_{b20}\sin\chi_{r2m} \\
v_{r2m}(\theta_2, z_{20}, \xi_2) &= q_{2m}(\theta_2, z_{20}, \xi_2)\sin\chi_{r2m} - R_{b20}\cos\chi_{r2m} \\
z_{r2m}(\theta_2, z_{20}, \xi_2) &= z_2(\theta_2, z_{20}) + R_{b20}\xi_2\cos\phi_{b20}\sin\phi_{b20} \\
&\mathbf{n}(\phi_{20} + \xi_2, \phi_{b20} ; O_{r2m})
\end{aligned}
$$

Making the coordinate system O_{r2m} rotate by ξ_2 in the opposite direction of θ_2 and superimposing it on the coordinate system O_2, the plane of action G_{20} moves to G_{2m} (inclination angle χ_{2m}), the point Q_m to P_m and the normal **n** to $\mathbf{n}_m$ on G_{2m}. The point Q_m in the coordinate system O_{r2m} and the point P_m in the coordinate system O_2 are the same one and have the same coordinate values in each coordinate system. Therefore, P_m $(u_{2m}, v_{2m}, z_{2m} ; O_2)$ and the inclination angle of its normal $\mathbf{n}_m$ are expressed as follows.

$$
\left.
\begin{aligned}
\chi_{2m} &= \chi_{20} - \xi_2 = \pi/2 - \phi_{20} - \xi_2 \\
q_{2m}(\theta_2, z_{20}, \xi_2) &= q_2(\theta_2, z_{20}) + R_{b20}\xi_2\cos^2\phi_{b20} \\
u_{2m}(\theta_2, z_{20}, \xi_2) &= q_{2m}(\theta_2, z_{20}, \xi_2)\cos\chi_{2m} + R_{b20}\sin\chi_{2m} \\
v_{2m}(\theta_2, z_{20}, \xi_2) &= q_{2m}(\theta_2, z_{20}, \xi_2)\sin\chi_{2m} - R_{b20}\cos\chi_{2m} \\
z_{2m}(\theta_2, z_{20}, \xi_2) &= z_2(\theta_2, z_{20}) + R_{b20}\xi_2\cos\phi_{b20}\sin\phi_{b20} \\
&\mathbf{n}_m(\phi_{20} + \xi_2, \phi_{b20} ; O_2)
\end{aligned}
\right\} \qquad (4.3\text{-}1)
$$

Equation (4.3-1) expresses an involute helicoid (tooth surface II) which passes P on the path of contact $\mathbf{g}_0$ at an angle of rotation θ_2 and its surface normals as functions of parameters z_{20} and ξ_2 in the coordinate system O_2, which means a group of involute helicoids (tooth surfaces II) when θ_2 varies. When θ_2 is fixed and $z_{20} = 0$, it expresses the tooth profile II through P and when θ_2 is fixed and $\xi_2 = 0$, it expresses the intersection w of the plane of action G_{20} and the involute helicoid (tooth surface II) through P.

When a group of tooth surfaces II are given by Eq. (4.3-1), the line of contact and the common contact normals at an angle of rotation θ_2 are obtained by eliminating the parameter z_{20} or ξ_2 through the fundamental requirement for contact.

4.3.2 Common contact normal $\mathbf{n}_m$ $(P_{m0}P_m)$

Figure 4.3-2 shows a line of contact PP_m and the common contact normal $\mathbf{n}_m$ $(P_{m0}P_m)$

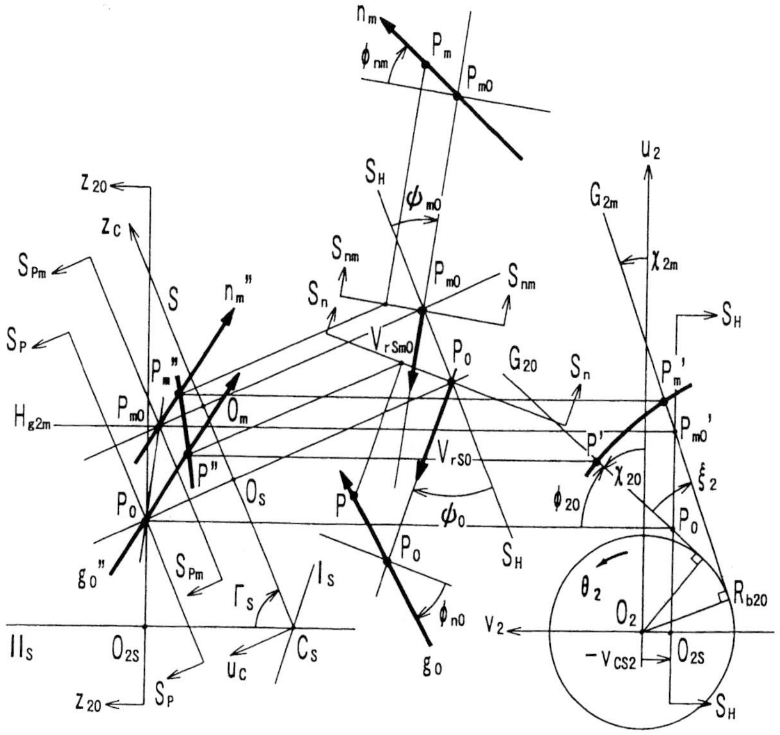

Fig.4.3-2 Line of contact PP_m and common contact normal $\mathbf{n}_m$

at a point of contact P_m. At an angle of rotation θ_2, gear II makes contact at P on the path of contact $\mathbf{g}_0$ and another point of contact P_m except P is chosen on tooth surface II at which the relative velocity is $\mathbf{V}_{rsm}$ and the common contact normal is $\mathbf{n}_m$.

When the intersection of $\mathbf{n}_m$ and the plane S_H is P_{m0} and the relative velocity at P_{m0} is $\mathbf{V}_{rsm0}$, the following equation is obtained because P_m is a point of contact.

$$\begin{aligned}\mathbf{n}_m \cdot \mathbf{V}_{rsm} &= \mathbf{n}_m \cdot (\mathbf{V}_{rsm0} + \omega_r \times |P_{m0}P_m| \cdot \mathbf{n}_m) \\ &= \mathbf{n}_m \cdot \mathbf{V}_{rsm0} = 0\end{aligned}$$

Namely it means that P_{m0} is also a point of contact.

The relative velocity $\mathbf{V}_{rsm0}$ through P_{m0} is on the plane S_{pm} parallel to the plane S_p, whose inclination angle to the plane S_H on the plane S_{pm} is defined as ϕ_{m0}. The plane S_{nm} which passes P_{m0} and is perpendicular to $\mathbf{V}_{rsm0}$ includes $\mathbf{n}_m$ ($P_{m0}P_m$) and is perpendicular to the plane S_{pm}, while the point of contact P_m and the common contact normal $\mathbf{n}_m$ are on the plane of action G_{2m} which inclines by ξ_2 from G_{20}, so that the inclination angle of $\mathbf{n}_m$ is given by $n_m(\phi_{20}+\xi_2, \phi_{b20}; O_2)$ in the coordinate system O_2. The common contact normal $\mathbf{n}_m$ is on the intersection of the planes S_{nm} and G_{2m}, therefore, the inclination angle ϕ_{m0} of the plane S_{nm} ($\mathbf{V}_{rsm0}$) is expressed by Eq. (2.1-20) as follows.

$$\tan\phi_{m0} = \tan(\phi_{20}+\xi_2)\sin\Gamma_s - \tan\phi_{b20}\cos\Gamma_s/\cos(\phi_{20}+\xi_2)$$

When P_{m0} is expressed by $P_{m0}(u_{cm0}, 0, z_{cm0}; C_s)$ in the coordinate system C_s, u_{cm0} is obtained from Eq. (2.2-8) as follows.

$$u_{cm0} = O_m P_{m0} = E\tan\phi_{m0}\sin(\Sigma-\Gamma_S)\sin\Gamma_S/\sin\Sigma$$
$$z_{cm0} = C_S O_m$$

Transforming P_{m0} into the coordinate sysem O_2, $P_{m0}(u_{2m0}, -v_{cs2}, z_{2m0}; O_2)$ is obtained as follows.

$$\left.\begin{aligned} u_{2m0} &= -u_{cm0}\cos\Gamma_S + z_{cm0}\sin\Gamma_S \\ v_{cs2} &= E\tan\Gamma_S/\{\tan(\Sigma-\Gamma_S)+\tan\Gamma_S\} \\ z_{2m0} &= -u_{cm0}\sin\Gamma_S - z_{cm0}\cos\Gamma_S - z_{2c0} \\ z_{2c0} &= -u_{c0}\sin\Gamma_S - z_{c0}\cos\Gamma_S \end{aligned}\right\} \quad (4.3\text{-}2)$$

where u_{c0} and z_{c0} : P_0 $(u_{c0}, 0, z_{c0}; C_S)$

Eliminating z_{cm0} from Eq. (4, 3-2), the following equation is obtained.

$$u_{2m0}\cos\Gamma_S + (z_{2m0} + z_{2c0})\sin\Gamma_S = -u_{cm0} \quad (4.3\text{-}3)$$

Equation (4.3-3) expresses the locus P_0P_{m0} of the intersection of a common contact normal $\mathbf{n}_m$ and the plane S_H. Because P_{m0} is also the intersection of the locus P_0P_{m0} and the intersection H_{g2m} of the plane of action G_{2m} and the plane S_H, it is expressed by the parameter ξ_2 in the coordinate system O_2 as follows.

$$\left.\begin{aligned} u_{2m0} &= R_{b20}/\cos(\phi_{20}+\xi_2) - v_{cs2}\tan(\phi_{20}+\xi_2) \\ v_{cs2} &= E\tan\Gamma_S/\{\tan(\Sigma-\Gamma_S)+\tan\Gamma_S\} \\ z_{2m0} &= -z_{2c0} - (u_{2m0}\cos\Gamma_S + u_{cm0})/\sin\Gamma_S \end{aligned}\right\} \quad (4.3\text{-}4)$$

Transforming P_{m0} into the coordinate system O_{q2}, q_{2m0} of $P_{m0}(q_{2m0}, -R_{b20}, z_{2m0}; O_{q2})$ is expressed as follows.

$$\left.\begin{aligned} q_{2m0} &= u_{2m0}\cos\chi_{2m} - v_{cs2}\sin\chi_{2m} \\ \chi_{2m} &= \pi/2 - \phi_{20} - \xi_2 = \chi_{20} - \xi_2 \end{aligned}\right\} \quad (4.3\text{-}5)$$

Equations (4.3-4) and (4.3-5) express the common contact normal $\mathbf{n}_m$ as a directed line which passes P_{m0} and has the inclination angle $\mathbf{n}_m(\phi_{20}+\xi_2, \phi_{b20}; O_2)$ with the parameter ξ_2 on the plane of action G_{2m}.

4. 3. 3 Equations of line of contact and surface of action

Figure 4.3-3 shows the relation among the points $Q_{m0}(P_{m0})$, $Q_m(P_m)$, Q and P on the plane of action G_{20}. When P_{m0} and P_m are given in the coordinate system O_{r2m} and the plane of action G_{2m} is rotated back by ξ_2 and superimposed on G_{20}, $\mathbf{n}_m$ rotates back to $\mathbf{n}$, P_{m0} to Q_{m0} and P_m to Q_m. When the intersection of G_{20} and tooth surface II is w and the intersection of w and $\mathbf{n}$ is Q, the following equation related among Q_{m0} $(q_{2m0}, -R_{b20}, z_{2m0}; O_{q2})$, P $(q_{2p}, -R_{b20}, z_{2p}; O_{q2})$ and an unknown point Q $\{q_2(\theta_2, \xi_2), -R_{b20}, z_2(\theta_2, \xi_2); O_{q2}\}$ is obtained in the coordinate system O_{q2}.

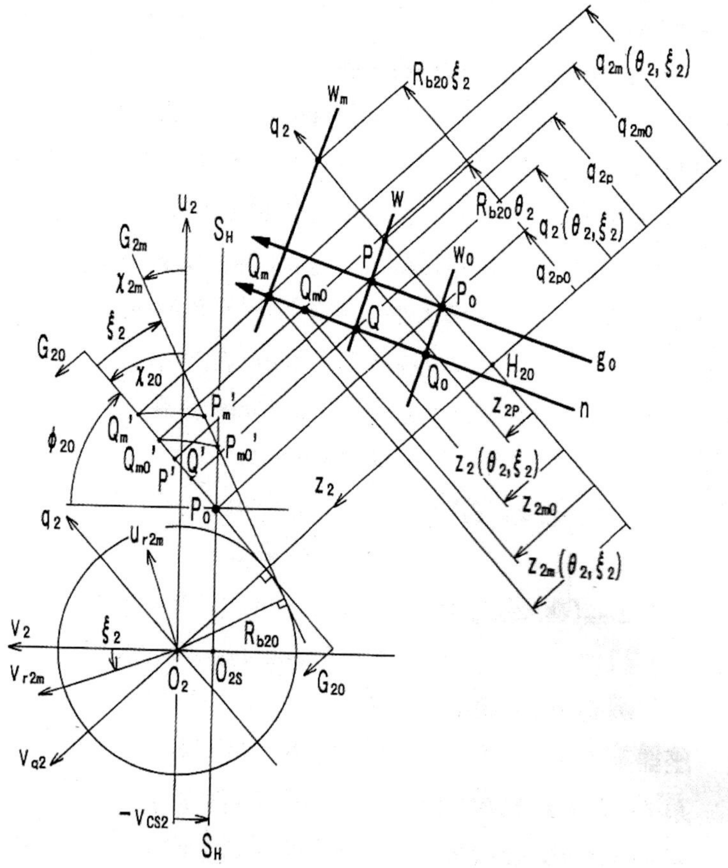

Fig.4.3-3 Relationship among Q_{m0} (P_{m0}), Q_m (P_m), Q and P on plane G_{20}

$$z_{2m0} = -\{q_2(\theta_2, \xi_2) - q_{2m0}\}\tan\phi_{b20} + \{q_{2p} - q_2(\theta_2, \xi_2)\}/\tan\phi_{b20} + z_{2p}$$

Therefore, Q is expressed in the coordinate system O_{q2} as follows.

$$q_2(\theta_2, \xi_2) = (q_{2m0}\tan\phi_{b20} + q_{2p}/\tan\phi_{b20} + z_{2p} - z_{2m0})/(\tan\phi_{b20} + 1/\tan\phi_{b20})$$
$$z_2(\theta_2, \xi_2) = z_{2p} + \{q_{2p} - q_2(\theta_2, \xi_2)\}/\tan\phi_{b20}$$

where $q_{2p} = q_{2p0} + R_{b20}\theta_2\cos^2\phi_{b20}$
$z_{2p} = R_{b20}\theta_2\cos\phi_{b20}\sin\phi_{b20}$

Q_m is expressed as follows with the parameter ξ_2 in the coordinate system O_{q2}.

$$q_{2m}(\theta_2, \xi_2) = q_2(\theta_2, \xi_2) + R_{b20}\xi_2\cos^2\phi_{b20}$$
$$z_{2m}(\theta_2, \xi_2) = z_2(\theta_2, \xi_2) + R_{b20}\xi_2\cos\phi_{b20}\sin\phi_{b20}$$

Therefore, a point of contact P_m (u_{2m}, v_{2m}, z_{2m} ; O_2) and the inclination angle of its common contact normal n_m are expressed with the parameter ξ_2 in the coordinate system O_2 as follows.

$$\left.\begin{aligned}
\chi_{2m} &= \chi_{20}-\xi_2 = \pi/2-\phi_{20}-\xi_2\\
q_{2m}(\theta_2,\xi_2) &= q_2(\theta_2,\xi_2)+R_{b20}\xi_2\cos^2\psi_{b20}\\
u_{2m}(\theta_2,\xi_2) &= q_{2m}(\theta_2,\xi_2)\cos\chi_{2m}+R_{b20}\sin\chi_{2m}\\
v_{2m}(\theta_2,\xi_2) &= q_{2m}(\theta_2,\xi_2)\sin\chi_{2m}-R_{b20}\cos\chi_{2m}\\
z_{2m}(\theta_2,\xi_2) &= z_2(\theta_2,\xi_2)+R_{b20}\xi_2\cos\psi_{b20}\sin\psi_{b20}\\
\mathbf{n}_m(\phi_{20}&+\xi_2,\ \psi_{b20};\ 0_2)
\end{aligned}\right\} \tag{4.3-6}$$

Equation (4.3-6) with the parameter ξ_2 expresses a line of contact (PP_m) and its common contact normal $\mathbf{n}_m$ at an angle of rotation θ_2 in the coordinate system O_2. Therefore, Eq. (4.3-6) expresses the surface of action as a set of lines of contact when θ_2 is varied. Equation (4.3-6) expresses a common contact normal $\mathbf{n}_m$ which is a path of contact when ξ_2 is given, therefore it also expresses the surface of action as a set of $\mathbf{n}_m$ when ξ_2 is varied.

Figure 4.3-2 shows schematically the surface of action represented by Eq. (4.3-6). The surface of action is a curved one drawn by a common contact normal $\mathbf{n}_m(P_{m0}P_m)$ which moves along the locus P_0P_{m0} on the plane S_H and changes the inclination angle $\mathbf{n}_m(\phi_{20}+\xi_2,\ \psi_{b20};\ 0_2)$ with variation of ξ_2. Generally speaking, the locus P_0P_{m0} varies according to tooth surface II and the location of the design point P_0, but it coincides with the instantaneous axis in cylindrical and bevel gears.

4．3．4 Equations of tooth surface I generated by an involute helicoid (tooth surface II)

Figure 4.3-4 shows the point of contact P_m and the common contact normal $\mathbf{n}_m$ ($P_{m0}P_m$) in the coordinate systems O_1 and O_{q1}. Transforming Eq. (4.3-6) into the coordinate systems O_1 and O_{q1} and moreover into the coordinate system O_{r1}, tooth surface I of gear I and the inclination angle of the normal $\mathbf{n}_m$ are obtained as follows.

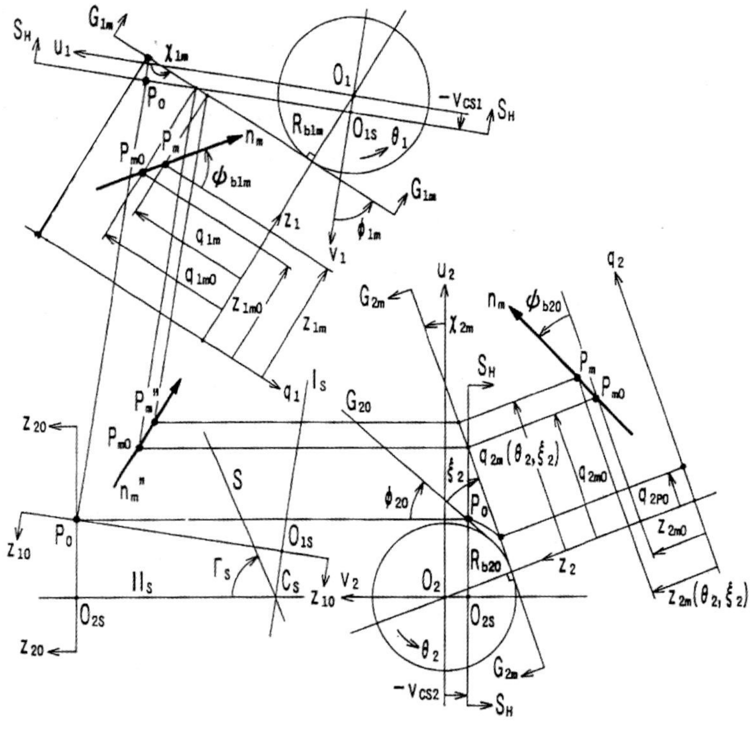

Fig.4.3-4 Point P_m and common normal $\mathbf{n}_m$

Transforming P_m $(u_{2m}, v_{2m}, z_{2m} ; O_2)$ in Eq. (4.3-6) into the coordinate system O_1, P_m $(u_{1m}, v_{1m}, z_{1m} ; O_1)$ is obtained by the equations between the coordinate systems O_2 and O_1 in 4.1(7) as follows.

$$
\begin{aligned}
u_{1m} &= -u_{2m}\cos\Sigma - (z_{2m} + z_{2c0})\sin\Sigma \\
v_{1m} &= v_{2m} + E \\
z_{1m} &= u_{2m}\sin\Sigma - (z_{2m} + z_{2c0})\cos\Sigma - z_{1c0} \\
z_{1c0} &= -u_{c0}\sin(\Sigma - \Gamma_s) + z_{c0}\cos(\Sigma - \Gamma_s)
\end{aligned}
$$

Transforming the inclination angle $n_m(\phi_{20} + \xi_2, \psi_{b20} ; O_2)$ of n_m into the coordinate system O_1, n_m $(\phi_{1m}, \psi_{b1m} ; O_1)$ is obtained by Eq. (2.1-5) as follows.

$$
\begin{aligned}
\tan\phi_{1m} &= -\tan(\phi_{20} + \xi_2)\cos\Sigma \ -\tan\psi_{b20}\sin\Sigma/\cos(\phi_{20} + \xi_2) \\
\sin\psi_{b1m} &= \cos\psi_{b20}\sin(\phi_{20} + \xi_2)\sin\Sigma - \sin\psi_{b20}\cos\Sigma
\end{aligned}
$$

When the plane of action which includes n_m and is parallel to the axis I is G_{1m}, $P_m(q_{1m}, -R_{b1m}, z_{1m} ; O_{q1})$ transformed from $P_m(u_{1m}, v_{1m}, z_{1m} ; O_1)$ is expressed as follows.

$$
\begin{aligned}
q_{1m} &= u_{1m}\cos\chi_{1m} + v_{1m}\sin\chi_{1m} \\
R_{b1m} &= u_{1m}\sin\chi_{1m} - v_{1m}\cos\chi_{1m} \\
\chi_{1m} &= \pi/2 - \phi_{1m}
\end{aligned}
$$

Transforming $P_m(u_{1m}, v_{1m}, z_{1m} ; O_1)$ into the coordinate system O_{r1}, $P_m(u_{r1m}, v_{r1m}, z_{r1m} ; O_{r1})$ of tooth surface I and the inclination angle of the normal n_m are obtained as follows.

$$
\left.
\begin{aligned}
\theta_1 &= i_0\theta_2 \quad (\text{when } \theta_2 = 0,\ \theta_1 = 0.) \\
\chi_{r1m} &= \pi/2 - \phi_{1m} - \theta_1 \\
u_{r1m} &= q_{1m}\cos\chi_{r1m} + R_{b1m}\sin\chi_{r1m} \\
v_{r1m} &= q_{1m}\sin\chi_{r1m} - R_{b1m}\cos\chi_{r1m} \\
z_{r1m} &= z_{1m} \\
& n_m(\phi_{1m} + \theta_1, \psi_{b1m} ; O_{r1})
\end{aligned}
\right\} \qquad (4.3\text{-}7)
$$

4. 3. 5 A family of gear pairs having the same involute helicoids for one member (a family of involute gear pairs) [44]

The axis of gear II, the same involute helicoid (base circle radius R_{b20} and helix angle ψ_{b20}) as tooth surface II and a point P_0 whose radius is R_{20} on the tooth surface II are given. The point P_0 and its normal n_0 are expressed in the coordinate system O_2 as follows.

$$
\begin{aligned}
& P_0(u_{2p0}, v_{2p0}, 0 ; O_2) \\
& n_0(\phi_{20}, \psi_{b20} ; O_2) \\
\phi_{20} + \varepsilon_{20} &= \cos^{-1}(R_{b20}/R_{20}) \\
\varepsilon_{20} &= \tan^{-1}(v_{2p0}/u_{2p0}) \\
R_{20} &= \sqrt{(u_{2p0}{}^2 + v_{2p0}{}^2)}
\end{aligned}
$$

In this theory, P_0 is normally chosen on the plane S_H and realizes $v_{2p0} = -v_{cs2}$, so that when the shaft angle Σ and the gear ratio i_0 (or the angle Γ_S of the instantaneous axis) are given, the offset E, the instantaneous axis S and the mating gear axis I are obtained and the gear pair is determined.

Figure 4.3-5 shows schematically a family of involute gear pairs, where P_0 and its normal n_0 are located adequately in the coordinate system O_2 by rotating gear II and are adopted as the design point and the path of contact of the gear pair. In Fig. 4.3-5, S, S_b and S_{hy} indicate the instantaneous axes of helical ($\Gamma_S = 0$ or π) and worm (or crossed helical), bevel and hypoid gears respectively.

The involute gear pair varies according to the location of P_0 as follows.

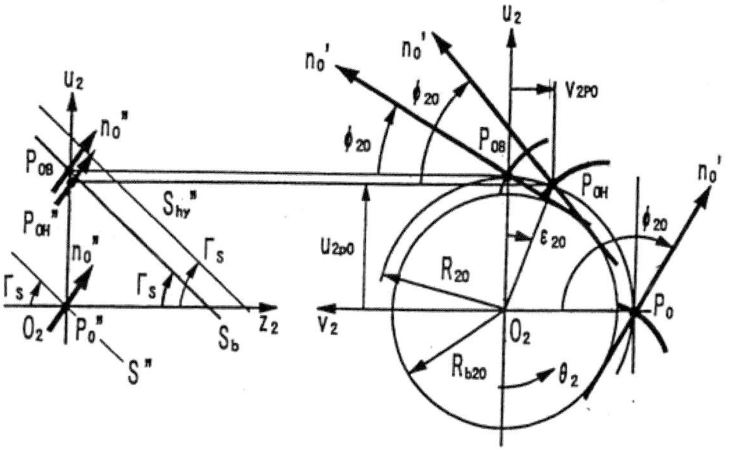

Fig.4.3-5 Family of gear pairs with the same involute helicoid for one member

(1) When P_0 is on the axis v_2

(a) $\Sigma = 0$ or π : external or internal helical gears

Because of $\Gamma_S = 0$ or π, the mating tooth surface I is an involute helicoid.

(b) $\Sigma \neq 0$: worm or crossed helical gears

Worm gears which adopt involute helicoids for either the worm member or the wheel one are realized. Crossed helical gears in point contact whose mating tooth surface I also has an involute helicoid corresponding to the normal n_0 (= the path of contact) are used practically.

(2) When P_0 (P_{0B}) is on the axis u_2 : bevel gears

Face gears and conical gears in point contact whose mating tooth surface I also has an involute helicoid corresponding to the normal n_0 (= the path of contact) are used practically.

(3) When P_0 (P_{0H}) is except mentioned above [42], [43]

When P_{0H} is chosen in the area where $|\varepsilon_{20}|$ is comparatively small (Fig. 4.3-5), it makes hypoid gears, face gears or conical gears near to bevel gears. When P_{0H} is chosen in the area where $|\varepsilon_{20}|$ nearly equals $\pi/2$, it makes a gear pair near to worm or crossed helical gears.

Giving an involute helicoid for one member, there is a family of involute gear pairs mentioned above, the line of contact, the surface of action and the mating conjugate tooth surface of which are expressed uniformly by Eqs. (4.3-6) and (4.3-7), which have been discussed separately in the present theories.

4.4 Design method and example of face gears having involute helicoids for gear I (pinion) [45]

In this section, using the results in 4.1 to 4.3, a design method for a pair of face gears which have involute helicoids for gear I (pinion) and the conjugate tooth surfaces for gear II (gear) and its design example are given.

4.4.1 Surface of action and conjugate tooth surfaces of a gear pair which has involute helicoids for a pinion

When the path of contact $\mathbf{g}_0$, the plane of action G_{10} and the involute helicoid for a pinion are given by Eqs. (4.2-4), (4.2-5) and (4.2-6) in the coordinate systems O_1 and O_{q1} respectively, a group of involute helicoids, the surface of action and the conjugate and the modified tooth surfaces of a gear are obtained in the coordinate systems O_1, O_{q1}, O_2, O_{q2} and O_{r2}, making reference to Eqs. (4.3-1), (4.3-6), (4.3-7) and (3.5-7) as follows.

(1) Equations of a group of involute helicoids for a pinion

Equations of a group of involute helicoids for the pinion are obtained by expressing Eq. (4.3-1) in the coordinate systems O_1 and O_{q1} as follows, where ξ_1 is a parameter.

$$\left.\begin{aligned}
\chi_{1m} &= \chi_{10}-\xi_1 = \pi/2-\phi_{10}-\xi_1 = \pi/2-\phi_{1m}\\
q_{1m}(\theta_1, z_{10}, \xi_1) &= q_1(\theta_1, z_{10})+R_{b10}\xi_1\cos^2\phi_{b10}\\
u_{1m}(\theta_1, z_{10}, \xi_1) &= q_{1m}(\theta_1, z_{10}, \xi_1)\cos\chi_{1m}+R_{b10}\sin\chi_{1m}\\
v_{1m}(\theta_1, z_{10}, \xi_1) &= q_{1m}(\theta_1, z_{10}, \xi_1)\sin\chi_{1m}-R_{b10}\cos\chi_{1m}\\
z_{1m}(\theta_1, z_{10}, \xi_1) &= z_1(\theta_1, z_{10})+R_{b10}\xi_1\cos\phi_{b10}\sin\phi_{b10}\\
&n_m(\phi_{10}+\xi_1, \phi_{b10}; 0_1)
\end{aligned}\right\} \qquad (4.4\text{-}1)$$

Equation (4.4-1) represents an involute helicoid which passes P on $\mathbf{g}_0$ at an angle of rotation θ_1, so that it represents a group of involute helicoids by varying θ_1. When θ_1 is fixed and $z_{10}=0$, the tooth profile through P is drawn by varying ξ_1. When θ_1 is fixed and $\xi_1 = 0$, the intersection of the plane of action G_{10} and the involute helicoid through P is drawn by varying z_{10}.

(2) Equations of surface of action

Making reference to 4.3.2 and 4.3.3, equations of surface of action are obtained by expressing Eq. (4.3-6) in the coordinate systems O_1 and O_{q1} as follows.

$$\left.\begin{aligned}
\chi_{1m} &= \chi_{10}-\xi_1 = \pi/2-\phi_{10}-\xi_1 = \pi/2-\phi_{1m}\\
q_{1m}(\theta_1, \xi_1) &= q_1(\theta_1, \xi_1)+R_{b10}\xi_1\cos^2\phi_{b10}\\
u_{1m}(\theta_1, \xi_1) &= q_{1m}(\theta_1, \xi_1)\cos\chi_{1m}+R_{b10}\sin\chi_{1m}\\
v_{1m}(\theta_1, \xi_1) &= q_{1m}(\theta_1, \xi_1)\sin\chi_{1m}-R_{b10}\cos\chi_{1m}\\
z_{1m}(\theta_1, \xi_1) &= z_1(\theta_1, \xi_1)+R_{b10}\xi_1\cos\phi_{b10}\sin\phi_{b10}
\end{aligned}\right\} \qquad (4.4\text{-}2)$$

where,

$$q_1(\theta_1, \xi_1) = (q_{1m0}\tan\phi_{b10} + q_{1p}/\tan\phi_{b10} + z_{1p} - z_{1m0})/(\tan\phi_{b10} + 1/\tan\phi_{b10})$$
$$z_1(\theta_1, \xi_1) = z_{1p} + (q_{1p} - q_1(\theta_1, \xi_1))/\tan\phi_{b10}$$
$$q_{1p} = R_{b10}\theta_1\cos^2\phi_{b10} + q_{1p0}$$
$$z_{1p} = R_{b10}\theta_1\cos\phi_{b10}\sin\phi_{b10}$$
$$q_{1m0} = u_{1m0}\cos\chi_{1m} - v_{cs1}\sin\chi_{1m}$$
$$u_{1m0} = R_{b10}/\cos(\phi_{10} + \xi_1) - v_{cs1}\tan(\phi_{10} + \xi_1)$$
$$z_{1m0} = -z_{1c0} + (u_{1m0}\cos(\Sigma - \Gamma_s) - u_{cm0})/\sin(\Sigma - \Gamma_s)$$
$$u_{cm0} = E\tan\phi_{m0}\sin(\Sigma - \Gamma_s)\sin\Gamma_s/\sin\Sigma$$
$$\tan\phi_{m0} = \tan(\phi_{10} + \xi_1)\sin(\Sigma - \Gamma_s) + \tan\phi_{b10}\cos(\Sigma - \Gamma_s)/\cos(\phi_{10} + \xi_1)$$

Equation (4.4-2) represents the surface of action which is a set of lines of contact or paths of contact. When θ_1 is fixed and ξ_1 varies, it is a line of contact. When ξ_1 is fixed and θ_1 varies, it is a path of contact (common contact normal).

(3) Equations of tooth surface for a pinion obtained through the surface of action

Transforming Eq. (4.4-2) into the coordinate system O_{r1}, a tooth surface for the pinion and its normal $\mathbf{n}_m(\pi/2 - \chi_{r1m}, \phi_{b10}; 0_{r1})$ are obtained as follows.

$$\left.\begin{aligned} \chi_{r1m} &= \chi_{1m} - \theta_1 \\ u_{r1m} &= q_{1m}\cos\chi_{r1m} + R_{b10}\sin\chi_{r1m} \\ v_{r1m} &= q_{1m}\sin\chi_{r1m} - R_{b10}\cos\chi_{r1m} \\ z_{r1m} &= z_{1m} \end{aligned}\right\} \qquad (4.4\text{-}3)$$

Equation (4.4-3) represents the involute helicoid through P which is one of the group given by Eq. (4.4-1) and expressed by θ_1 and ξ_1. When θ_1 is fixed, it expresses the line of contact through P and when ξ_1 is fixed, it expresses the tooth profile through P.

(4) Equations of conjugate tooth surface of a gear

Transforming Eq. (4.4-2) into the coordinate systems O_2 and O_{q2}, the surface of action is expressed by the parameters θ_1 and ξ_1 as follows.

$$\left.\begin{aligned} u_{2m} &= -u_{1m}\cos\Sigma + (z_{1m} + z_{1c0})\sin\Sigma \\ v_{2m} &= v_{1m} - E \\ z_{2m} &= -u_{1m}\sin\Sigma - (z_{1m} + z_{1c0})\cos\Sigma - z_{2c0} \\ \tan\phi_{2m} &= -\tan(\phi_{10} + \xi_1)\cos\Sigma + \tan\phi_{b10}\sin\Sigma/\cos(\phi_{10} + \xi_1) \\ \sin\phi_{b2m} &= -\cos\phi_{b10}\sin(\phi_{10} + \xi_1)\sin\Sigma - \sin\phi_{b10}\cos\Sigma \\ \chi_{2m} &= \pi/2 - \phi_{2m} \\ q_{2m} &= u_{2m}\cos\chi_{2m} + v_{2m}\sin\chi_{2m} \\ R_{b2m} &= u_{2m}\sin\chi_{2m} - v_{2m}\cos\chi_{2m} \end{aligned}\right\} \qquad (4.4\text{-}4)$$

Therefore, transforming Eq. (4.4-4) into the coordinate system O_{r2}, a point $P_m(u_{r2m}, v_{r2m}, z_{r2m}; 0_{r2})$ on a conjugate tooth surface of the gear and its normal $\mathbf{n}_m(\pi/2 - \chi_{r2m}, \phi_{b2m}; 0_{r2})$ are obtained as follows, where the coordinate system O_{r2} coincides with the coordinate system O_2 when $\theta_2 = 0$.

4. Involute helicoid and its conjugate tooth surface

$$\left.\begin{aligned} \theta_2 &= \theta_1 / i_0 \\ \chi_{r2m} &= \chi_{2m} - \theta_2 \\ u_{r2m} &= q_{2m}\cos\chi_{r2m} + R_{b2m}\sin\chi_{r2m} \\ v_{r2m} &= q_{2m}\sin\chi_{r2m} - R_{b2m}\cos\chi_{r2m} \\ z_{r2m} &= z_{2m} \end{aligned}\right\} \tag{4.4-5}$$

(5) Modified tooth surfaces

Making reference to Eq. (3.5-7), the modified tooth surface of a gear is defined as a set of point P_{2mf} which corresponds to a point P_m chosen at will on the conjugate tooth surface of the gear and has an amount of modification $\triangle p_{2nm}(\theta_1, \xi_1)$ in the direction of its normal $\mathbf{n}_m$. Therefore, $P_{2mf}(u_{r2mf}, v_{r2mf}, z_{r2mf};\ O_{r2})$ is obtained as follows.

$$\left.\begin{aligned} u_{r2mf} &= u_{r2m} + \triangle p_{2nm}(\theta_1, \xi_1)\cos\phi_{b2m}\cos\chi_{r2m} \\ v_{r2mf} &= v_{r2m} + \triangle p_{2nm}(\theta_1, \xi_1)\cos\phi_{b2m}\sin\chi_{r2m} \\ z_{r2mf} &= z_{r2m} + \triangle p_{2nm}(\theta_1, \xi_1)\sin\phi_{b2m} \end{aligned}\right\} \tag{4.4-6}$$

$\triangle p_{2nm}(\theta_1, \xi_1)$ is given generally by a quadratic surface of θ_1 and ξ_1.

A modified tooth surface of a pinion is obtained in the same way.

4.4.2 Flow of design and selection of variables

Figure 4.4-1 shows the flow of the design of face gears.

(1) The shaft angle Σ, the offset $E \geqq 0$ of the two axes and the gear ratio i_0 are given and the basic coordinate systems are determined by the two axes and their common perpendicular as shown in 2.1.

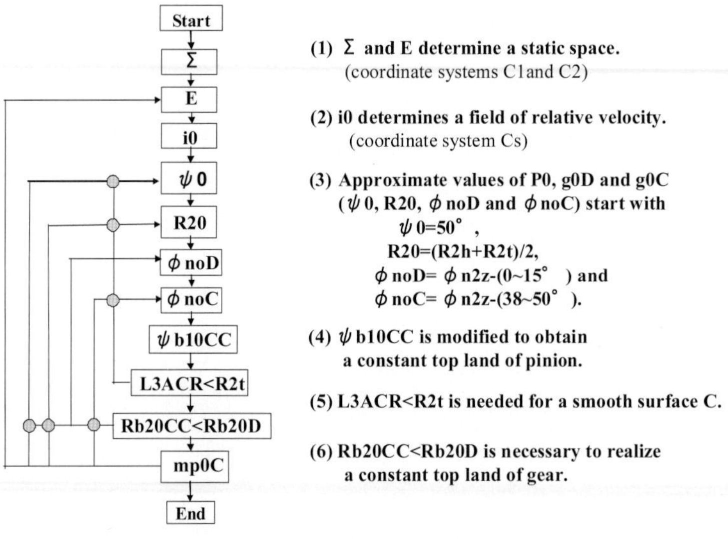

Start
Σ
E
i0
ψ0
R20
φnoD
φnoC
ψb10CC
L3ACR<R2t
Rb20CC<Rb20D
mp0C
End
(1) Σ and E determine a static space.
(coordinate systems C1and C2)
(2) i0 determines a field of relative velocity.
(coordinate system Cs)
(3) Approximate values of P0, g0D and g0C
(ψ0, R20, φnoD and φnoC) start with
ψ0=50°,
R20=(R2h+R2t)/2,
φnoD= φn2z-(0~15°) and
φnoC= φn2z-(38~50°).
(4) ψb10CC is modified to obtain
a constant top land of pinion.
(5) L3ACR<R2t is needed for a smooth surface C.
(6) Rb20CC<Rb20D is necessary to realize
a constant top land of gear.

Fig.4.4-1 Flow of design procedure of involute face gears

(2) According to 3.2 and 3.3, the design point $P_0(u_{c0}, 0, z_{c0};\ C_S)$ and a pair of paths of contact through P_0, $\mathbf{g}_{0D}(\psi_0, \phi_{n0D};\ C_S)$ and $\mathbf{g}_{0C}(\psi_0, \phi_{n0C};\ C_S)$ are selected. Using the fundamental requirement for contact, Eq. (3.2-1) or Eq. (3.2-2), the design of a pair of paths of contact leads to the selection of four independent variables, ψ_0 (or u_{c0}), R_{20} (or z_{c0}), ϕ_{n0D} and ϕ_{n0C}.

(3) The involute helicoids which have $\mathbf{g}_{0D}$ and $\mathbf{g}_{0C}$ for a pinion are given and the four variables are corrected so as to realize the smooth conjugate tooth surfaces of a gear all over the facewidth that have no cusp at the tip and sufficient contact ratios on both the drive

and coast sides, which leads to the end of the design procedure. A design example selected according to Fig. 4. 4-1 is shown in the following.

4. 4. 3 Selection of design point P_0 and paths of contact g_{0D} and g_{0C}

(1) Basic dimensions Σ, E and i_0

They are given as shown in Table 4. 4-1.

(2) Calculation of design point P_0, design reference line g_t and limit path of contact g_{2S}

Face gears are intended to be used for automobiles, therefore $\phi_0 = 60°$ is given and $R_{20} = 77$ mm is selected from the given R_{2t} and R_{2h}. From ϕ_0 and R_{20} selected above, the design point P_0, the design reference line g_t and the limit path of contact g_{2S} are calculated by Eqs. (3. 2-1), (3. 2-7), (3. 3-2) and (3. 4-3), the results of which are shown in Table 4. 4-1.

$$P_0(u_{c0},\ 0,\ z_{c0};\ C_S)$$
$$g_t(\phi_0,\ \phi_{nt};\ C_S),\quad g_{2s}(\phi_0,\ \phi_{n2S};\ C_S)$$
$$u_{c0} = E\sin(\Sigma-\Gamma_S)\sin\Gamma_S\tan\phi_0/\sin\Sigma$$
$$z_{c0} = (\sqrt{(R_{20}^2 - v_{cs2}^2)} + u_{c0}\cos\Gamma_S)/\sin\Gamma_S$$
$$\tan\phi_{nt} = \cos\Gamma_S/(-\cos\phi_0/\tan\varepsilon_{20}+\sin\Gamma_S\sin\phi_0)$$
$$\tan\phi_{n2S} = -\tan\varepsilon_{20}\cos\phi_0\cos\Gamma_S$$
$$\tan\varepsilon_{20} = -v_{cs2}/u_{2p0}$$

(3) Involute helicoid for a pinion which has limit path of contact g_{2S} through P_0

Substituting the limit path of contact g_{2S} for the path of contact (g_0), $P_0(q_{1p0},\ -R_{b10},\ 0;\ 0_{q1})$ and $g_{2S}(\phi_{10},\ \phi_{b10};\ 0_1)$ are obtained as follows.

$$P_0(10.91,\ -29.43,\ 0;\ 0_{q1})\ \text{(mm)},\qquad g_{2S}(16.76°,\ 58.58°;\ 0_1)$$

Substituting these P_0 and g_{2S} for Eq. (4. 2-6), the involute helicoid for a pinion which has the limit path of contact g_{2S} is obtained.

Figure 4. 4-2 shows the involute helicoid for the pinion, where (a) is a bird's eye view and (b) is a view from the positive direction of the axis z_1. g_{r2S1} is the tooth profile of the pinion corresponding to the limit path of contact g_{2S} and ξ_{r2S1} is the intersection of the surface of action (G_{10}) and the involute helicoid.

4. Involute helicoid and its conjugate tooth surface

Table 4.4-1 Dimensions of face gears

Shaft angle Σ (°)	90	
Ratio of angular velocities $i_0 (N_2/N_1)$	4.1 (41/10)	
Angle of instantaneous axis Γ_s (°)	76.29	
Offset E (mm)	35 (below center)	
Dispositions of shafts I and II to instantaneous axis		
v_{cs1} (mm)	-1.97	
v_{cs2} (mm)	33.03	
Outside radius of gear R_{2h} (mm)	95	
Inside radius of gear R_{2t} (mm)	67	
Radius of design cylinder of gear R_{20} (mm)	77	
Inclination angle of plane S_n ϕ_0 (°)	60.0	
Design point in coordinate system C_s		
$P_0(u_{c0}, 0, z_{c0}; C_S)$ (mm)	(13.96, 0, 75.00)	
Dimensions of equivalent rack		
$\mathbf{g}_t(\phi_0, \phi_{nt}; C_S)$ (°)	(60.0, 7.13)	
$\mathbf{g}_{2s}(\phi_0, \phi_{n2s}; C_S)$ (°)	(60.0, 3.22)	
$\mathbf{g}_{0D}(\phi_0, \phi_{n0D}; C_S)$ (°)	(60.0, 13.93)	
$\mathbf{g}_{0C}(\phi_0, \phi_{n0C}; C_S)$ (°)	(60.0, -24.07)	
Reference pitch p_{gt} (mm)	9.664	
Normal pitches p_{g0D} (mm)	9.596	
p_{g0C} (mm)	8.266	
Gear addendum A_{d2} (mm)	0.0	
Working depth h_k (mm)	3.813	
(where cutter top land = 2mm and clearance = 2mm are assumed.)		
Phase angle of tooth surface C θ_{2wsc} (°)	-3.13	
Dimensions of pinion		
Radius of design cylinder R_{10} (mm)	31.391	
Outside radius R_{1h} (mm)	35.17	
Root radius R_{1t} (mm)	29.39	
Facewidth $z_{1ch} - z_{1ct}$ (mm)	32	
Dimensions of tooth surface C_C		
$P_{0CC}(u_{c0CC}, 0, z_{c0CC}; C_S)$ (mm)	(15.34, 0, 69.30)	
$\mathbf{g}_{0CC}(\phi_{0CC}, \phi_{n0CC}; C_S)$ (°)	(62.30, -21.60)	
Phase angle of tooth surface C_C θ_{1wscc} (°)	6.37	
Dimensions of involute helicoids		
	Surface D	Surface C_C
Lead (mm)	109.83	
Radii of base cylinder (mm)	31.389	17.754
Helix angles on design cylinder (°)	60.89	45.44
Transverse circular thickness on design cylinder (mm)	12.685	
Transverse top land (mm)	6.556	
Contact ratios along path of contact through the middle of facewidth (total contact ratios)	m_{0D} = 2.10 (4.21)	m_{0C} = 0.88 (2.10)

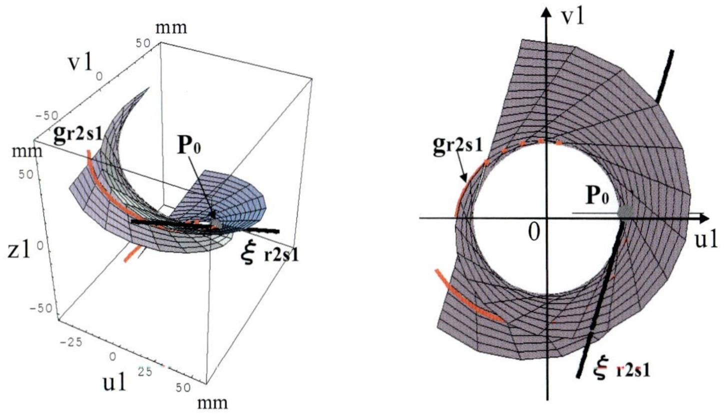

(a) Bird's-eye view **(b) View from direction z1**

Fig.4.4-2 Involute helicoid with g_{2S}

(4) Surface of action of the involute helicoid having the limit path of contact g_{2S}

Figure 4.4-3 shows the surface of action drawn by Eq. (4.4-2) where (a) is a bird's eye view, (b) and (c) are Figure (a) seen from the positive directions of the axes v_1 and z_1 respectively.

The surface of action of the involute helicoid is a complicated one composed of paths of contact (common contact normals) expressed by ξ_1 = constant (g_{2S} when $\xi_1 = 0$) and lines of contact expressed by θ_1 = constant as shown in Fig. 4.4-3.

$\phi_{1m} = 0$ in Fig. 4.4-3(c) means that the common contact normal is on the plane of rotation of the gear in this example. Generally, a pair of tooth surfaces D (drive side) and C (coast side) of the gear are designed for each tooth normal to be inclined inversely to the plane of rotation of the gear, therefore, the area represented by $\phi_{1m} < 0$ is named D and the area by $\phi_{1m} > 0$ is C, which correspond to the tooth surfaces D and C.

(5) Limit lines of action on the surface of action

The limit line of action on the surface of action for the gear is obtained by Eq. (2.4-4) as a set of points tangent to intersections of the surface of action and planes of rotation of the gear. For the pinion it is obtained in the same way.

Figure 4.4-3 shows the limit lines of action, where there are L_{2D} and L_{2C} of the gear and L_{1D} and L_{1C} of the pinion which are tangent to the base cylinder, corresponding to the areas D ($\phi_{1m} < 0$) and C ($\phi_{1m} > 0$) respectively. L_{1CD} is another limit line of action of the pinion which is not tangent to the base cylinder in the area C.

The effective surface of action D must not include the limit lines of action L_{2D} and L_{1D} as mentioned in 3.6-1 to realize a smooth tooth surface D of the gear. For the same reason, the effective surface of action C must not include L_{2C}, L_{1C} and L_{1CD}.

Figure 4.4-4 shows variations of radii R_{2SZD} and R_{2SZC} of the limit lines of action of the gear L_{2D} and L_{2C} along the axis z_2. In the region of mesh of the gear shown by the shaded rectangle, the radius R_{2SZD} of L_{2D} is smaller than the radius R_{2SZC} of L_{2C}. Adopting the same involute helicoid which has the limit path of contact g_{2S} as the path of contact for both the D and C sides, the D side has a smooth tooth surface in the region bigger than the radius R_{2SZD}, while the C side has no tooth surface in the region smaller than the radius

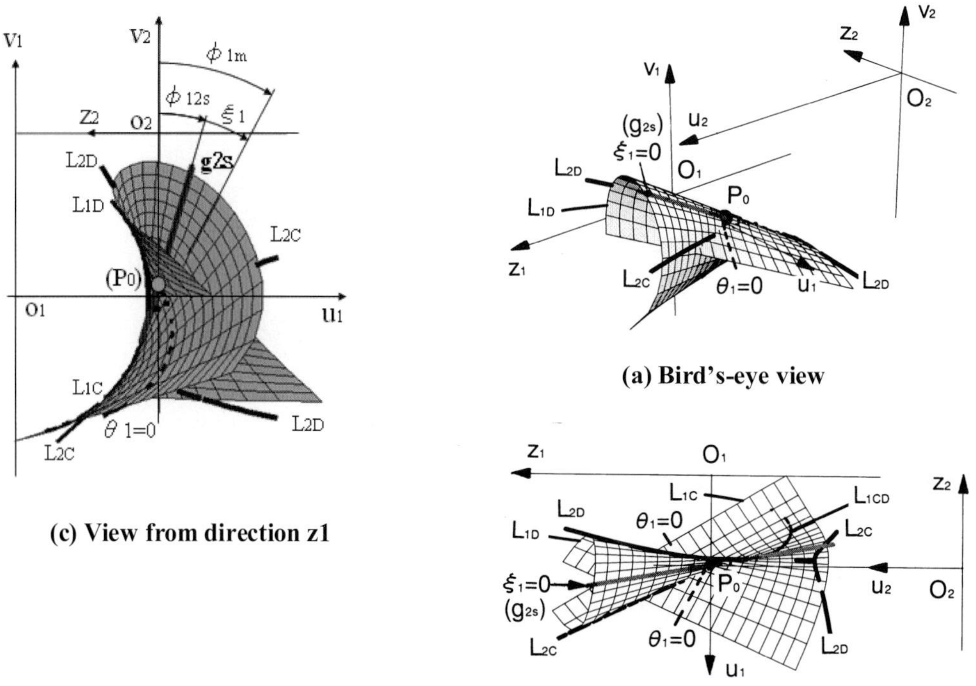

(c) View from direction z1

(a) Bird's-eye view

(b) View from direction v1

Fig.4.4-3 Surface of action of involute helicoid with $\mathbf{g}_{2S}$

R_{2SZC}, so that the tooth surfaces of the gear have a big difference in the small ends. This is often experienced in the design of face gears, which indicates that the selection of the facewidth of face gears has little freedom because the smooth tooth surfaces and the top land are severely limited generally when the same involute helicoid is given to both tooth surfaces D and C.

(6) Selection of paths of contact $\mathbf{g}_{0D}$ and $\mathbf{g}_{0C}$

In Fig. 4. 4-4, in order to make R_{2SZC} smaller and spread the effective smooth tooth surface C in the small end, the inclination angle of $\mathbf{g}_{0C}$ is changed to $\phi_{n0C} = \phi_{n2S} - 38°$ and the radius R_{2SZCC} of the limit line of action of the gear in the surface of action C is calculated and shown, where R_{2SZCC} is nearly equal to R_{2SZD} of $\phi_{n0D} = \phi_{n2S}$. It means that by selecting the inclination angles of $\mathbf{g}_{0D}$ and $\mathbf{g}_{0C}$ differently like $\phi_{n0D} = \phi_{n2S}$ and $\phi_{n0C} = \phi_{n2S}$ $-38°$ and giving the different radii of base cylinder for the pinion in the D and C sides respectively, the difference between the radii of the limit lines of action of the gear becomes smaller and sufficiently smooth tooth surfaces of the gear are realized in both the D and C sides. Finally, the inclination angles of $\mathbf{g}_{0D}$ and $\mathbf{g}_{0C}$ are compared with $\mathbf{g}_t$ from the view point of symmetry of the tooth shape and are determined as shown in Table 4. 4-1.

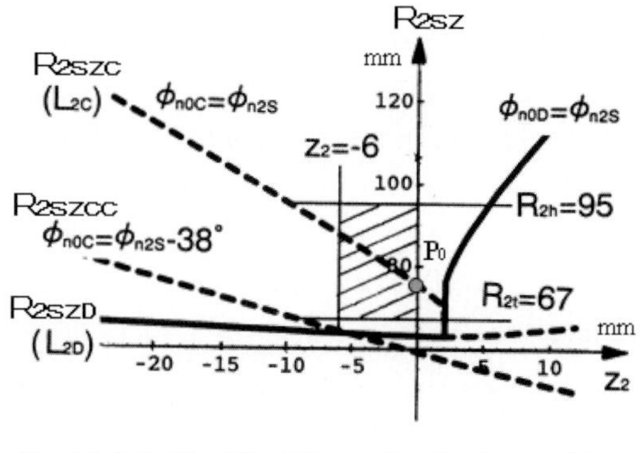

Fig.4.4-4 Radii of limit lines of action L_{2D} and L_{2C}

4．4．4 Modification of involute helicoids of the pinion

The involute helicoids of the pinion which have $\mathbf{g}_{0D}$ and $\mathbf{g}_{0C}$ selected above have different leads on the D and C sides, therefore the top land is not constant when the pinion is cylindrical such as in this example. To make the leads of the D and C sides equal, the involute helicoids are modified as follows. There are some methods to make the leads equal, but in this example, the lead of the D side is fixed and that of the C side is modified by the helix angle ψ_{b10C} only on the same plane of action G_{10C} with the same radius R_{b10C} of base cylinder to make a new tooth surface C_C.

Figure 4.4-5 shows the relation between the design point P_0 and the intersection P_{wSCC} of a new path of contact $\mathbf{g}_{0CC}$ and the new tooth surface C_C (which is indicated by the intersection w_{SCC} of G_{10C} and the tangent plane) at an angle of rotation $\theta_1 = 0$.

(1) Helix angle ψ_{b10CC} of tooth surface C_C

Because the lead of the tooth surface C_C is that of the D, the helix angle ψ_{b10CC} is obtained as follows.

$$\tan\psi_{b10CC} = R_{b10C}\tan\psi_{b10D}/R_{b10D} \tag{4.4-7}$$

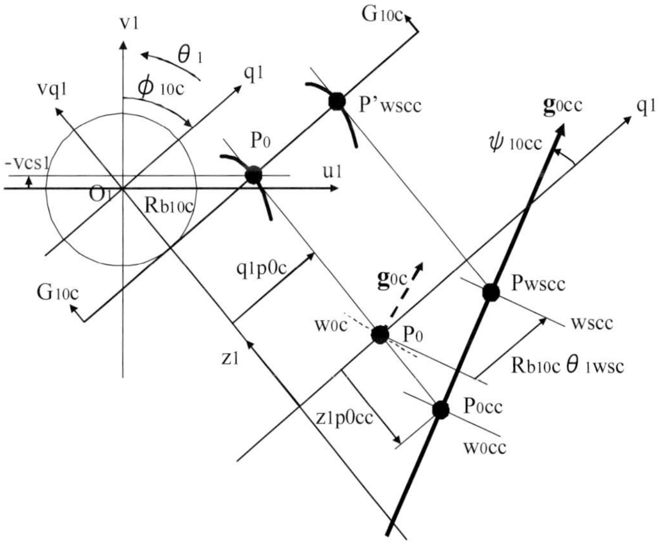

Fig.4.4-5 Tooth surface C_C and point of contact P_{wscc} at θ_1=0

(2) Path of contact $\mathbf{g}_{0CC}$ of tooth surface C_C

When the inclination angle of the new path of contact (common contact normal) $\mathbf{g}_{0CC}$ of the tooth surface C_C on the plane of action G_{10C} is expressed by $\mathbf{g}_{0CC}(\phi_{10C}, \psi_{b10CC}; O_1)$, the intersection $P_{0CC}(u_{c0CC}, 0, z_{c0CC}; C_S)$ of the plane S_H and $\mathbf{g}_{0CC}(\varphi_{0CC}, \phi_{n0CC}; C_S)$ is obtained by Eq. (2.1-21) in the coordinate system C_S as follows, where $\Sigma = \pi/2$.

$$\left.\begin{aligned}
\sin\phi_{n0CC} &= -\cos\psi_{b10CC}\sin\phi_{10C}\sin\Gamma_S+\sin\psi_{b10CC}\cos\Gamma_S\\
\tan\varphi_{0CC} &= \tan\phi_{10C}\cos\Gamma_S+\tan\psi_{b10CC}\sin\Gamma_S/\cos\phi_{10C}\\
u_{c0CC} &= E\tan\varphi_{0CC}\cos\Gamma_S\sin\Gamma_S\\
z_{c0CC} &= (u_{1p0}-u_{c0CC}\sin\Gamma_S)/\cos\Gamma_S\\
u_{1p0} &= \sqrt{(R_{10}^2-v_{cs1}^2)}
\end{aligned}\right\} \tag{4.4-8}$$

Expressed in the coordinate systems O_1 and O_{q1}, they are obtained as follows.

$$\left.\begin{aligned}
&P_{0CC}(u_{1p0}, -v_{cs1}, z_{1p0CC}; O_1),\quad P_{0CC}(q_{1p0C}, -R_{b10C}, z_{1p0CC}; O_{q1})\\
&z_{1p0CC} = -u_{c0CC}\cos\Gamma_S+z_{c0CC}\sin\Gamma_S-z_{1c0}
\end{aligned}\right\} \tag{4.4-9}$$

(3) Phase angle of tooth surface C_C from P_{0CC}

When $\theta_1 = 0$, the tooth surface C_C rotates by the phase angle θ_{1WSC} from P_0 and intersects g_{0CC} at P_{WSCC}. Therefore, the phase angle θ_{1WSCC} of P_{WSCC} from P_{0CC} is obtained as follows.

$$\left.\begin{aligned} P_{0CC}P_{WSCC} &= R_{b10C}\,\theta_{1WSC}\cos\phi_{b10CC} - z_{1p0CC}\sin\phi_{b10CC} \\ \theta_{1WSCC} &= 2\theta_{1p}(P_{0CC}P_{WSCC})/p_{g0CC} \\ p_{g0CC} &= 2\pi R_{b10C}\cos\phi_{b10CC}/N_1 \end{aligned}\right\} \qquad (4.4\text{-}10)$$

where $2\theta_{1p}$: angular pitch of pinion and N_1 : number of teeth of pinion.

(4) Equations of path of contact g_{0CC}

Using Eqs. (4.4-7) to (4.4-10), the path of contact g_{0CC} is expressed in the coordinate systems O_1 and O_{q1} as follows.

$$\left.\begin{aligned} q_1(\theta_1) &= R_{b10C}(\theta_1+\theta_{1WSCC})\cos^2\phi_{b10CC} + q_{1p0C} \\ \chi_1(\theta_1) &= \chi_{10C} = \pi/2-\phi_{10C} \\ u_1(\theta_1) &= q_1\cos\chi_{10C} + R_{b10C}\sin\chi_{10C} \\ v_1(\theta_1) &= q_1\sin\chi_{10C} - R_{b10C}\cos\chi_{10C} \\ z_1(\theta_1) &= R_{b10C}(\theta_1+\theta_{1WSCC})\cos\phi_{b10CC}\sin\phi_{b10CC} + z_{1p0CC} \end{aligned}\right\} \qquad (4.4\text{-}11)$$

The calculation results of g_{0CC} are shown in Table 4.4-1.

4.4.5 Conjugate tooth surfaces of the gear

The surfaces of action of the involute helicoids of the pinion which have the paths of contact g_{0D} and g_{0CC} respectively are obtained by Eq. (4.4-4), therefore transforming them into the coordinate system O_{r2} by Eq. (4.4-5), the conjugate tooth surfaces of the gear are obtained.

Figure 4.4-6(a) shows the conjugate tooth surface C_C of the gear. It is drawn by the parameters θ_1 and ξ_1 and the line of contact at $\theta_1 = 0$ intersects g_{0CC} corresponding to $\xi_1 = 0$ at P_{WSCC}. Because the limit line of action of the gear L_{r2C} passes in the part bigger than the inside radius of gear R_{2t}, the gear tooth surface C_C lacks a part in the small end.

Figure 4.4-6(b) shows the conjugate tooth surface D of the gear in the same way, where the line of contact at $\theta_1 = 0$ and the path of contact $\xi_1 = 0$ corre-

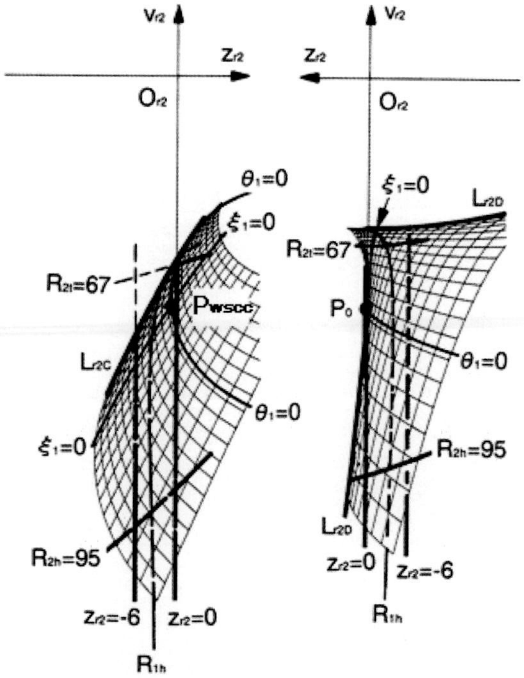

Fig.4.4-6 Tooth surfaces C_C and D of gear

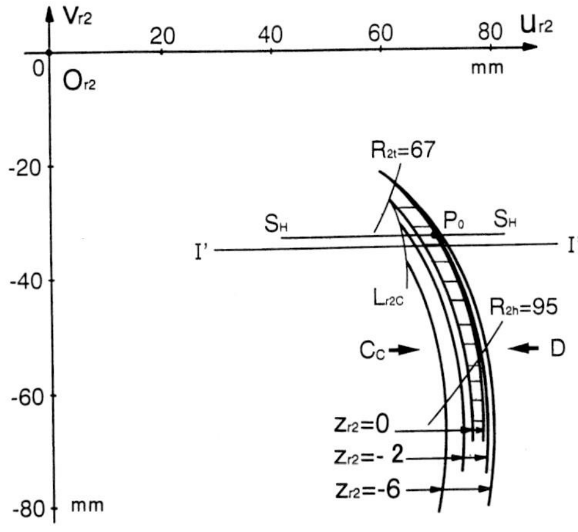

Fig.4.4-7 Gear tooth curves in planes of rotation

Fig.4.4-8 A trial product, which has pinion cut by a conventional machine and gear cut by a ball end mill controlled numerically

sponding to g_{0D} are drawn. Because the limit line of action of the gear L_{r2D} passes in the part smaller than the inside radius of gear R_{2t}, the tooth surface D of the gear is smooth all over the effective area.

In order to realize no variation of bearing loads, giving the amount of modification $\Delta p_{2nm}(\theta_1, \xi_1)$ in Eq. (4.4-6) as a function of ξ_1 only, the tooth bearing must be formed along a path of contact (ξ_1= constant) chosen adequately in Fig. 4.4-6.

Figure 4.4-7 shows the intersections of the tooth surfaces D and C_C of the gear and the planes of rotation $Z_{r2}=0$ (tip plane), $Z_{r2}=-2$ mm and $Z_{r2}=-6$ mm, where the acceptable top land (shaded part) is realized, which varies slightly by being narrower toward the large end.

Figure 4.4-8 shows a trial product of the example, the pinion is cut by a conventional gear shaper and the gear is cut by a ball end mill controlled numerically. Good tooth bearings and engagement are obtained.

4．4．6　Contact ratios

Contact ratios are calculated directly from the tooth surfaces of the gear as the ratio of the amount of angle of rotation which is necessary for a line of contact to pass the effective tooth surface from the small to large ends and the angular pitch, where the effective tooth surface is surrounded by the tip curves, the inside and outside radii and the limit lines of action of the pinion and gear. The calculation results are shown in Table 4.4-1.

Because the face angle of the gear is always 90° having no relation to Γ_S, face gears have a substantial difference between the contact ratios in the drive and the coast sides. In this example, it is a big design problem to give a sufficient contact ratio in the coast side.

4．5　Design example of face gears having an involute spur gear for gear Ⅰ(pinion)

Another pair of face gears which have the same shaft angle, offset, gear ratio, inside and outside radii of the gear as shown in Table 4.4-1 and have an involute spur gear for the pinion is calculated, and the surface of action and the conjugate tooth surfaces of the gear are shown.

(1) Given conditions

(a) The radius of the design cylinder of the gear R_{20} = 77 mm is the same.

(b) Because the pinion is an ordinary spur gear, the involute helicoids have the same base cylinder in the drive and the coast sides and the helix angle equals 0.

$$\left.\begin{aligned} \psi_{b10D} &= \psi_{b10C} = 0 \\ R_{b10D} &= R_{b10C} = R_{b10} \end{aligned}\right\} \tag{4.5-1}$$

Present design methods of face gears having a spur gear for the pinion means how to choose the base cylinder of the pinion and the facewidth of the gear so as to have sufficiently smooth tooth surfaces of the gear with no cusp at the tip under Eq. (4.5-1).

(2) Selection of the base cylinder of the pinion

The relation between the inclination angles $g_0(\phi_{10}, \psi_{b10}; 0_1)$ and $g_0(\phi_{20}, \psi_{b20}; 0_2)$ of $\mathbf{g}_0$ is given by Eq. (2.1-7) as follows, where $\Sigma = \pi/2$.

$$\left.\begin{aligned} \tan\phi_{20} &= \tan\psi_{b10}/\cos\phi_{10} \\ \sin\psi_{b20} &= -\cos\psi_{b10}\sin\phi_{10} \end{aligned}\right\} \tag{4.5-2}$$

Substituting Eq. (4.5-1) for Eq. (4.5-2) and simplifying,

$$\left.\begin{aligned} &\phi_{20D} = \phi_{20C} = 0 \\ &\psi_{b20D} = -\phi_{10D}, \quad \psi_{b20C} = -\phi_{10C} \end{aligned}\right\} \tag{4.5-3}$$

Eq. (4.5-3) means that the planes of action of the gear G_{20D} and G_{20C} are always perpendicular to the axis of the pinion, so that the planes of action of the pinion G_{10D} and G_{10C} are perpendicular to G_{20D} and G_{20C} and their intersections are the paths of contact $\mathbf{g}_{0D}$ and $\mathbf{g}_{0C}$. Therefore, the plane of rotation of the pinion through the design point P_0 is the plane of action $G_{20D} = G_{20C}$, so the radii of base cylinder of the gear $R_{b20D} = R_{b20C}$ are obtained as follows.

$$R_{b20D} = R_{b20C} = \sqrt{(R_{20}^2 - V_{cs2}^2)} = R_{b20} \tag{4.5-4}$$

The fundamental requirement for contact (2.2-6) is transformed by Eqs. (4.5-1), (4.5-3) and (4.5-4) as follows.

$$\begin{aligned} &\cos\phi_{10D} = i_0 R_{b10}/R_{b20} \\ &\cos\phi_{10C} = i_0 R_{b10}/R_{b20} \\ &\therefore\ \phi_{10D} = -\phi_{10C} \ (\text{where } \phi_{10D} \neq \phi_{10C}) \end{aligned} \tag{4.5-5}$$

In ordinary face gears, the radii of base cylinder $R_{b10D} = R_{b10C} = R_{b10}$ are chosen for the pressure angles ($\phi_{10C} = -\phi_{10D}$) at the design point P_0 to be 20° to 25° so as to realize smooth tooth surfaces of the gear around P_0 as much as possible.

(3) Design point $P_{0D}(=P_0)$ and P_{0C}

The intersections P_{0D} (u_{1c0D}, $-v_{cs1}$, z_{1c0D}; C_1) and P_{0C} (u_{1c0C}, $-v_{cs1}$, z_{1c0C}; C_1) of the plane S_H and the paths of contact $\mathbf{g}_{0D}$ and $\mathbf{g}_{0C}$ are obtained as follows, because $\mathbf{g}_{0D}$ and $\mathbf{g}_{0C}$ are on the plane of rotation of the pinion.

$$\left.\begin{aligned} u_{1c0D} &= R_{b10}/\cos\phi_{10D} - v_{cs1}\tan\phi_{10D} \\ z_{1c0D} &= R_{b20} \\ u_{1c0C} &= R_{b10}/\cos\phi_{10C} - v_{cs1}\tan\phi_{10C} \\ z_{1c0C} &= R_{b20} \end{aligned}\right\} \quad (4.5\text{-}6)$$

When $v_{cs1} \neq 0$ (the offset $E \neq 0$), P_{0D} does not coincide with P_{0C} because of Eq. (4.5-6), while $\mathbf{g}_{0D}$ and $\mathbf{g}_{0C}$ intersect on the plane $v_1 = 0$.

(4) Calculation results of surface of action and conjugate tooth surfaces of the gear

When $\phi_{10D} = -\phi_{10C} = -20°$ ($R_{b10D} = R_{b10C} = 15.94$ mm) are given, they are obtained as follows.

Figure 4.5-1 shows the surface of action which corresponds to Fig. 4.4-3(a), where the paths of contact $\mathbf{g}_{0D}$ and $\mathbf{g}_{0C}$, the line of contact at $\theta_1 = 0$, the limit lines of action of the gear L_{2D} and L_{2C} and the limit lines of action of the pinion L_{1D} and L_{1C} are drawn.

Figure 4.5-2 shows the variations of the radii R_{2SZD} and R_{2SZC} of the limit lines of action L_{2D} and L_{2C} along the axis z_2. The region of mesh is effective all over the tooth surface C, while it is not effective inside the design point P_0 in the tooth surface D.

Figures 4.5-3(a) and (b) show the tooth surface C_C which has a phase angle different from the C and the tooth surface D of the gear seen from the negative or the positive direction of the axis v_{r2} respectively, where $\mathbf{g}_{r20D}$ and $\mathbf{g}_{r20C}$ are the $\mathbf{g}_{0D}$ and $\mathbf{g}_{0C}$ expressed in the coordinate system O_{r2}, $R_{1h} = 21$ mm is the tip circle radius of the pinion and L_{r21D} and L_{r21C} are the limit lines of action of the pinion drawn on the tooth surfaces of the gear. The tooth surface D lacks a smooth one in the small end, while the tooth surface C_C is smooth from the small to large ends. However, the engagements in both tooth surfaces D and C_C are limited by the tip circle of the pinion.

Figure 4.5-4 shows the intersections of the tooth surfaces D and C_C of the gear and the planes of rotation $Z_{r2} = 0$ (tip plane), $Z_{r2} = -3$ mm and $Z_{r2} = -6$ mm and the limit lines of action of the gear L_{r2D} and L_{r2C}, corresponding to Fig. 4.4-7. The facewidth of the gear generated by the pinion with this same base cylinder has an acceptable area only from $R_{2t} \fallingdotseq 80$ mm limited by L_{r2D} to the large end limited by the top land ($R_{2h} = 95$ mm). In present design methods for face gears, there are few variables which can be chosen at will because Eq. (4.5-1) is given, therefore the facewidth is forced to be selected so as to avoid the limit lines of action in the small end and to ensure a sufficient top land in the large end.

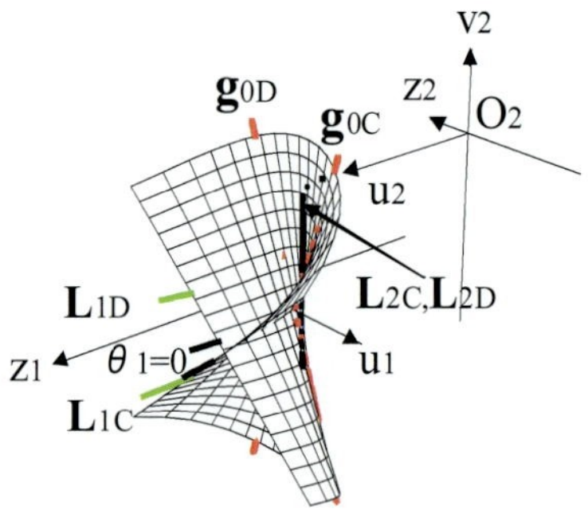

Bird's-eye view

Fig.4.5-1 Surface of action

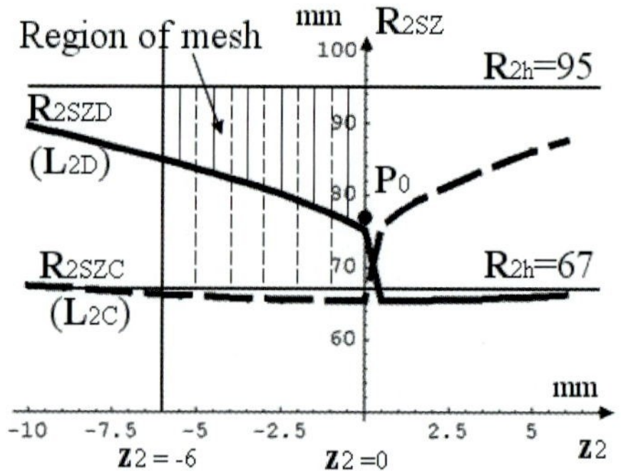

Fig.4.5-2 Radii of limit lines of action L_{2D} and L_{2C}

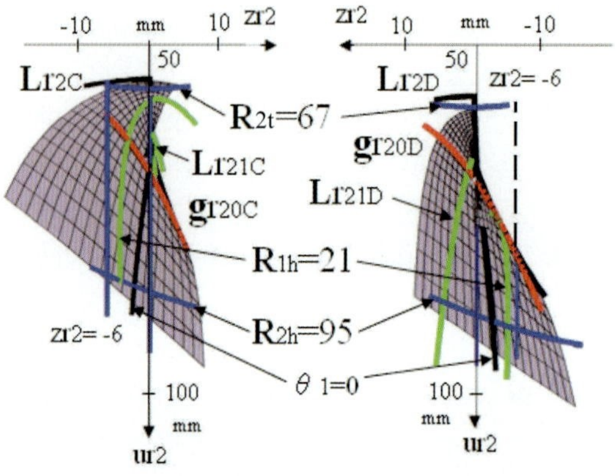

(a) Surface Cc (b) Surface D

Fig.4.5-3 Tooth surfaces C_C and D of gear

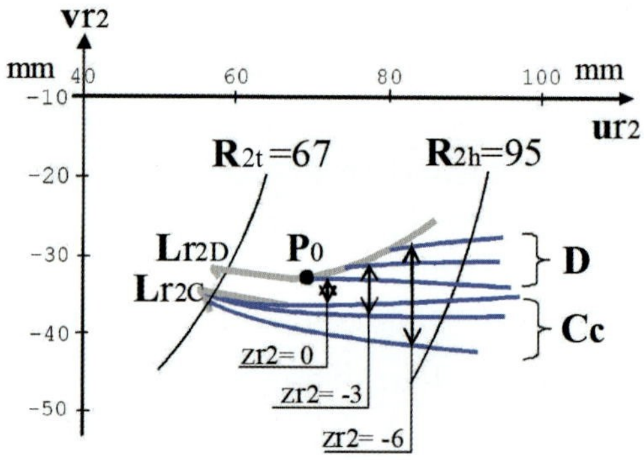

Fig.4.5-4 Gear tooth curves in planes of rotation

4. 6 Summary and references

When the shaft angle and the offset of the two axes, the gear ratio (constant) and the path of contact $\mathbf{g}_0$ which realizes no variation of bearing loads are given, gear pairs which have the involute helicoid corresponding to $\mathbf{g}_0$ for one member and its conjugate tooth surface for the other are analyzed and the following results are obtained.

(1) Equations of the plane of action having the path of contact $\mathbf{g}_0$ and the corresponding involute helicoids are obtained. Giving these involute helicoids corresponding to $\mathbf{g}_0$ to gears I (pinion) and II (gear), a pair of involute gears in point contact are obtained.

(2) Equations of the line of contact, the surface of action and the conjugate tooth surface are obtained when an involute helicoid is given for one member of gears.

(3) Helical, crossed helical, worm, conical and face gears which have involute helicoids for one member and have been treated separately are the gear pairs determined by where a point P_0 and its normal $\mathbf{n}_0$ on the involute helicoid are located as the design point and the path of contact in the static space. The surfaces of action and the mating tooth surfaces are expressed by the equations of items (1) and (2).

(4) Examples of face gears which have involute helicoids for gear I (pinion) and the conjugate tooth surfaces for gear II (gear) are shown, the results of which are summarized as follows.

(a) The radius of the design reference cylinder of the gear (gear II) R_{20} and the inclination

angle ϕ_0 of the plane S_n are chosen from the inside and outside radii R_{2t} and R_{2h} of the gear and the design point P_0 is determined.
(b) Using the surface of action of the involute helicoid which has the limit path of contact $\mathbf{g}_{2S}$ through P_0, the paths of contact $\mathbf{g}_{0D}$ (drive side) and $\mathbf{g}_{0C}$ (coast side) are selected for the radii of the limit lines of action of the gear to be smaller than the inside radius of the gear R_{2t}, which are realized by giving the involute helicoids whose radii of base cylinder of the pinion are different on the drive and the coast sides.
(c) In order to realize a cylindrical pinion, the involute helicoids of the drive and the coast sides are modified so as to have the same lead.
(d) The conjugate tooth surfaces of the gear are calculated and the results are shown.
(e) The surface of action and the conjugate tooth surfaces of a gear pair which have an involute spur gear for the pinion are calculated and shown for reference.

Selecting the design point P_0 and the paths of contact $\mathbf{g}_{0D}$ and $\mathbf{g}_{0C}$ mentioned above, the problem of the present face gears, namely no smooth tooth surface in the small end and a cusp at the tip in the large end, is overcome and the face gears designed according to the new method can be used for power transmission. The new method is applicable to cylindrical gears when $\Sigma \to 0$ and bevel gears when $E \to 0$, so that it is effective for all kinds of gears from cylindrical to hypoid gears.

References

(40) 本多捷，歯面の接触と動荷重の基礎理論（第３報），機論，62-603，C(1996)，4349-4356.
(41) 本多捷，歯面の接触と動荷重の基礎理論（第４報），機論，62-603，C(1996)，4357-4362.
(42) 荻野修作，インボリュートねじ歯車の理論（第１報），機論，23-134，第４部(1957)，726-732.
(43) 三留謙一，円すい形インボリュート歯車の研究（第１報，設計および製作法），機論，48-430，C(1982)，852-859.
(44) Roth, K., *Zahnradtechnik Evolventen-Sonderverzahnungen zur Getriebeverbesserung*, Springer, 1998, 435-440.
(45) Honda, S., "A unified designing method applicable to all kinds of gears for power transmission", Proceedings of JSME Int. Conf. MPT2001-Fukuoka, 506-512.

5. Variation of tooth bearing and engagement error caused by assembly errors

In this chapter, when gear pairs designed according to Section 4.4 have assembly errors, it is discussed how the tooth bearing and the engagement error are varied by them. Gear I (pinion) has involute helicoids and gear II (gear) has tooth surfaces modified from the conjugate ones generated by gear I.

5.1 Movement of a point of contact and engagement error

5.1.1 Definition of assembly errors

Figure 5.1-1 shows the relation among the coordinate systems O_3, O_4, O_5 and O_2, the first three of which have assembly errors to the last, where the origins O_3, O_4 and O_5 are the same.

The coordinate system O_5 is obtained by transferring and rotating the coordinate system O_2 according to the following five items which are infinitesimal of the first order, while the pinion coordinate system O_1 is fixed.

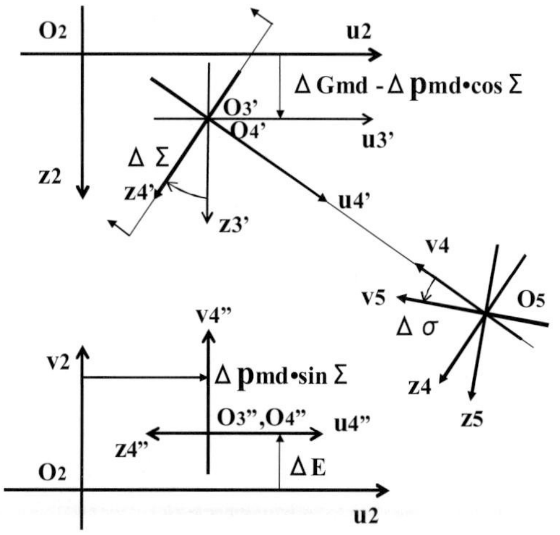

Fig.5.1-1 Relationship among coordinate systems O_2, O_3, O_4 and O_5

(1) $\triangle Gmd$: error in the direction of the gear axis (positive in the direction of z_2)
(2) $\triangle E$: error in the direction of the offset (positive in the direction of v_2)
(3) $\triangle Pmd$: error in the direction of the pinion axis (positive in the direction of z_1)
(4) $\triangle \Sigma$: error in rotation around the axis of v_2 (positive in clockwise seen from the positive direction of v_2 which means positive in increase of the shaft angle Σ)
(5) $\triangle \sigma$: error in rotation around the axis of u_2 (positive in counterclockwise seen from the positive direction of u_2)

The rotation around the axis z_2 does not result in an error because the point of contact moves along the true path of contact according to the corresponding rotation of the pinion, therefore it has no influence on the tooth bearing and the engagement error.

5.1.2 Relation between coordinate systems O_5 and O_2

When the coordinate system $O_3(u_3, v_3, z_3)$ is obtained by transferring the coordinate system O_2 by the assembly errors $\triangle Gmd$, $\triangle Pmd$ and $\triangle E$, the relation between the coordinate systems O_3 and O_2 is as follows.

$$u_3 = u_2 - \triangle Pmd \sin\Sigma$$
$$v_3 = v_2 - \triangle E$$
$$z_3 = z_2 - (\triangle Gmd - \triangle Pmd \cos\Sigma)$$

When the coordinate system $O_4(u_4, v_4, z_4)$ is obtained by rotating the coordinate system O_3 by the error $\triangle\Sigma$, the relation between the coordinate systems O_4 and O_3 is as follows.

$$u_4 = u_3 + z_3\triangle\Sigma$$
$$v_4 = v_3$$
$$z_4 = z_3 - u_3\triangle\Sigma$$

In the same way, when the coordinate system $O_5(u_5, v_5, z_5)$ is obtained by rotating the coordinate system O_4 by the error $\triangle\sigma$, the relation between the coordinate systems O_5 and O_4 is as follows.

$$u_5 = u_4$$
$$v_5 = v_4 + z_4\triangle\sigma$$
$$z_5 = z_4 - v_4\triangle\sigma$$

Eliminating the infinitesimal items of the second order, the relation between the coordinate systems O_5 and O_2 is obtained as follows.

$$u_5 = u_2 - \triangle u_2$$
$$v_5 = v_2 - \triangle v_2$$
$$z_5 = z_2 - \triangle z_2$$

where,

$$\left.\begin{aligned}\triangle u_2 &= \triangle Pmd \sin\Sigma - z_2\triangle\Sigma\\ \triangle v_2 &= \triangle E - z_2\triangle\sigma\\ \triangle z_2 &= \triangle Gmd - \triangle Pmd \cos\Sigma + u_2\triangle\Sigma + v_2\triangle\sigma\end{aligned}\right\} \quad (5.1\text{-}1)$$

Five assembly errors $\triangle Pmd$, $\triangle Gmd$, $\triangle E$, $\triangle\Sigma$ and $\triangle\sigma$ are replaced by $\triangle u_2$, $\triangle v_2$ and $\triangle z_2$ in the coordinate system O_2, where the normal inclines by $\triangle\Sigma$ and $\triangle\sigma$.

5．1．3　Expression of modified tooth surface of a gear in coordinate system O_2

Based on Fig. 3. 5-3, Fig. 5. 1-2 shows schematically the relation among a point of contact P_m, its common contact normal $\mathbf{n}_m$ and the line of contact on the conjugate tooth surface (expressed by $\theta_1 - \xi_1$ plane) of an involute helicoid explained in Chapter 4.

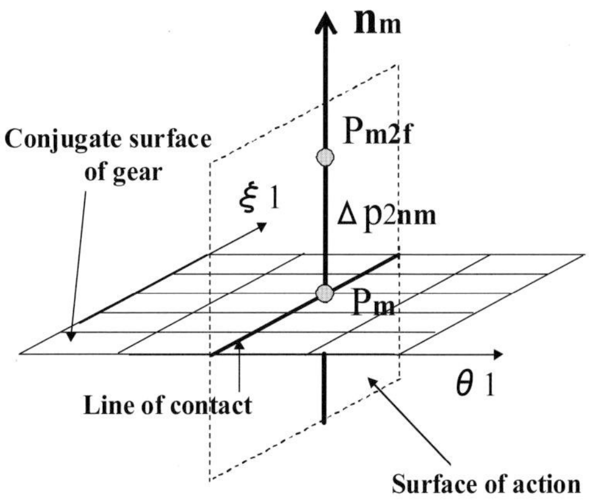

Fig.5.1-2 Definition of modified surface P_{m2f} of gear

A point $P_{m2f}(u_{r2m2f}, v_{r2m2f}, z_{r2m2f}; 0_{r2})$ on the modified tooth surface of a gear is expressed by Eq. (4. 4-6), which has an amount of modification $\triangle p_{2nm}$ in the direction of $\mathbf{n}_m$ corresponding to the point $P_m(u_{r2m}, v_{r2m}, z_{r2m}; O_{r2})$ on the conjugate tooth surface of the gear.

Because the point $P_m(u_{2m}, v_{2m}, z_{2m}; O_2)$ given by Eq. (4.4-4) is also on the surface of action, $P_{m2f}(u_{2m2f}, v_{2m2f}, z_{2m2f}; O_2)$ can be expressed as a point which has an amount of modification $\triangle p_{2nm}(\theta_1, \xi_1)$ on the common contact normal $\mathbf{n}_m$ from P_m in the coordinate system O_2 as follows, where $\triangle p_{2nm}(\theta_1, \xi_1)$ can be replaced by $\triangle p_{2nm}(\theta_2, \xi_2)$.

$$\left.\begin{aligned} u_{2m2f} &= u_{2m} + \triangle p_{2nm}(\theta_1, \xi_1)\cos\phi_{b2m}\cos\chi_{2m} \\ v_{2m2f} &= v_{2m} + \triangle p_{2nm}(\theta_1, \xi_1)\cos\phi_{b2m}\sin\chi_{2m} \\ z_{2m2f} &= z_{2m} + \triangle p_{2nm}(\theta_1, \xi_1)\sin\phi_{b2m} \end{aligned}\right\} \quad (5.1\text{-}2)$$

Strictly speaking, the common contact normal $\mathbf{n}_m$ is not the normal of the modified tooth surface at P_{m2f}, but in the following discussion, $\mathbf{n}_m$ is regarded as the normal of the modified tooth surface at P_{m2f} because $\triangle p_{2nm}$ is infinitesimal of the first order.

Because P_m is also on the tooth surface of the pinion, the modified tooth surface of the gear given by Eq. (5.1-2) can be considered as the tooth surface which keeps the distance $\triangle p_{2nm}(\theta_1, \xi_1)$ from the corresponding tooth surface of the pinion, where any surface represented by $\triangle p_{2nm}(\theta_1, \xi_1)$ can be given.

5.1.4 Normal gap δLm of modified tooth surface of a gear caused by assembly errors

Figure 5.1-3 shows the relation between the normal distance δ Lz2m of the modified tooth surface of the gear and the assembly error $\triangle z_{2m}$ in the direction of z_2 at a point of contact P_{m2f}. A point of contact P_m is P_{m2} on the conjugate tooth surface of the gear and P_{m1} on the tooth surface of the pinion, while the common contact normal $\mathbf{n}_m$ is $\mathbf{n}_{m2}$ as the normal of the tooth surface of the gear and $\mathbf{n}_{m1}$ as the normal of the pinion. It is supposed that the tooth surfaces of the pinion and gear in the vicinity of the point of contact P_m can be approximated by the common tangent plane W_m which is W_{m1} for the pinion and W_{m2} for the gear. Transferring from the coordinate system O_2 to O_5 through the assembly error $\triangle z_{2m}$, P_{m2f} moves to P_{m5}, $\mathbf{n}_{m2}$ to $\mathbf{n}_{m5}$ which inclines by $\triangle\Sigma$ and $\triangle\sigma$ and the modified tooth surface of the gear W_{m2f} to W_{m5}. Because the tooth surface of the pinion is fixed, the point of contact on the tooth surface of the pinion ($P_{m1} = P_m$) and the common contact normal ($\mathbf{n}_{m1} = \mathbf{n}_m$) are not changed as shown in Fig. 5.1-3, therefore it is assumed that the common contact normal at W_{m5} remains $\mathbf{n}_{m1} = \mathbf{n}_m$. When the intersection of the common contact normal $\mathbf{n}_m$ and the modified tooth surface of the gear W_{m5} through P_{m5} is P_{m6}, the normal distance δ Lz2m from P_{m2f} to W_{m5} can be expressed as follows, because $\triangle\Sigma$ and $\triangle\sigma$ are supposed to be infinitesimal of the first order.

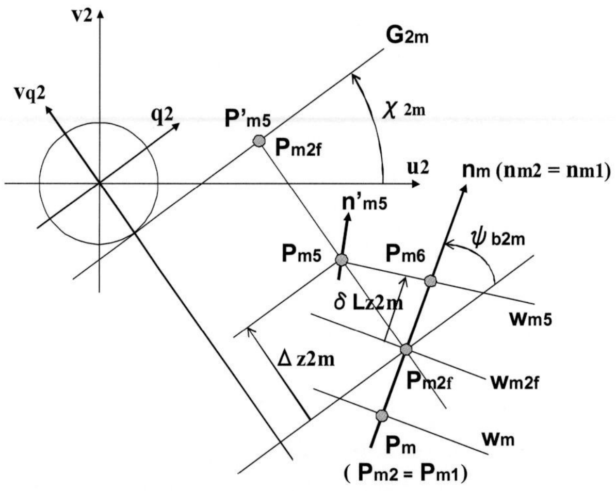

Fig.5.1-3 Normal distance δ Lz2m between tangent planes W_{m2f} and W_{m5}

$$\delta Lz2m = P_{m2f}P_{m6}\cos\triangle\Sigma\cos\triangle\sigma$$
$$= P_{m2f}P_{m6} = \triangle z_{2m}\sin\phi_{b2m} \quad (5.1\text{-}3)$$

The normal distances $\delta Lu2m$ and $\delta Lv2m$ caused by $\triangle u_{2m}$ and $\triangle v_{2m}$ are obtained in the same way.

$$\left.\begin{aligned}\delta Lu2m &= \triangle u_{2m}\cos\phi_{b2m}\cos\chi_{2m}\\ \delta Lv2m &= \triangle v_{2m}\cos\phi_{b2m}\sin\chi_{2m}\end{aligned}\right\} \quad (5.1\text{-}4)$$

The normal gap δLm caused by the assembly errors is regarded approximately as the sum of the normal distances, so that it is expressed as follows.

$$\delta Lm = \delta Lu2m + \delta Lv2m + \delta Lz2m \quad (5.1\text{-}5)$$

Figure 5.1-4 shows schematically the relation among P_m, P_{m2f}, P_{m6}, P_{m7} and δLm, where P_{m7} is the intersection of the modified tooth surface of the gear moved by the assembly errors $\triangle u_{2m}$, $\triangle v_{2m}$ and $\triangle z_{2m}$ and the common contact normal $\mathbf{n}_m$ and $P_{m2f}P_{m7}$ equals δLm. Because a point of contact P_m and its common contact normal $\mathbf{n}_m$ are given as functions of θ_1 and ξ_1, the normal gap δLm caused by the assembly errors in the vicinity of the point of contact chosen at will is obtained by giving θ_1 and ξ_1.

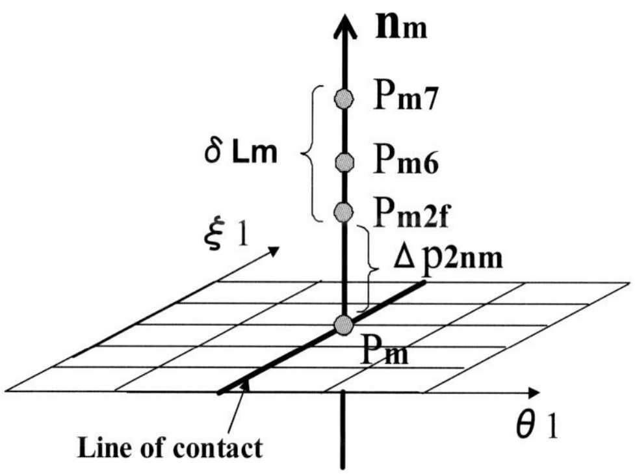

Fig.5.1-4 Schematic view of S_{knm}

5．1．5　Normal clearance $S_{knm}(\theta_1, \xi_1)$ between the tooth surface of a pinion and the modified one of a gear

The normal clearance $P_mP_{m7} = S_{knm}$ between the tooth surface of the pinion and the modified one of the gear in the vicinity of P_m is obtained from Fig.5.1-4 as follows.

$$S_{knm}(\theta_1, \xi_1) = \triangle p_{2nm}(\theta_1, \xi_1) + \delta Lm(\theta_1, \xi_1) \quad (5.1\text{-}6)$$

The point P_{m2} on the conjugate tooth surface of the gear moves to P_{m2f} due to the amount of modification of the gear $\triangle p_{2nm}$ and moreover to P_{m7} due to the normal gap δLm caused by the assembly errors.

5．1．6　Contact (interference) point P_{m8} between the tooth surface of a pinion and the modified one of a gear with assembly errors

Because $S_{knm}(\theta_1, \xi_1)$ represents the clearance between the tooth surface of the pinion and the modified one of the gear in the vicinity of the common contact normal at an angle of rotation θ_1, a contact or interference point P_{m8} is obtained as follows.

(1) When material part of tooth surface of a gear is in the negative direction of normal n_{m2} at P_{m2} (tooth surface D)

Because $S_{knm}(\theta_1, \xi_1) \geqq 0$ means that there is an interference between tooth surfaces, the location $\xi_1 = \xi_{18}$ at which $S_{knm}(\theta_1, \xi_1)$ becomes maximum algebraically at an angle of rotation θ_1 is the contact (interference) point $P_{m8}(\theta_1, \xi_{18})$ between the tooth surface of the pinion and the modified one of the gear.

(2) When material part of tooth surface of a gear is in the positive direction of normal n_{m2} at P_{m2} (tooth surface C)

Because $S_{knm}(\theta_1, \xi_1) \geqq 0$ means that there is a clearance between tooth surfaces, the location $\xi_1 = \xi_{18}$ at which $S_{knm}(\theta_1, \xi_1)$ becomes minimum algebraically at an angle of rotation θ_1 is the contact (interference) point $P_{m8}(\theta_1, \xi_{18})$ between the tooth surface of the pinion and the modified one of the gear.

Varying θ_1, $P_{m8}(\theta_1, \xi_{18})$ draws a path of contact on the modified tooth surface of the gear, therefore the tooth bearing under no loads is represented by one pitch on the path of contact, $S_{knm}(\theta_1, \xi_{18})$ of which is larger (or smaller) than the other part.

5.1.7 Engagement errors

When $\triangle p_{ep}(\theta_1, \xi_{18})$ and $\triangle p_{eG}(\theta_1, \xi_{18})$ are the other expressions of $S_{knm}(\theta_1, \xi_{18})$ calculated in the directions of the axes q_1 and q_2 in the coordinate systems O_{q1} and O_{q2} respectively, they are expressed as follows.

$$\left.\begin{aligned} \triangle p_{eP}(\theta_1, \xi_{18}) &= S_{knm}(\theta_1, \xi_{18})/\cos\phi_{b10} \\ \triangle p_{eG}(\theta_1, \xi_{18}) &= S_{knm}(\theta_1, \xi_{18})/\cos\phi_{b2m}(\theta_1, \xi_{18}) \end{aligned}\right\} \quad (5.1\text{-}7)$$

The engagement errors $\triangle\theta_{1k}$ and $\triangle\theta_{2k}$ of the pinion and the gear at θ_1 are obtained as follows.

$$\left.\begin{aligned} \triangle\theta_{1k}(\theta_1, \xi_{18}) &= \triangle p_{ep}(\theta_1, \xi_{18})/R_{b10} \\ \triangle\theta_{2k}(\theta_1, \xi_{18}) &= \triangle p_{eG}(\theta_1, \xi_{18})/R_{b2m} = \triangle\theta_{1k}/i_0 \end{aligned}\right\} \quad (5.1\text{-}8)$$

5.1.8 A pair of tooth surfaces insensitive to assembly errors

According to Eqs. (5.1-3) and (5.1-4), the normal gap δLm caused by the assembly errors $\triangle Gmd$, $\triangle Pmd$ and $\triangle E$ depends only on the inclination angle of the common contact normal $n_m(\chi_{2m}, \phi_{b2m}; C_2)$, so that a pair of tooth surfaces having the plane surface of action mentioned in 3.5 is feasible to realize tooth surfaces insensitive to the assembly errors because χ_{2m} and ϕ_{b2m} are constant. This hypothesis is already valid for involute helical gears and is reinforced by the fact that the tooth bearing of the face gears having involute helicoids for a pinion has no relation to the movement in the direction of the axis of the pinion as mentioned in 5.2. In addition, it suggests also the hypothesis that the tooth bearing of the formate gears which give conical tooth surfaces for the ring gear are insensitive to movement in the direction of the axis of the ring gear. It must be verified by experiments hereafter.

5．2　Calculated examples of variation of tooth bearing and engagement error

In this section, the variation of tooth bearing and the engagement error on the drive side are calculated. On the coast sides, they are obtained in the same way.

5．2．1　Dimensions of face gears and the conjugate tooth surface of a gear

Table 5.2-1 shows the dimensions of the face gears having involute helicoids for a pinion designed by the method mentioned in Chapter 4.

Table 5.2-1　Dimensions of face gears

Shaft angle Σ(°)	90	
Ratio of angular velocities i_0 (N_2/N_1)	4.1(41/10)	
Angle of instantaneous axis Γ_S(°)	76.29	
Offset E(mm)	31.75(below center)	
Inclination angle of plane S_n ϕ_0(°)	55.84	
Dimensions of gear	Pinion	Gear
Radii of design cylinder R_{10}, R_{20}(mm)	26.24	70
Outside radii R_{1h}, R_{2h}(mm)	30.8	85
Inside(Root)radii R_{1t}, R_{2t}(mm)	25.23	62
Face and root angles (°)	0	90
Facewidth of pinion $z_{1ch}-z_{1ct}$(mm)	28	
Dimensions of tooth surface D		
$P_0(u_{c0}, 0, z_{c0}; C_S)$ (mm)	(10.44, 0, 67.66)	
$\mathbf{g}_{0D}(\phi_0, \phi_{n0D}; C_S)$ (°)	(55.0, 11.30)	
Dimensions of tooth surface C_C		
$P_{0CC}(u_{c0CC}, 0, z_{c0CC}; C_S)$ (mm)	(11.65, 0, 62.68)	
$\mathbf{g}_{0CC}(\phi_{0CC}, \phi_{n0CC}; C_S)$ (°)	(57.91, -23.96)	
Phase angle of tooth surface C_C θ_{1WSCC}(°)	5.87	
Dimensions of involute helicoids		
	Surface D	Surface C_C
Lead(mm)	111.874	
Radii of base cylinder(mm)	26.174	15.471
Helix angles on design cylinder(°)	55.774	40.987
Transverse circular thickness on design cylinder(mm)	11.617	
Transverse top land(mm)	4.01	

Contact ratios along path of contact through the middle of facewidth (total contact ratios)

$m_{0D} = 1.75$ (4.21), $m_{0C} = 1.23$ (2.28)

Figure 5.2-1 shows the calculated example of the conjugate tooth surface D of the gear with the parameters θ_1 and ξ_1, where the boundary curves, namely, R_{1h} (outside radius of pinion), R_{2h} (outside radius of gear), R_{2t} (inside radius of gear) and z_{2h} (tip plane of gear), the limit line of action L_{r2}, the lines of contact at $\theta_1 = 0$ and 36° and the paths of contact at $\xi_1 = -16.0°$ and $-26.3°$ are drawn and where P_{mZ2h} and P_{mr1h} are the intersections of the paths of contact and the tip plane of the gear and the tip cylinder of the pinion respectively.

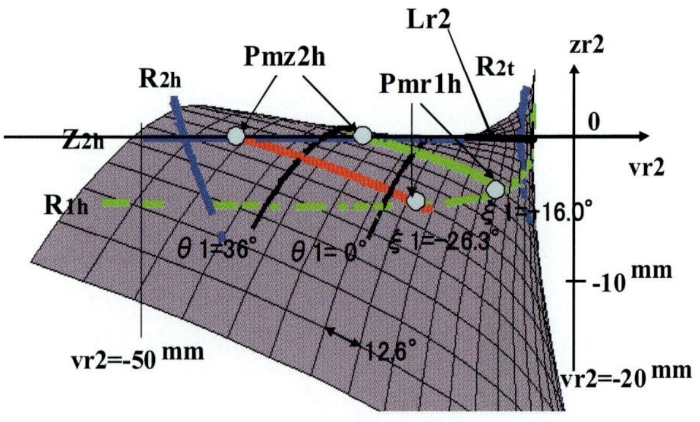

Fig.5.2-1 Conjugate tooth surface D of gear

5. 2. 2 Normal gap $\delta Lm(\theta_1, \xi_1)$ caused by assembly errors

The following five assembly errors defined in 5.1.1 are given and handled individually.

(1) $\triangle Gmd = \pm 0.1$mm (gear axis)
(2) $\triangle Pmd = \pm 0.1$mm (pinion axis)
(3) $\triangle E = \pm 0.1$mm (offset)
(4) $\triangle \Sigma = \pm 0.1°$ (around axis v_2)
(5) $\triangle \sigma = \pm 0.1°$ (around axis u_2)

Because of $\Sigma = \pi/2$, Eq. (5.1-1) is simplified as follows.

$$\left.\begin{aligned} \triangle u_{2m} &= \triangle Pmd - z_{2m} \triangle \Sigma \\ \triangle v_{2m} &= \triangle E - z_{2m} \triangle \sigma \\ \triangle z_{2m} &= \triangle Gmd + u_{2m} \triangle \Sigma + v_{2m} \triangle \sigma \end{aligned}\right\} \quad (5.2\text{-}1)$$

where $P_m(u_{2m}, v_{2m}, z_{2m}; O_2)$ is given by Eq. (4.4-4).

Figures 5.2-2 from (a) to (e) show the normal gaps δLmG, δLmP, δLmE, $\delta Lm\Sigma$ and $\delta Lm\sigma$ corresponding to each assembly error, where the boundary curves R_{2h}, R_{2t}, z_{2h} and R_{1h}, the path of contact $\xi_1 = -26.3°$ and the limit line of action L_{r2} are drawn. The influences on the tooth bearing caused by the assembly errors in these examples are as follows.

(1) In all the normal gaps $\delta Lm(\theta_1, \xi_1)$, $\partial(\delta Lm)/\partial\theta_1 = 0$, so that they are the surface on which $\partial(\delta Lm)/\partial\xi_1$ varies only and their differences from (a) to (e) caused by the assembly errors are nothing but those of $\partial(\delta Lm)/\partial\xi_1$. $\delta Lm(\theta_1, \xi_1)$ depends entirely on the inclination angle of $\mathbf{n}_m(\pi/2-\chi_{2m}, \psi_{b2m}; C_2)$ which is a function of ξ_1 only expressed by Eq. (4.4-4). $\delta Lm\Sigma$ and $\delta Lm\sigma$ are also functions of ξ_1 only, because the increments of $\delta Lm\Sigma$ and $\delta Lm\sigma$ caused by those of (u_{2m}, v_{2m}, z_{2m}) due to θ_1 result in 0.

(2) In the normal gaps, |δLm|s of △Gmd (Fig. 5. 2-2(a)) and △Σ(Fig. 5. 2-2(d)) are larger compared with those of △E (Fig. 5. 2-2(c)) and △σ (Fig. 5. 2-2(e)).

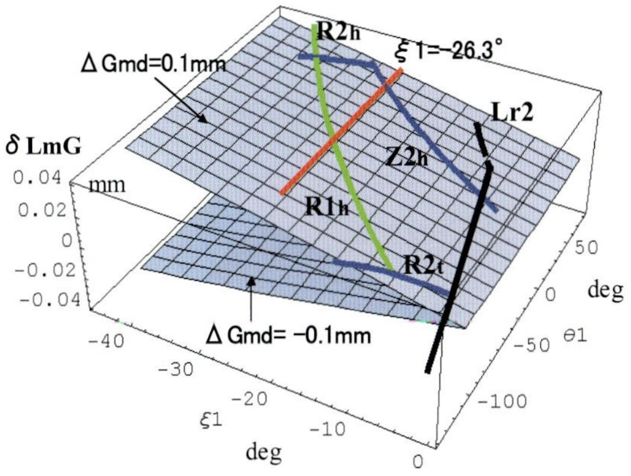

Fig.5.2-2(a) Normal gap δLmG caused by △Gmd in Drive side

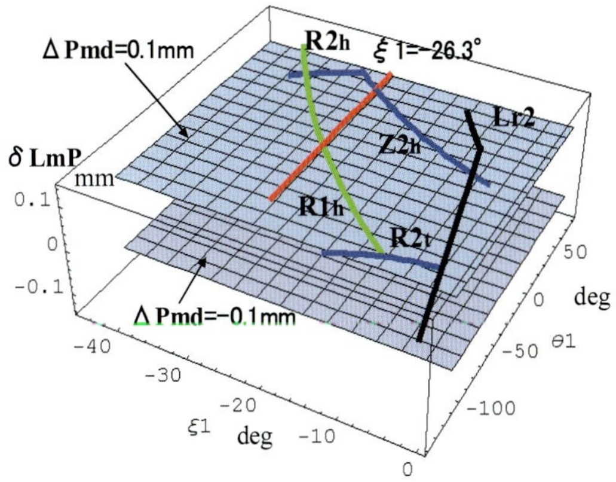

Fig.5.2-2(b) Normal gap δLmP caused by △Pmd in Drive side

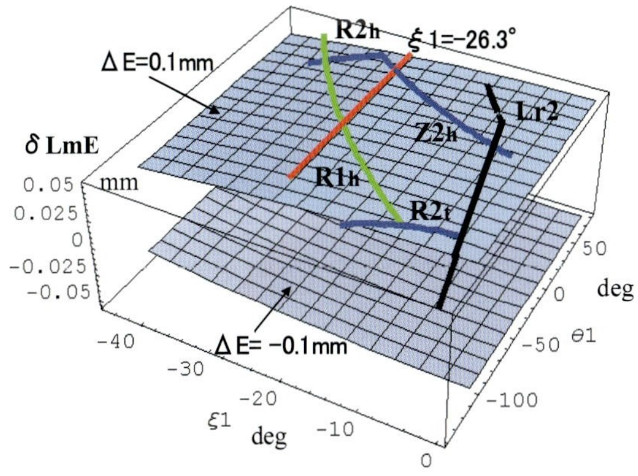

Fig.5.2-2(c) Normal gap δLmE caused by △E in Drive side

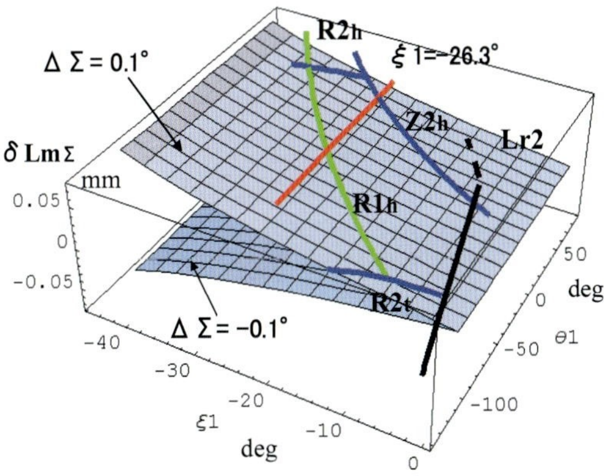

Fig.5.2-2(d) Normal gap δLmΣ caused by △Σ in Drive side

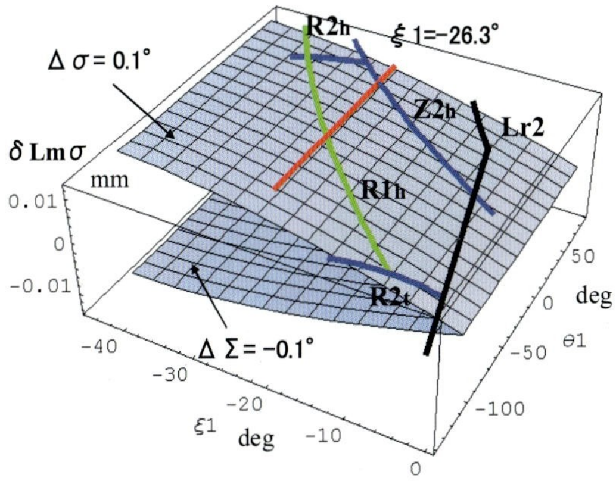

Fig.5.2-2(e) Normal gap δLmσ caused by △σ in Drive side

(3) In addition, △Pmd (Fig. 5.2-2(b)) has no variation of the tooth bearing because $\partial(\delta Lm)/\partial \xi_1 = 0$. This is because the relation between the involute helicoid of the pinion and the conjugate tooth surface of the gear has no change even if the phase angle of the involute helicoid is changed by △Pmd. There is another more direct explanation. Eq. (5.1-3) is expressed as follows in the coordinate system C_1.

$$\delta Lz1m = \Delta z_{1m} \sin\phi_{b1m} = \Delta Pmd \sin\phi_{b10}$$

Namely, δLz1m caused by △Pmd has no variation due to ξ_1.

(4) The conjugate tooth surface of the gear having the assembly errors except △Pmd makes contact at one (P_{m8}) of the two intersections of the line of contact at an angle of θ_1 and the boundaries of the tooth surface including the limit line of action, which has the algebraically larger δLm in the case of the tooth surface D. For examples, in the case of △Gmd (Fig. 5.2-2(a)),

(a) When △Gmd ≧ 0, P_{m8} is on the line of contact around the bottom in the large end,
(b) When △Gmd ≦ 0, P_{m8} is on the line of contact around the tip in the small end.

5.2.3 Amount of tooth surface modification of gear $\Delta p_{2nm}(\theta_1, \xi_1)$

Figure 5.2-3 shows the amount of tooth surface modification of the gear $\Delta p_{2nm}(\theta_1, \xi_1)$.

The modified tooth surface of the gear is designed as follows.

(1) In all the normal gaps $\delta Lm(\theta_1, \xi_1)$, $\partial(\delta Lm)/\partial\theta_1 = 0$ and they are the surfaces on which $\partial(\delta Lm)/\partial\xi_1$ varies only, therefore $\partial(\Delta p_{2nm})/\partial\theta_1 = 0$ is supposed, which means that $\Delta p_{2nm}(\theta_1, \xi_1)$ is a function of ξ_1 only and the variation of the engagement error along a path of contact (in the direction of θ_1) is always 0.

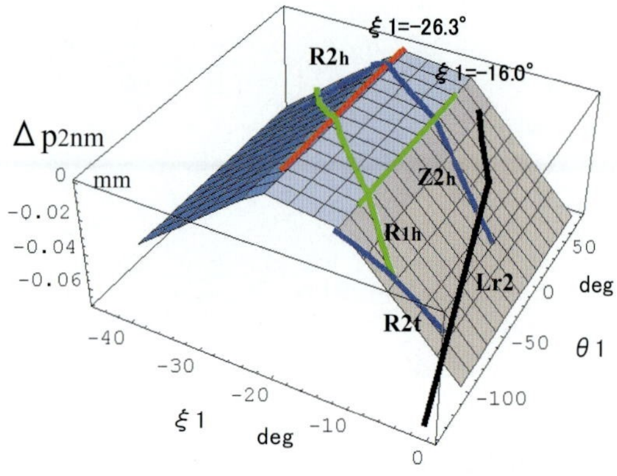

Fig.5.2-3 Amount of surface modification $\Delta p_{2nm}(\xi_1, \theta_1)$

(2) The path of contact $\xi_1 = -26.3°$ is the tooth bearing with no assembly errors and under no loads.

(3) When △Gmd = ±0.1mm, $\partial(\delta LmG)/\partial\xi_1$ of which is comparatively large, is given, the tooth surface is modified for the path of contact $\xi_1 = -26.3°$ to be the limit in the large end side and for the path of contact $\xi_1 = -16.0°$ to be the limit in the small end side.

According to the design policy mentioned above, $\Delta p_{2nm}(\theta_1, \xi_1)$ is given by a linear function of ξ_1 to simplify the problem as follows, where Δp_{2nm} is expressed in mm.

$$\begin{aligned}\triangle p_{2nm}(\theta_1, \xi_1) &= 0.003(\xi_1 + 26.3^\circ) && (\xi_1 < -26.3^\circ)\\ &= -0.0005(\xi_1 + 26.3^\circ) && (-26.3^\circ \leqq \xi_1 < -16.0^\circ)\\ &= -0.004(\xi_1 + 16.0^\circ) - 0.0052 && (-16.0^\circ \leqq \xi_1)\end{aligned}$$

5．2．4 Normal clearance $S_{knm}(\theta_1, \xi_1)$ and contact point $P_{m8}(\theta_1, \xi_{18})$

Based on the results of 5.2.2, the influences of $\triangle Gmd$ and $\triangle E$ are examined as follows: the former represents the case with a large $|\partial(\delta Lm)/\partial \xi_1|$ and the latter represents the case with a small one.

(1) When $\triangle Gmd = \pm 0.1$mm (Figs. 5.2-4(a) and (b))

Figure 5.2-4(a) shows the normal clearance $S_{knmG}(\theta_1, \xi_1)$ and the path of $P_{m8}(\theta_1, \xi_{18})$ (tooth bearing) in the case of $\triangle Gmd = 0.1$mm. The path of P_{m8} is a polygonal line which connects the tip of the gear (z_{2h}), the line $\xi_1 = -26.3^\circ$ and the tip of the pinion (R_{1h}). $S_{knmG}(\theta_1, \xi_1)$ is maximum and constant along the line $\xi_1 = -26.3^\circ$ (from P_{mZ2h} to P_{mr1h}) and the tooth bearing under no loads remains the same as that under $\triangle Gmd = 0$.

Figure 5.2-4(b) shows the case of $\triangle Gmd = -0.1$mm. $S_{knmG}(\theta_1, \xi_1)$ is maximum and constant along the line $\xi_1 = -16.0^\circ$ (from P_{mZ2h} to P_{mr1h}). The tooth bearing under no loads moves from the line $\xi_1 = -26.3^\circ$ to $\xi_1 = -16.0^\circ$ toward the small end (Fig. 5.2-1).

(2) When $\triangle E = \pm 0.1$mm (Figs. 5.2-4(c) and (d))

In the same way, Figs. 5.2-4(c) and (d) show those in the cases of $\triangle E = \pm 0.1$mm respectively. Because the $|\partial(\delta LmE)/\partial \xi_1|$s are smaller compared with those of $\triangle Gmd = \pm 0.1$ mm, the tooth bearing under no loads (from P_{mZ2h} to P_{mr1h}) remains the same as that under $\triangle E = 0$.

5．2．5 Curves of engagement error $\Delta\theta_{1k}(\theta_1, \xi_{18})$

Figure 5.2-5(a) shows the curves of engagement error $\triangle\theta_{1k}(\theta_1, \xi_{18})$ caused by $\triangle Gmd = \pm 0.1$mm, where the variations of $\triangle\theta_{1k}(\theta_1, \xi_{18})$ equal 0 in the regions from P_{mZ2h} to P_{mr1h} and occur in other regions by engaging along the tips of the pinion and gear. The difference of δLmG caused by $\triangle Gmd$ makes the tooth bearing move and gives the different curves of $\triangle\theta_{1k}(\theta_1, \xi_{18})$.

Figure 5.2-5(b) shows those in the case of $\triangle E = \pm 0.1$mm in the same way.

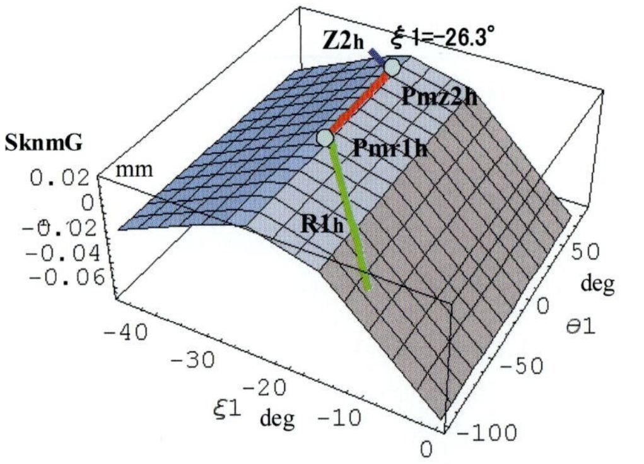

Fig.5.2-4(a) S_{knmG} and the path of contact in the case of △Gmd=0.1mm on Drive side

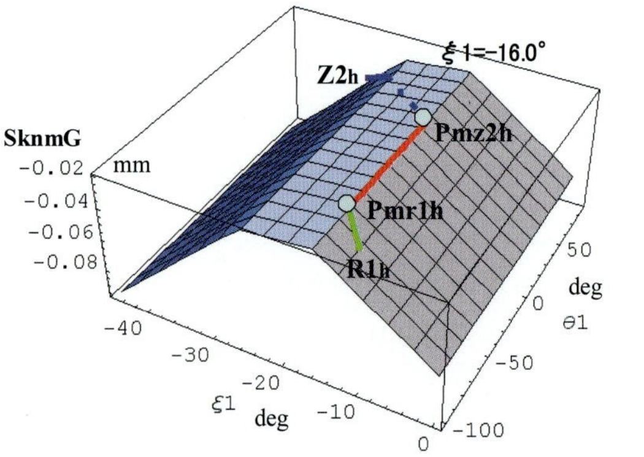

Fig.5.2-4(b) S_{knmG} and the path of contact in the case of △Gmd= −0.1mm on Drive side

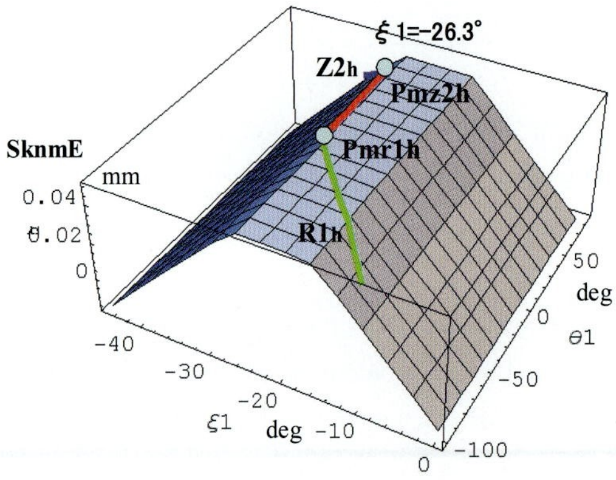

Fig.5.2-4(c) S_{knmE} and the path of contact in the case of △E=0.1mm on Drive side

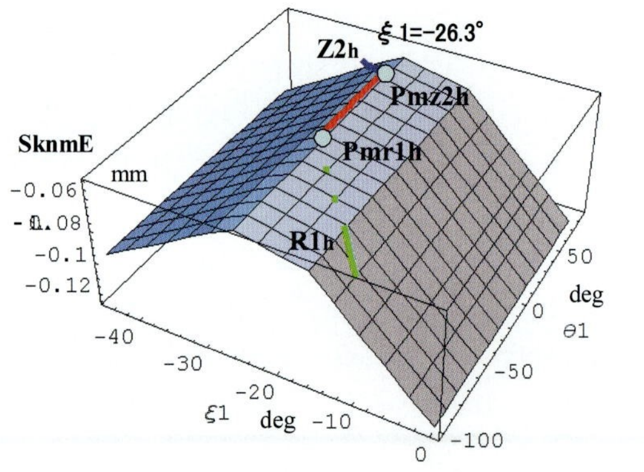

Fig.5.2-4(d) S_{knmE} and the path of contact in the case of △E= −0.1mm on Drive side

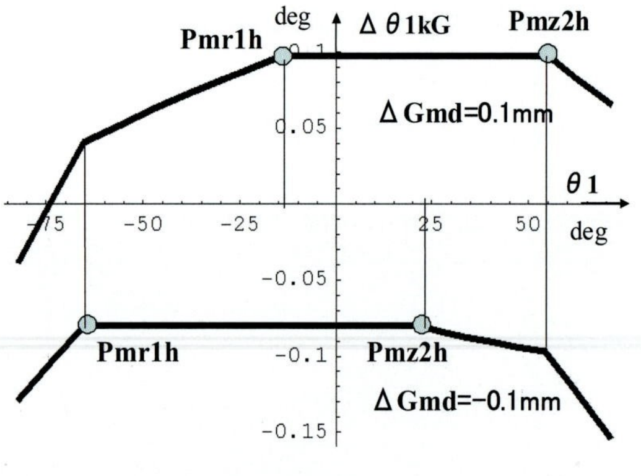

Fig.5.2-5(a) Engagement error $\triangle\theta_{1k}$ caused by △Gmd on Drive side

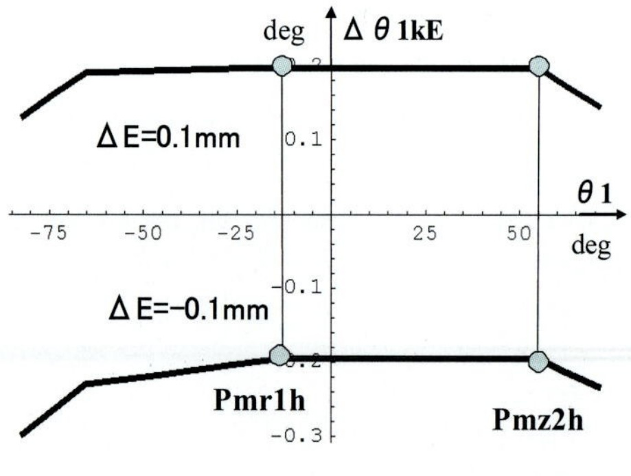

Fig.5.2-5(b) Engagement error $\triangle\theta_{1k}$ caused by △E on Drive side

5．3 Summary

In a gear pair which has involute helicoids for the pinion and their conjugate tooth surfaces for the gear, it is clarified how to calculate the amount of movement of the tooth bearing and the variation of engagement error caused by the five assembly errors (△Pmd, △Gmd, △E, △Σ and △σ). The results are summarized as follows.

(1) The assembly errors are replaced by the amount of movement ($\triangle u_2$, $\triangle v_2$, $\triangle z_2$) in the direction of each axis of the coordinate system O_2.

(2) The normal gap $\delta Lm(\theta_1, \xi_1)$ is obtained, which occurs between the points of contact on tooth surfaces of the pinion and gear caused by the movement ($\triangle u_2$, $\triangle v_2$, $\triangle z_2$).

(3) In the gear pair, $\delta Lm(\theta_1, \xi_1)$ depends on the inclination angle of the common contact normal $\mathbf{n}_m(\pi/2-\chi_{2m}, \phi_{b2m}; C_2)$, so that it is a function of ξ_1 only. The five assembly errors are represented by the amount of the normal gap δLm and the larger $|\delta Lm|$ makes for a larger movement of the tooth bearing.

(4) Therefore, the tooth surfaces which have the plane surface of action mentioned in 3.5 are expected to be insensitive to assembly errors, because the inclination angle of $\mathbf{n}_m$ has no variations.

(5) The amount and the shape of modification of the tooth surface must be designed so as to compensate for $\delta Lm(\theta_1, \xi_1)$.

(6) According to the idea mentioned above, the modified tooth surface of a gear is designed and the movement of the tooth bearing and the curve of the engagement error are calculated for examples.

These calculated results explain the observations of the tooth bearing in the trial products mentioned in 4.4.5.

6. Rotational motion of gears and dynamic increment of tooth load

In gear pairs which have involute helicoids (tooth surfaces II) and their conjugate tooth surfaces (tooth surfaces I) discussed in Section 4.3, when the tooth surfaces II are modified smoothly and engage with the tooth surfaces I under no loads and under load, the conditions of contact of tooth surfaces, the rotational motion (equations of motion) and the dynamic increment of tooth load are discussed in this chapter.

6.1 Path of contact with its common contact normal at each point of a gear pair having modified tooth surfaces II and tooth surfaces I under no loads (61), (63)

6.1.1 Point of contact P_m and its path of contact h_m between a tooth surface II (involute helicoid) and its conjugate tooth surface I

Figure 6.1-1 shows a line of contact PP_m at an angle of rotation θ_2 chosen at will and a path of contact h_m (P_0P_m) with its common contact normal at each point between the tooth surfaces II and I in the coordinate systems O_2 and O_{q2}. $P_0(\theta_2 = 0)$ is the design point, $\mathbf{n}_0$ ($= \mathbf{g}_0$) is the common contact normal at P_0 and G_{20} is the plane of action including $\mathbf{n}_0$. P_m is a point chosen at will on the line of contact PP_m, $\mathbf{n}_m$ is the common contact normal at P_m and G_{2m} is the plane of action including $\mathbf{n}_m$, which inclines to the plane G_{20} by ξ_{2m}. When the plane of action G_{2m} rotates with gear II by ξ_{2m} and is superimposed on the plane of action G_{20}, P_m moves to Q_m, and $\mathbf{n}_m$ to $\mathbf{n}$. w_m, w and w_0 are the intersection lines of the plane G_{20} and the tangent planes W_m at Q_m, W at P and W_0 at P_0 respectively. In Fig. 6.1-1, the signs (' and ") indicate the orthographic projections of points and vectors. In other figures hereafter, they represent the same.

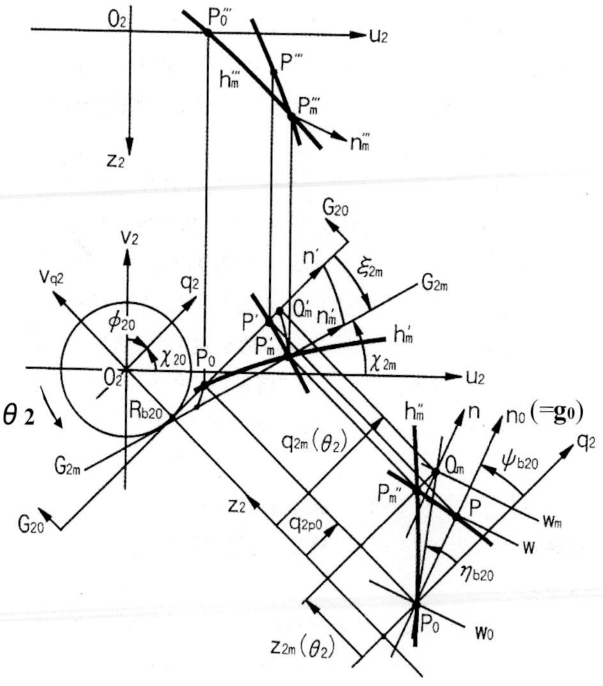

Fig.6.1-1 A line of contact PP_m and a path of contact h_m

Because the surface of action of tooth surfaces II and I is given by Eq. (4.3-6), when the path of contact through the design point $P_0(\theta_2 = 0)$ is supposed to be the curve which satisfies the equation of $dz_{2m}/dq_{2m} = \tan\eta_{b20}$, which means that the inclination angle on each plane of action is always η_{b20} (constant), the following equation is obtained at the intersection P_m of the path of contact and the line of contact at θ_2.

6.1 Path of contact with its common contact normal at each point of a gear pair having modified tooth surfaces II and tooth surfaces I under no loads

$$z_{2m}(\theta_2, \xi_2) = \{q_{2m}(\theta_2, \xi_2) - q_{2p0}\}\tan\eta_{b20} \tag{6.1-1}$$

Substituting the solution $\xi_{2m}(\theta_2)$ of Eq. (6.1-1) for Eq. (4.3-6), the path of contact with its common contact normal at each point satisfying Eq. (6.1-1) is expressed by the combination of the parameters $(\theta_2, \xi_{2m}(\theta_2))$ on the surface of action given by Eq. (4.3-6).

Substituting the solution $\xi_{2m}(\theta_2)$ of Eq. (6.1-1) for Eq. (2.3-2) and considering that the tooth surface II is an involute helicoid, a point of contact $P_m(q_{2m}, -R_{b20}, z_{2m}; O_{q2})$ between the tooth surfaces II and I and the inclination angle $n_m(\phi_{20}+\xi_{2m}, \phi_{b20}; O_2)$ of its common contact normal are obtained as follows in the coordinate systems O_2 and O_{q2}.

$$\left.\begin{aligned}
q_{2m}(\theta_2) &= R_{b20}(\theta_2+\xi_{2m}(\theta_2))/(\tan\phi_{b20}\tan\eta_{b20}+1)+q_{2p0}\\
z_{2m}(\theta_2) &= \{q_{2m}(\theta_2)-q_{2p0}\}\tan\eta_{b20}\\
R_{b2m}(\theta_2) &= R_{b20}\\
\eta_{b2m}(\theta_2) &= \eta_{b20}\\
\phi_{b2m}(\theta_2) &= \phi_{b20}\\
\chi_{2m}(\theta_2) &= \chi_{20}-\xi_{2m}(\theta_2) = \pi/2-\phi_{20}-\xi_{2m}(\theta_2)
\end{aligned}\right\} \tag{6.1-2}$$

When $\eta_{b20}=\phi_{b20}$, Eq. (6.1-2) represents the path of contact $\mathbf{g}_0$ because of $\xi_{2m}=0$.

The path of contact $h_m(u_{2m}, v_{2m}, z_{2m}; O_2)$ with the inclination angle of its common contact normal $\mathbf{n}_m$ at each point is expressed in the coordinate systems O_2 as follows.

$$\left.\begin{aligned}
u_{2m} &= q_{2m}(\theta_2)\cos\chi_{2m}+R_{b20}\sin\chi_{2m}\\
v_{2m} &= q_{2m}(\theta_2)\sin\chi_{2m}-R_{b20}\cos\chi_{2m}\\
z_{2m} &= \{q_{2m}(\theta_2)-q_{2p0}\}\tan\eta_{b20}\\
&n_m(\phi_{20}+\xi_{2m}, \phi_{b20}; O_2)
\end{aligned}\right\} \tag{6.1-3}$$

The tooth profile II and the inclination angle of its normal $\mathbf{n}_{rm}$ at each point are obtained as follows by transforming Eq. (6.1-3) into the coordinate system O_{r2} which rotates with gear II and coincides with the coordinate system O_2 when $\theta_2=0$.

$$\left.\begin{aligned}
\chi_{r2m} &= \pi/2-\phi_{20}-\xi_{2m}-\theta_2\\
u_{r2m} &= q_{2m}(\theta_2)\cos\chi_{r2m}+R_{b20}\sin\chi_{r2m}\\
v_{r2m} &= q_{2m}(\theta_2)\sin\chi_{r2m}-R_{b20}\cos\chi_{r2m}\\
z_{r2m} &= z_{2m}(\theta_2)\\
&n_{rm}(\phi_{20}+\xi_{2m}+\theta_2, \phi_{b20}; O_{r2})
\end{aligned}\right\} \tag{6.1-4}$$

Equation (6.1-4) is drawn as the curve B_2 in Fig. 6.1-4.

6.1.2 Point of contact P_{me} and its path of contact h_{me} between the modified tooth surface II and the tooth surface I

Figure 6.1-2 shows a point of contact P_{me} at an angle of rotation θ_2 and its path of contact h_{me} between the modified tooth surface II and the tooth surface I corresponding to the point P_m and the path of contact h_m. Because of the tooth surface modification, the point of contact P_m moves to P_{me}, the common contact normal $\mathbf{n}_m$ to $\mathbf{n}_{me}$ and the plane of action G_{2m} to G_{2me}, while gear I rotates by $\theta_{1e}=\theta_1+\triangle\theta_1$.

The path of contact h_{me} whose variables are indicated by subscript e is assumed to be given by Eq. (2.3-2) as follows.

(1) Because the modified tooth surface II is assumed to have the design point P_0 and its common contact normal $\mathbf{n}_0$ in common with the tooth surface II, four initial values, q_{2p0}, R_{b20}, χ_{20} and ψ_{b20} are determined.
(2) The modified tooth surface II is assumed to be modified smoothly to make contact along the tooth profile II, therefore

$$\left.\begin{aligned} \eta_{b2me}(\theta_2) &= \eta_{b20} \\ \psi_{b2me}(\theta_2) &= \psi_{b20} \end{aligned}\right\} \quad (6.1\text{-}5)$$

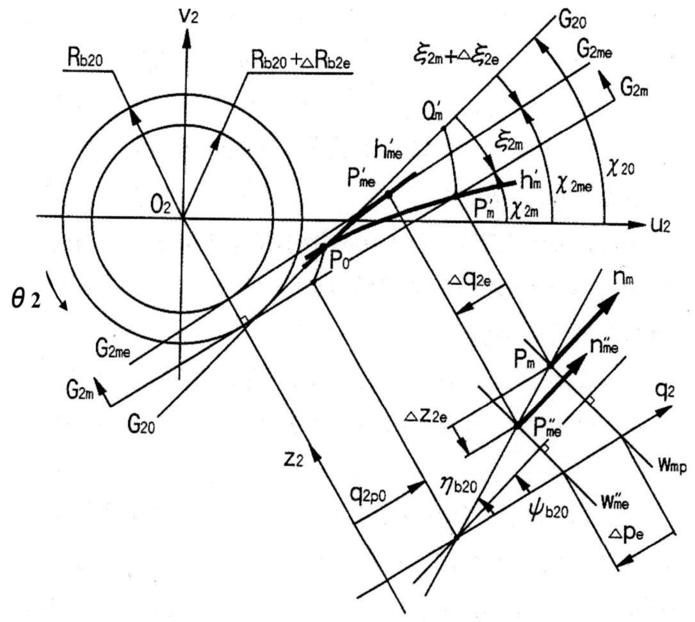

Fig.6.1-2 A point of contact P_{me} and its path of contact h_{me} in the case of $\triangle R_{b2e} < 0$

(3) The radius of base cylinder of gear II which is tangent to the plane of action G_{2me} including the normal $\mathbf{n}_{me}$ is assumed to be given by Eq. (2.3-2) as follows, where the variation $\triangle R_{b2e}$ of the radius of base cylinder of gear II is infinitesimal of the first order. (Note:6.1-1)

$$\left.\begin{aligned} R_{b2me}(\theta_2) &= R_{b20}+\triangle R_{b2e}(\theta_2) \qquad (\triangle R_{b2e} \ll R_{b20}) \\ \triangle R_{b2e}(\theta_2) &= \int_0^{\theta 2} (dR_{b2me}/d\theta_2)\,d\theta_2 \end{aligned}\right\} \quad (6.1\text{-}6)$$

(4) The inclination angle of the plane of action G_{2me} is expressed as follows.

$$\chi_{2me}(\theta_2) = \chi_{2m}(\theta_2) - \triangle\xi_{2me} = \chi_{20} - \xi_{2m}(\theta_2) - \triangle\xi_{2me}$$

where $\triangle\xi_{2me}$: inclination angle of the plane G_{2me} to the plane G_{2m}.

In the gear pairs in this theory, the difference between η_{b20} and ψ_{b20} is assumed to be sufficiently small, so that $\xi_{2m}(\theta_2)$ is also considered to be sufficiently small, because $\xi_{2m}(\theta_2) = 0$ when $\eta_{b20} = \psi_{b20}$. Moreover the amount of modification along the tooth profile II corresponding to $\triangle R_{b2e}$ is sufficiently small, so that $\triangle\xi_{2me}$ can be neglected because it is infinitesimal of the second order. Therefore, the inclination angle χ_{2me} of the plane of action G_{2me} is obtained as follows.

$$\chi_{2me}(\theta_2) = \chi_{2m}(\theta_2) = \pi/2 - \phi_{20} - \xi_{2m}(\theta_2) \qquad (6.1\text{-}7)$$

Namely, the common contact normal $\mathbf{n}_{me}$ is supposed to be parallel to $\mathbf{n}_m$ in the result.

6. 1 Path of contact with its common contact normal at each point of a gear pair having modified tooth surfaces II and tooth surfaces I under no loads

Substituting Eqs. (6. 1-5), (6. 1-6) and (6. 1-7) for Eq. (2. 3-2) and simplifying, the point of contact $P_{me}(q_{2me}, -R_{b2me}, z_{2me}\ ;\ 0_{q2})$ and the inclination angle of its common contact normal $n_{me}(\phi_{20}+\xi_{2m}, \phi_{b20}\ ;\ 0_2)$ between the modified tooth surface II and the tooth surface I at θ_2 are expressed as follows in the coordinate systems O_2 and O_{q2}, where $\triangle R_{b2e}\cdot(d\chi_{2me}/d\theta_2)$ is neglected because it is infinitesimal of the second order.

$$q_{2me}(\theta_2) = q_{2m}(\theta_2) + \int_0^{\theta 2} \triangle R_{b2e} d\theta_2/(\tan\phi_{b20}\tan\eta_{b20}+1)$$

$$R_{b2me}(\theta_2) = R_{b20}+\triangle R_{b2e}$$

$$z_{2me}(\theta_2) = z_{2m}(\theta_2) + \int_0^{\theta 2} \triangle R_{b2e} d\theta_2/(\tan\phi_{b20}+1/\tan\eta_{b20})$$

$$\chi_{2me}(\theta_2) = \chi_{2m}(\theta_2) = \chi_{20}-\xi_{2m}(\theta_2) = \pi/2-\phi_{20}-\xi_{2m}(\theta_2)$$

Therefore, the point P_{me} and its common contact normal $\mathbf{n}_{me}$ can be represented by the deviation from the corresponding point P_m in the coordinate system O_{q2} and by the inclination angle of the common contact normal $\mathbf{n}_m$ as follows.

$$\left.\begin{aligned}
q_{2me}(\theta_2) &= q_{2m}(\theta_2)+\triangle q_{2e}(\theta_2)\\
z_{2me}(\theta_2) &= z_{2m}(\theta_2)+\triangle z_{2e}(\theta_2)\\
R_{b2me}(\theta_2) &= R_{b20}+\triangle R_{b2e}(\theta_2)\\
\triangle q_{2e}(\theta_2) &= \triangle p_e(\theta_2)/(\tan\phi_{b20}\tan\eta_{b20}+1)\\
\triangle z_{2e}(\theta_2) &= \triangle p_e(\theta_2)/(\tan\phi_{b20}+1/\tan\eta_{b20})\\
\triangle R_{b2e}(\theta_2) &= d\triangle p_e(\theta_2)/d\theta_2\\
\triangle p_e(\theta_2) &= \int_0^{\theta 2} \triangle R_{b2e}(\theta_2)d\theta_2\\
&(\triangle p_e(0)= 0,\ \text{because } P_0 \text{ is the point of contact at } \theta_2 = 0.)\\
n_{me}(\phi_{20}+\xi_{2m}(\theta_2), \phi_{b20}\ ;\ 0_2) &= n_m
\end{aligned}\right\} \quad (6.1\text{-}8)$$

The path of contact $h_{me}(u_{2me}, v_{2me}, z_{2me}\ ;\ 0_2)$ with the inclination angle of the common contact normal $\mathbf{n}_{me}$ at each point is expressed by the deviation from corresponding $h_m(u_{2m}, v_{2m}, z_{2m}\ ;\ 0_2)$ and the inclination angle of $\mathbf{n}_m$ in the coordinate system O_2 as follows.

$$\left.\begin{aligned}
u_{2me} &= u_{2m}+\triangle q_{2e}\cos\chi_{2me}+\triangle R_{b2e}\sin\chi_{2me}\\
v_{2me} &= v_{2m}+\triangle q_{2e}\sin\chi_{2me}-\triangle R_{b2e}\cos\chi_{2me}\\
z_{2me} &= z_{2m}+\triangle z_{2e}\\
n_{me}(\phi_{20}+\xi_{2m}, \phi_{b20}; 0_2) &= n_m
\end{aligned}\right\} \quad (6.1\text{-}9)$$

Transforming h_{me} and $\mathbf{n}_{me}$ into the coordinate system O_{r2}, the modified tooth profile II with its normal at each point is obtained, which means the path of contact on the corresponding modified tooth surface II and which is also used to represent the modified tooth surface II from now on in this chapter.

The deviation of the point of contact P_{me} from P_m is shown in Fig. 6. 1-2. When the tangent plane W_{mp} at P_m intersects the plane of action G_{2m} along the line w_{mp} and the tangent plane W_{me} at P_{me} intersects the plane of action G_{2me} along the line w_{me} shown by the orthographic projection w_{me}" on the plane G_{2m}, $\triangle p_e$ represents the deviation of

w_{me}" from w_{mp} in the direction of the axis q_2 and is called the deviation of tangent plane. The line w_{mp} in Fig. 6.1-2 coincides with the line w_m in Fig. 6.1-1 by rotating the plane G_{2m} together with gear II by ξ_{2m} and superimposing it on the plane G_{20}. P_{me} is the point of contact when the gear pair has only one pair of the modified tooth surface II and the tooth surface I and it is expressed in Eq. (6.1-8) by the deviation from the point P_m, namely $\triangle q_{2e}$, $\triangle R_{b2e}$ and $\triangle z_{2e}$ which are represented by two parameters, the deviation of tangent plane $\triangle p_e$ and the inclination angle of path of contact η_{b20}. In Fig. 6.1-2, the ordinary case where the tooth surface II is modified convexly ($\triangle p_e<0$, $\triangle R_{b2e}<0$) is drawn.

6.1.3 Determination of the path of contact h_{me} between the modified tooth surface II and the tooth surface I

When the modified tooth surface II is the modified involute helicoid defined in 6.1.2, the path of contact h_{me} can be determined by measuring the single flank error and the helix form deviation as follows.

Gear I rotates by $\theta_{1e}=\theta_1+\triangle\theta_1$ at an angle of rotation θ_2 of gear II and the common contact normal $\mathbf{n}_{me}$ at P_{me} is supposed to be parallel to $\mathbf{n}_m$ at P_m, so that the fundamental requirement for contact is as follows.

$$R_{b1me}(d\theta_{1e}/dt)\cos\phi_{b1me} = (R_{b20}+\triangle R_{b2e})(d\theta_2/dt)\cos\phi_{b20} \tag{6.1-10}$$

where R_{b1me} : radius of base cylinder of the tooth surface I

ϕ_{b1me} : inclination angle of $\mathbf{n}_{me}$ on the plane of action G_{1me} at P_{me} and

$\triangle R_{b2e}$: amount of variation of the radius of base cylinder of gear II.

P_{me} is also the point on the tooth surface I, so that the following equation is obtained by the fundamental requirement for contact.

$$R_{b1me}\cos\phi_{b1me} = R_{b20}\cos\phi_{b20}/i_0 \tag{6.1-11}$$

Substituting Eq. (6.1-11) for Eq. (6.1-10),

$$\left.\begin{aligned} &(R_{b20}/i_0)(d\theta_{1e}/dt) = (R_{b20}+\triangle R_{b2e})(d\theta_2/dt) \\ &i = (d\theta_{1e}/dt)/(d\theta_2/dt) = i_0(1+\triangle R_{b2e}/R_{b20}) \end{aligned}\right\} \tag{6.1-12}$$

Equation (6.1-12) is the fundamental requirement for contact between the modified tooth surface II and the tooth surface I and the ratio i of angular velocities varies by $i_0\triangle R_{b2e}/R_{b20}$.

When the angles of rotation of gears II and I are θ_2 and θ_{1e} at time t, integrating Eq. (6.1-12) from 0 to t, Eq. (6.1-13) is obtained, where $\theta_2 = \theta_{1e} = 0$, when t = 0.

$$\left.\begin{aligned} \int_0^{\theta 1e}(R_{b20}/i_0)d\theta_{1e} &= \int_0^{\theta 2}(R_{b20}+\triangle R_{b2e})d\theta_2 \\ R_{b20}(\triangle\theta_1/i_0) &= \int_0^{\theta 2}\triangle R_{b2e}d\theta_2 \\ R_{b20}\triangle\theta_{2k} &= \triangle p_e(\theta_2) \end{aligned}\right\} \tag{6.1-13}$$

Substituting the single flank error $\triangle\theta_{2k}$ measured between the modified tooth surface II and the tooth surface I for Eq. (6.1-13), $\triangle p_e$ and $\triangle R_{b2e}$ can be obtained.

Figure 6.1-3 shows examples of the deviation of tangent plane $\triangle p_e(\theta_2)$ between the modified tooth surface II and the tooth surface I and the variation $\triangle R_{b2e}(\theta_2)$ of the radius of base cylinder of gear II. When $\triangle R_{b2e}(\theta_2)$ is supposed to be given by a polynomial expression of θ_2, $\triangle p_e(\theta_2)$ is expressed by Eq. (6.1-8) as follows.

$$\left.\begin{aligned}\triangle R_{b2e}(\theta_2) &= a_0+a_1\theta_2+a_2\theta_2^2+\cdots\\ \triangle p_e(\theta_2) &= a_0\theta_2+a_1\theta_2^2/2+a_2\theta_2^3/3+\cdots\end{aligned}\right\}\quad(6.1\text{-}14)$$

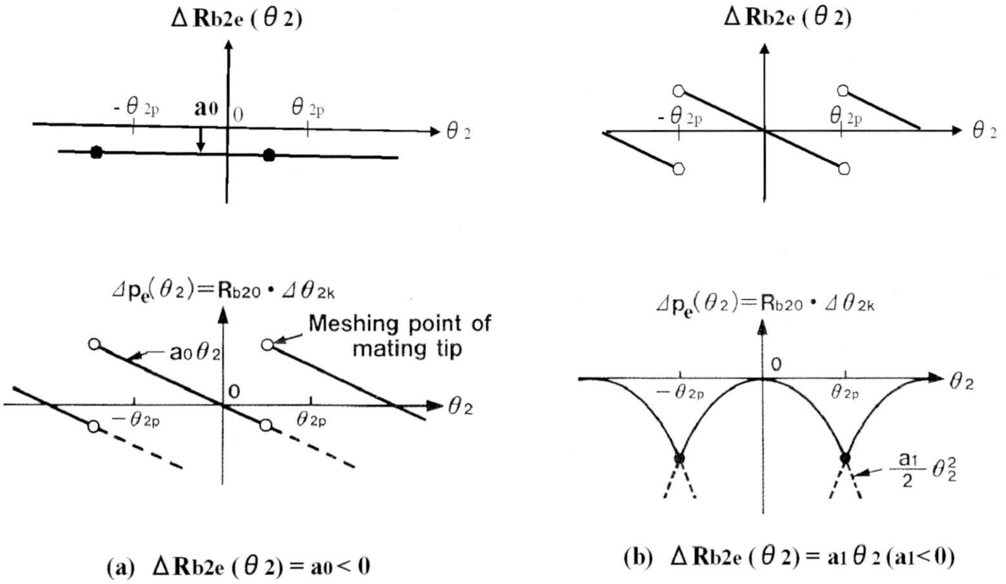

Fig.6.1-3 Examples of $\triangle p_e$ and $\triangle R_{b2e}$

Figure 6.1-3(a) shows tooth surface II with some pressure angle error.
Figure 6.1-3(b) shows tooth surface II modified symmetrically and convexly. When tooth surface II is modified concavely, it is expressed by $a_1 > 0$.
In the following discussion, the modified tooth surface II whose measured $\triangle p_e(\theta_2)$ is supposed to be approximated by a quadratic symmetrical expression of θ_2 is dealt with, which means that $\triangle R_{b2e}$ is a linear expression of θ_2.

The inclination angle η_{b20} of the path of contact h_{me} can be found by measuring the helix form deviation of the modified tooth surface II on several cylinders with different radii, because the modified tooth profile II corresponding to h_{me} means the row of the points on the modified tooth surface II whose normals have the same inclination angle ψ_{b20} on their planes of action.

Substituting those results of $\triangle p_e$ and η_{b20} mentioned above for Eqs. (6.1-8) and (6.1-9), the point of contact P_{me} and the path of contact h_{me} between the modified tooth surface II and the tooth surface I are determined in the coordinate systems O_2 and O_{q2}.

The present methods approximate a modified involute helicoid by giving the profile error, the helix form deviation, the pressure angle error and the bias and calculate numerically the points of contact between the modified involute helicoid and the tooth surface I, while the method in this theory can easily determine analytically the path of contact with its common contact normal at each point between them by giving the single flank error and the helix form deviation.

6. 1. 4 Path of contact of a gear pair with the modified tooth surfaces II and the tooth surfaces I

(1) Tooth profiles II and simultaneous points of contact

Figure 6.1-4 shows the path of contact h_m and four adjacent tooth profiles II, A_2, B_2, C_2 and D_2, whose intersections are simultaneous points of contact shown as P_{mA}, P_{mB}, P_{mC} and P_{mD} at an angle of rotation θ_2, where their common contact normals are shown as n_{mA}, n_{mB}, n_{mC} and n_{mD} in the coordinate system O_2. Planes of action G_{2mA}, G_{2mB}, G_{2mC} and G_{2mD} including the common contact normals n_{mA}, n_{mB}, n_{mC} and n_{mD} at the simultaneous points of contact P_{mA}, P_{mB}, P_{mC} and P_{mD} are inclined to the plane G_{20} by angles ξ_{2mA}, ξ_{2mB}, ξ_{2mC} and ξ_{2mD} respectively. When $2\theta_{2p}$ is the angular pitch of gear II and the phase angle of the point P_{mB} is assumed to be 0, the phase angles of points P_{mA}, P_{mB}, P_{mC} and P_{mD} can be represented as $+2\theta_{2p}$, 0, $-2\theta_{2p}$ and $-4\theta_{2p}$ on the h_m respectively, so that the simultaneous points of contact and their contact normals can be represented by Eq. (6.1-2) as follows.

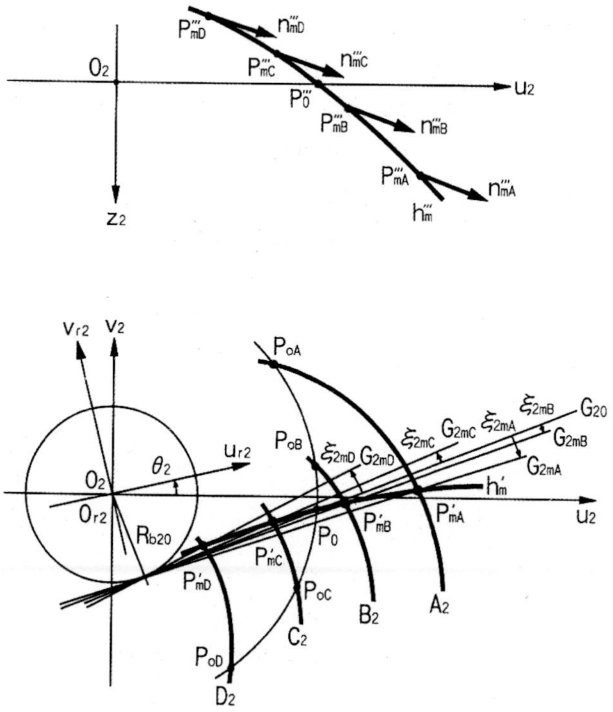

Fig.6.1-4 A path of contact and adjacent tooth profiles II

$$
\left.
\begin{array}{ll}
P_{mA}: P_{mA}\{q_{2m}(\theta_2+2\theta_{2p}), -R_{b20}, z_{2m}(\theta_2+2\theta_{2p});\ 0_{q2}\}, & n_{mA}(\phi_{20}+\xi_{2mA}, \phi_{b20}) \\
P_{mB}: P_{mB}\{q_{2m}(\theta_2), -R_{b20}, z_{2m}(\theta_2);\ 0_{q2}\}, & n_{mB}(\phi_{20}+\xi_{2mB}, \phi_{b20}) \\
P_{mC}: P_{mC}\{q_{2m}(\theta_2-2\theta_{2p}), -R_{b20}, z_{2m}(\theta_2-2\theta_{2p});\ 0_{q2}\}, & n_{mC}(\phi_{20}+\xi_{2mC}, \phi_{b20}) \\
P_{mD}: P_{mD}\{q_{2m}(\theta_2-4\theta_{2p}), -R_{b20}, z_{2m}(\theta_2-4\theta_{2p});\ 0_{q2}\}, & n_{mD}(\phi_{20}+\xi_{2mD}, \phi_{b20})
\end{array}
\right\} \quad (6.1\text{-}15)
$$

Substituting Eq. (6.1-15) for Eq. (6.1-3) and transforming them into the coordinate system O_{r2}, the tooth profiles II, A_2, B_2, C_2 and D_2, and their normals are obtained. In Fig. 6.1-4, points P_{0A}, P_{0B}, P_{0C} and P_{0D} on the tooth profiles II correspond to the design point P_0 respectively.

(2) Modified tooth profiles Ⅱ and fundamental requirements for multiple tooth pair contact

When the gear pair has the modified tooth profiles Ⅱ, A_{2e}, B_{2e}, C_{2e} and D_{2e}, corresponding to the tooth profiles Ⅱ, A_2, B_2, C_2 and D_2 adjacent by the angular pitch $2\theta_{2p}$, points P_{mAe}, P_{mBe}, P_{mCe} and P_{mDe} on the modified tooth profiles Ⅱ at θ_2 corresponding to the points P_{mA}, P_{mB}, P_{mC} and P_{mD} are given by Eq. (6.1-8), so that the deviations of tangent plane and the variations of the radius of base cylinder are expressed as follows.

$$\left.\begin{aligned}
P_{mAe}&: \triangle p_{Ae} = \triangle p_e(\theta_2+2\theta_{2p}), & \triangle R_{b2Ae} &= d\triangle p_{Ae}/d\theta_2\\
P_{mBe}&: \triangle p_{Be} = \triangle p_e(\theta_2), & \triangle R_{b2Be} &= d\triangle p_{Be}/d\theta_2\\
P_{mCe}&: \triangle p_{Ce} = \triangle p_e(\theta_2-2\theta_{2p}), & \triangle R_{b2Ce} &= d\triangle p_{Ce}/d\theta_2\\
P_{mDe}&: \triangle p_{De} = \triangle p_e(\theta_2-4\theta_{2p}), & \triangle R_{b2De} &= d\triangle p_{De}/d\theta_2
\end{aligned}\right\} \qquad (6.1\text{-}16)$$

where,

$\triangle p_{Ae}$, $\triangle p_{Be}$, $\triangle p_{Ce}$ and $\triangle p_{De}$: deviations of tangent plane at points of contact P_{mAe}, P_{mBe}, P_{mCe} and P_{mDe} and

$\triangle R_{b2Ae}$, $\triangle R_{b2Be}$, $\triangle R_{b2Ce}$ and $\triangle R_{b2De}$: variations of the radius of base cylinder at points of contact P_{mAe}, P_{mBe}, P_{mCe} and P_{mDe}.

Figure 6.1-5 shows the deviations of tangent plane $\triangle p_{Ae}$, $\triangle p_{Be}$, $\triangle p_{Ce}$ and $\triangle p_{De}$ and the radii of base cylinder R_{b2mAe}, R_{b2mBe}, R_{b2mCe} and R_{b2mDe} of the modified tooth profiles Ⅱ based on Figs. 6.1-3(b) and 6.1-4 and Eq. (6.1-16) at an angle of rotation θ_2, where P_{mAe}, P_{mBe}, P_{mCe} and P_{mDe} mean the points at which the modified tooth profiles Ⅱ, A_{2e}, B_{2e}, C_{2e} and D_{2e}, are feasible to be in contact simultaneously at θ_2.

For the points P_{mAe}, P_{mBe}, P_{mCe} and P_{mDe} to be in contact simultaneously, their deviations of tangent plane from each point, P_{mA}, P_{mB}, P_{mC} and P_{mD} respectively must be equal, therefore, Eq. (6.1-17) is obtained.

$$\triangle p_{Ae} = \triangle p_{Be} = \triangle p_{Ce} = \triangle p_{De} \qquad (6.1\text{-}17)$$

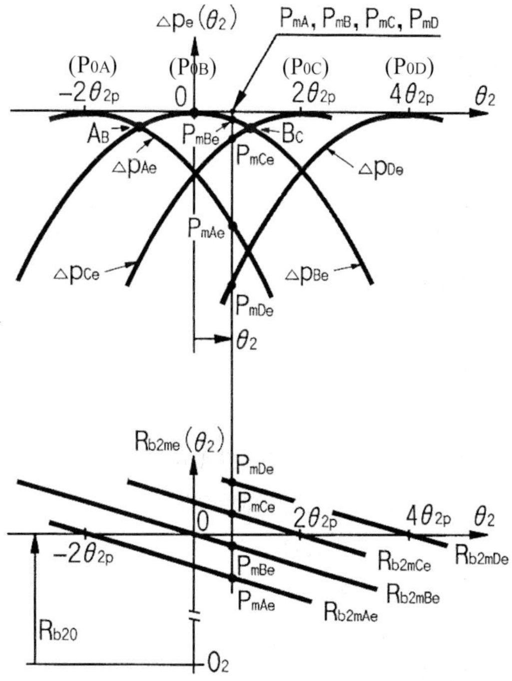

Fig.6.1-5 Examples of $\triangle p_e$ and R_{b2me} of modified tooth profiles Ⅱ

In addition, for Eq. (6.1-12) to be valid at the points P_{mAe}, P_{mBe}, P_{mCe} and P_{mDe} at the same time, the radii of base cylinder must be equal, because $(d\theta_{1e}/dt)$ and $(d\theta_2/dt)$ are the same at the simultaneous points of contact. Therefore, Eq. (6.1-18) is obtained.

$$\triangle R_{b2Ae} = \triangle R_{b2Be} = \triangle R_{b2Ce} = \triangle R_{b2De} \qquad (6.1\text{-}18)$$

Equations (6.1-17) and (6.1-18) represent the fundamental requirements for multiple tooth pair contact.

When P_{mBe} and P_{mCe} are the simultaneous points of contact at θ_2, the following equations are obtained from Eqs. (6.1-16), (6.1-17) and (6.1-18).

$$\triangle p_{Be} = \triangle p_e(\theta_2) = \triangle p_e(\theta_2 - 2\theta_{2p}) = \triangle p_{Ce}$$
$$\triangle R_{b2Be} = d\triangle p_{Be}/d\theta_2 = d\triangle p_e(\theta_2)/d\theta_2 = d\triangle p_e(\theta_2 - 2\theta_{2p})/d\theta_2 = d\triangle p_{Ce}/d\theta_2 = \triangle R_{b2Ce}$$

In order that the modified tooth profiles II are in contact simultaneously, the deviation of tangent plane $\triangle p_e(\theta_2)$ must be a periodic function of $2\theta_{2p}$ including the first-order derivative. Since it is difficult to give $\triangle p_e(\theta_2)$ a periodic function, the modified tooth profiles II can hardly be in contact simultaneously under no loads.

When the modified tooth profile II has $\triangle p_e(\theta_2)$ which is not a periodic function but a symmetrically convex one as shown in Fig. 6.1-5, there is no multiple tooth pair contact.

(3) Path of contact of the gear pair with the modified tooth surfaces II and the tooth surfaces I

In the example shown in Fig. 6.1-5, because the tooth pair which has the most proceeding $\triangle p_e$ at θ_2 in the direction of rotation results in contact, the point of contact of the gear pair at θ_2 in Fig. 6.1-5 is determined as follows according to the regions of θ_2, where the point A_B is located at $-\theta_{2p}$ and the point B_C at θ_{2p}, because $\triangle p_e$ is symmetrically convex.

$$-3\theta_{2p} < \theta_2 \leqq -\theta_{2p} : P_{mAe} : \triangle p_e(\theta_2 + 2\theta_{2p})$$
$$-\theta_{2p} < \theta_2 \leqq \theta_{2p} : P_{mBe} : \triangle p_e(\theta_2)$$
$$\theta_{2p} < \theta_2 \leqq 3\theta_{2p} : P_{mCe} : \triangle p_e(\theta_2 - 2\theta_{2p})$$
. . . .

The location of the point of contact on the path of contact h_{me} is expressed by Eq. (6.1-9) as follows.

$$-3\theta_{2p} < \theta_2 \leqq -\theta_{2p} : h_{me}(\theta_2 + 2\theta_{2p})$$
$$-\theta_{2p} < \theta_2 \leqq \theta_{2p} : h_{me}(\theta_2)$$
$$\theta_{2p} < \theta_2 \leqq 3\theta_{2p} : h_{me}(\theta_2 - 2\theta_{2p})$$
. . . .

The equations above indicate that the path of contact in actual contact of the gear pair with the modified tooth surfaces II and the tooth surfaces I is always represented by the region from $h_{me}(-\theta_{2p})$ to $h_{me}(\theta_{2p})$, which corresponds to the curve from the point A_B to B_C shown in Fig. 6.1-5 and includes just one pitch of the gear pair. The deviations of tangent plane of the modified tooth profiles II, A_{2e} and B_{2e}, at the point A_B are equal but their radii of base cylinder are different, so that they are not in contact simultaneously at A_B. It is also the same at B_C.

6.1.5 Summary

In gear pairs which have involute helicoids for the tooth surfaces II and their conjugate rigid tooth surfaces I generated through a ratio i_0 of angular velocities, the modified tooth surfaces II are defined by giving a single flank error and a helix form deviation and the conditions of contact between the modified tooth surfaces II and the tooth surfaces I are analyzed under no loads, which is summarized as follows.

(1) Equations of a path of contact with its common contact normal at each point and the corresponding tooth profile II of the gear pair having involute helicoids (tooth surfaces II) and their conjugate tooth surfaces I are obtained under no loads.

(2) When the modified tooth surface II makes contact in the vicinity of the tooth profile II and its tangent plane at θ_2 is supposed to be parallel to that of the tooth surface II, the path of contact between the modified tooth surface II and the tooth surface I is expressed by giving the single flank error (deviation of tangent plane) and the helix form deviation (inclination angle of path of contact) in the coordinate system O_2.

(3) The fundamental requirements for multiple tooth pair contact in the gear pair having the modified tooth surfaces II and the conjugate tooth surfaces I are clarified. There is no multiple tooth pair contact in the gear pair having the modified tooth surfaces II which are modified symmetrically and convexly under no loads.

(Note 6.1-1) : the sign of $\triangle R_{b2e}$ is defined inversely to references (61) and (63), so that it is changed in the same way in the corresponding Eqs. (6.1-8 to 10) and (6.1-12 to 14).

6．2 Paths of contact and their common contact normal at each point of a gear pair under load [61]

In this section, the paths of contact and their common contact normal at each point of a gear pair having the modified involute helicoids for tooth surfaces II and the tooth surfaces I are obtained under load, where the tooth surfaces I are the conjugate rigid surfaces of the involute helicoids mentioned in 6.1.

6．2．1 Point of contact P_{mt} and its common contact normal n_{mt} between the modified tooth surface II and the tooth surface I under load

Figure 6.2-1 shows a point of contact P_{mt} and its common contact normal n_{mt} between the modified tooth surface II assumed to be elastic and the tooth surface I assumed to be rigid under torque T_2 in the coordinate system O_{q2}, together with the corresponding points P_m and P_{me} and their common contact normals n_m and n_{me} under no loads respectively. The load on the modified tooth surface II moves the point P_{me} to P_{mt} and the common contact normal n_{me} (parallel to n_m) to n_{mt} where the inclination angle $\triangle\xi_{2t}$ of n_{mt} to n_m is assumed to be negligibly small, which means that n_{mt} is parallel to n_m and the tangent plane W_{mt} at P_{mt} is parallel to the plane W_{mp} at P_m. The variation of the angle η_{b20} is also assumed to be negligibly small. When the tangent plane W_{mp} at P_m intersects the plane of action G_{2m} along the line w_{mp} and the tangent plane W_{mt} at P_{mt} intersects the plane of action G_{2mt} along the line w_{mt} shown by the orthographic projection w_{mt}" on the plane G_{2m}, $\triangle p_t$ represents the deviation of w_{mt}" from w_{mp} in the direction of the axis q_2 and is called the deviation of tangent plane under load. The point of contact under load $P_{mt}(q_{2mt}, -R_{b2mt}, z_{2mt}\ ;\ O_{q2})$ and the inclination angle of its common contact normal $n_{mt}(\pi/2-\chi_{2mt}, \phi_{b20}\ ;\ O_2)$ are obtained as follows by substituting subscript t for e in Eq. (6.1-8) in the same way as P_{me} and n_{me}.

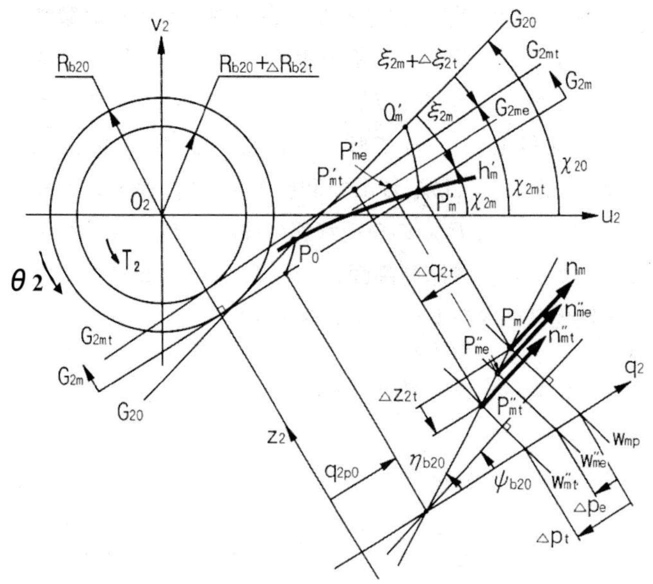

Fig.6.2-1 Relationship among points of contact P_m, P_{me} and P_{mt}

$$\left.\begin{aligned}
q_{2mt}(\theta_2) &= q_{2m}(\theta_2)+\triangle q_{2t}(\theta_2)\\
z_{2mt}(\theta_2) &= z_{2m}(\theta_2)+\triangle z_{2t}(\theta_2)\\
R_{b2mt}(\theta_2) &= R_{b20}+\triangle R_{b2t}(\theta_2)\\
\triangle q_{2t}(\theta_2) &= \triangle p_t(\theta_2)/(\tan\psi_{b20}\tan\eta_{b20}+1)\\
\triangle z_{2t}(\theta_2) &= \triangle p_t(\theta_2)/(\tan\psi_{b20}+1/\tan\eta_{b20})\\
\triangle R_{b2t}(\theta_2) &= d\triangle p_t(\theta_2)/d\theta_2\\
\triangle p_t(\theta_2) &= \int_0^{\theta 2} \triangle R_{b2t}(\theta_2)d\theta_2+\triangle p_t(0)\\
n_{mt}(\pi/2-\chi_{2mt} &= \phi_{20}+\xi_{2m},\ \psi_{b20};\ 0_2) = n_m
\end{aligned}\right\} \quad (6.2\text{-}1)$$

where $\triangle p_t(0)$: deviation of tangent plane at $\theta_2 = 0$ under load.

Figure 6.2-2 shows the relationship between the deviations of tangent plane $\triangle p_e$ and $\triangle p_t$ at the points P_m, P_{me} and P_{mt} respectively. In this theory, the modified tooth surface II is assumed to be given by $\triangle p_e(\theta_2) = a_1\theta_2{}^2/2$ ($a_1 < 0$), which means that it is symmetrical and convex, so that the deflection $\triangle p_t - \triangle p_e$ is always negative when $T_2 \geqq 0$. In the case of $T_2 < 0$, when the modified tooth surface is symmetrical and convex, it is dealt with in the same way because $\triangle p_t$ and $\triangle p_e$ are symmetrical with regard to the axis θ_2.

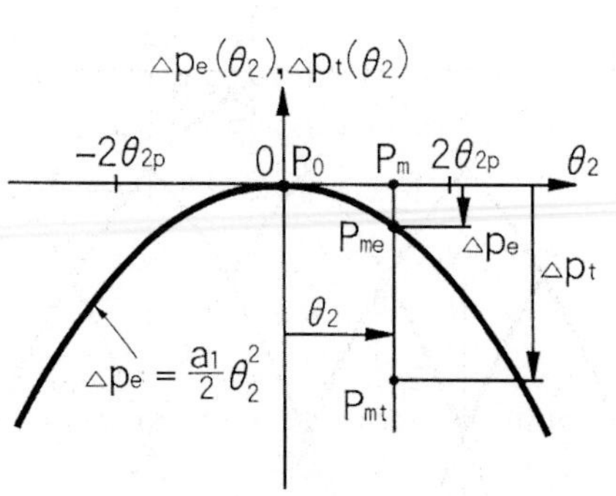

Fig.6.2-2 Relationship between deviations of tangent plane $\triangle p_e$ and $\triangle p_t$

The point $P_{mt}(u_{2mt}, v_{2mt}, z_{2mt}\ ;\ 0_2)$ is expressed in the coordinate system O_2 as follows.

$$\left.\begin{aligned} \chi_{2mt} &= \chi_{2m} = \chi_{20} - \xi_{2m} = \pi/2 - \phi_{20} - \xi_{2m} \\ u_{2mt} &= u_{2m} + \triangle q_{2t}\cos\chi_{2mt} + \triangle R_{b2t}\sin\chi_{2mt} \\ v_{2mt} &= v_{2m} + \triangle q_{2t}\sin\chi_{2mt} - \triangle R_{b2t}\cos\chi_{2mt} \\ z_{2mt} &= z_{2m} + \triangle z_{2t} \end{aligned}\right\} \qquad (6.2\text{-}2)$$

6. 2. 2 Deviations of tangent plane between the modified tooth surfaces II and the tooth surfaces I under load

(1) Regions having the same number of tooth pairs in mesh under load

Figure 6.2-3 shows the deviations of tangent plane $\triangle p_{Ae}$, $\triangle p_{Be}$, $\triangle p_{Ce}$, • •, and $\triangle p_{Ke}$ of the modified tooth surfaces II which are called by the corresponding modified tooth profiles, A_{2e}, B_{2e}, C_{2e}, • • and K_{2e} adjacent to each other by the angular pitch $2\theta_{2p}$, based on Figs. 6.1-5 and 6.2-2.

The deviations of tangent plane of the modified tooth surfaces II at an angle of rotation θ_2 are given by Eq. (6.1-16) as follows.

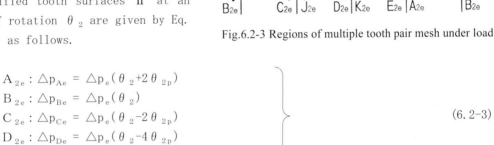

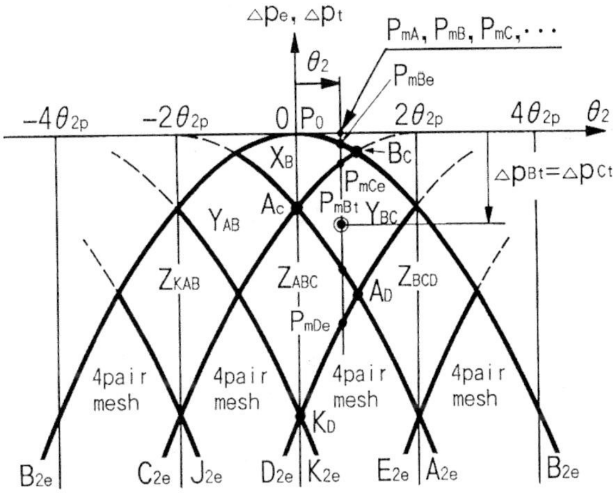

Fig.6.2-3 Regions of multiple tooth pair mesh under load

$$\left.\begin{aligned} A_{2e} &: \triangle p_{Ae} = \triangle p_e(\theta_2 + 2\theta_{2p}) \\ B_{2e} &: \triangle p_{Be} = \triangle p_e(\theta_2) \\ C_{2e} &: \triangle p_{Ce} = \triangle p_e(\theta_2 - 2\theta_{2p}) \\ D_{2e} &: \triangle p_{De} = \triangle p_e(\theta_2 - 4\theta_{2p}) \\ &\cdots\cdots\cdots \end{aligned}\right\} \qquad (6.2\text{-}3)$$

When the modified tooth surfaces II are under load, the point P_{mBe} on the B_{2e} makes contact at the point P_{mBt} and the deviation of tangent plane at P_{mBt} is $\triangle p_{Bt}$, P_{mBt} is in the region of $\triangle p_{Bt} \leqq \triangle p_{Be}$ in Fig. 6.2-3 and its location depends on θ_2 and the load, so that the number of tooth pairs in mesh varies as follows according to which region in Fig. 6.2-3 P_{mBt} belongs to.

X_B : region of one pair mesh of the modified tooth surface B_{2e}.

Y_{AB} or Y_{BC} : region of two pair mesh of the modified tooth surfaces A_{2e} and B_{2e} or B_{2e} and C_{2e}.

Z_{KAB}, Z_{ABC} or Z_{BCD} : region of three pair mesh of the modified tooth surfaces K_{2e}, A_{2e} and B_{2e} or A_{2e}, B_{2e} and C_{2e} or B_{2e}, C_{2e} and D_{2e}.

Regions of more than four pair mesh are determined in the same way.

(2) Fundamental requirements for multiple tooth pair contact under load

The fundamental requirements for multiple tooth pair contact of the modified tooth surfaces II under load are obtained by making reference to Eqs. (6.1-17) and (6.1-18) under no loads as follows.

When the load deflects the modified tooth surfaces II and makes them contact simultaneously at points P_{mBt}, P_{mCt}, P_{mAt}, P_{mDt}, $\cdots$ corresponding to the points P_{mBe}, P_{mCe}, P_{mAe}, P_{mDe}, $\cdots$ at an angle of rotation θ_2 in Fig. 6.2-3, all the deviations of tangent plane must be equal.

$$\triangle p_{Bt} = \triangle p_{Ct} = \triangle p_{At} = \triangle p_{Dt} = \cdots \qquad (6.2\text{-}4)$$

where $\triangle p_{Bt}$, $\triangle p_{Ct}$, $\triangle p_{At}$ and $\triangle p_{Dt}$: deviations of tangent plane under load.

In the same way, all the radii of base cylinder must be equal, because $(d\theta_{1e}/dt)$ and $(d\theta_2/dt)$ are the same at the simultaneous points of contact.

$$R_{b2mBt} = R_{b2mCt} = R_{b2mAt} = R_{b2mDt} = \cdots \qquad (6.2\text{-}5)$$

where R_{b2mBt}, R_{b2mCt}, R_{b2mAt} and R_{b2mDt} : radii of base cylinder at simultaneous
points of contact P_{mBt}, P_{mCt}, P_{mAt} and P_{mDt} under load.

Eqs. (6.2-4) and (6.2-5) are the fundamental requirements for multiple tooth pair contact of the modified tooth surfaces II under load. In Fig. 6.2-3, the point P_{mCe} moves and makes contact at P_{mCt} at the same time which is superimposed on the point of contact P_{mBt}.

(3) Equivalent tooth load F_{q2}

When the modified tooth surfaces II are in contact simultaneously at the points P_{mAt}, P_{mBt}, P_{mCt}, $\cdots$, total torque T_{2d} of gear II transmitted by the tooth loads shared at the points of contact is given as follows because the radii of base cylinder are the same by Eq. (6.2-5).

$$\begin{aligned}
T_{2d} &= F_{q2A}R_{b2mAt} + F_{q2B}R_{b2mBt} + F_{q2C}R_{b2mCt} + \cdots \\
&= (F_{q2A} + F_{q2B} + F_{q2C} + \cdots)R_{b2mBt} \\
&= F_{q2}R_{b2mBt} \\
F_{q2} &= F_{q2A} + F_{q2B} + F_{q2C} + \cdots
\end{aligned} \qquad (6.2\text{-}6)$$

where F_{q2A}, F_{q2B}, F_{q2C}, $\cdots$: tooth loads shared at the points of contact P_{mAt}, P_{mBt}, P_{mCt}, $\cdots$, in the direction of the axis q_2.

Equivalent tooth load F_{q2} defined by Eq. (6.2-6) is the scalar sum of the tooth loads F_{q2A}, F_{q2B} and F_{q2C}, $\cdots$, which can be expressed as follows.

$$F_{q2} = F_{q2S} + F_{q2d} \quad (|F_{q2d}| \ll |F_{q2S}|)$$
$$F_{q2S} = -T_2/R_{b20} \quad (T_2 \text{ is positive in the direction indicated in Fig. 6.2-1})$$

where

F_{q2S} : static equivalent tooth load transmitted by the true involute helicoid with the radius R_{b20} of base cylinder and

F_{q2d} : increment of equivalent tooth load F_{q2} caused by both the variation of base cylinder due to the surface modification and deflection and the variation of rotation.

When the static deviation of tangent plane (static engagement error) is only required, $F_{q2} = F_{q2S}$ is satisfactory.

Figure 6.2-4 shows the tooth loads F_{q2B} and F_{q2C} in the region of two pair mesh Y_{BC} where the points P_{mBt} and P_{mCt} are in contact simultaneously. The tooth loads F_{q2B} and F_{q2C} are on the planes of action G_{2mBt} and G_{2mCt} which are parallel to the planes G_{2mB} and G_{2mC} respectively and are tangent to the same base cylinder whose radius is $R_{b2mBt} = R_{b2mCt}$, so that they incline to the axis u_2 by the inclination angles χ_{2mB} and χ_{2mC} of the planes G_{2mBt} and G_{2mCt} respectively.

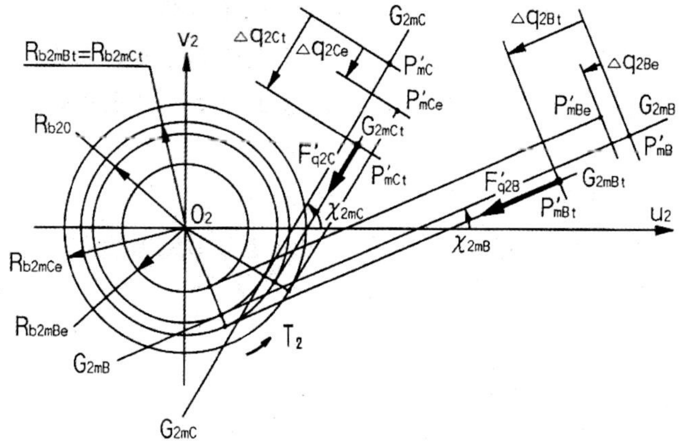

Fig.6.2-4 Relationship between tooth loads at simultaneous points of contact P_{mBt} and P_{mCt}

(4) Deviations of tangent plane $\Delta p_{Bt}(\theta_2)$ at P_{mBt} under load

Deviations of tangent plane Δp_{Bt} at P_{mBt} under load are expressed as follows according to where the point P_{mBt} exists in the regions of mesh shown in Fig. 6. 2-3.

$$\left.\begin{aligned}\Delta p_{Bt} &= \Delta p_{XB} && \text{(when } P_{mBt} \text{ is in the region of } X_B\text{)}\\ &= \Delta p_{YBC} \text{ or } \Delta p_{YAB} && \text{(when } P_{mBt} \text{ is in the region of } Y_{BC} \text{ or } Y_{AB}\text{)}\\ &= \Delta p_{ZABC},\ \Delta p_{ZKAB} \text{ or } \Delta p_{ZBCD} && \text{(when } P_{mBt} \text{ is in the region of } Z_{ABC},\ Z_{KAB} \text{ or } Z_{BCD}\text{)}\\ &\cdots\cdots\cdots\end{aligned}\right\} \quad (6.2\text{-}7)$$

Each $\Delta p_{Bt}(\theta_2)$ is obtained as follows.

(a) $\Delta p_{Bt}(\theta_2)$ in the region of one pair mesh X_B

When P_{mBt} is in the region of one pair mesh X_B, $\Delta p_{Bt}(\theta_2) = \Delta p_{XB}$ is obtained as follows.

$$F_{q2} = F_{q2B} = K(\theta_2)\{\Delta p_{XB} - \Delta p_e(\theta_2)\}$$

Substituting $\Delta p_X(\theta_2)$ for Δp_{XB},

$$\begin{aligned}\Delta p_{Bt}(\theta_2) &= \Delta p_{XB} = \Delta p_X(\theta_2)\\ &= \{K(\theta_2)\Delta p_e(\theta_2) + F_{q2}\}/K(\theta_2)\end{aligned} \quad (6.2\text{-}8)$$

where $K(\theta_2)$: tooth-pair spring constant in the direction of F_{q2}.

Although $K(\theta_2)$ differs according to the inclination angle η_{b20} of the path of contact in the same gear pair, $K(\theta_2)$ is assumed to be that of the gear pair which has the path of contact $\mathbf{g}_0$ because η_{b20} is taken to be close to ϕ_{b20} in this theory and to be approximated by an even function of θ_2. The surface modification is considered to have little influence on $K(\theta_2)$.

Δp_{Bt} corresponding to F_{q2B} inclines to the axis u_2 by the inclination angle χ_{2mB} of the plane G_{2mBt} at P_{mBt} because it is defined in the direction of the axis q_2 on the plane G_{2mBt} parallel to the plane G_{2mB}.

(b) $\Delta p_{Bt}(\theta_2)$ in the region of two pair mesh Y_{BC} or Y_{AB} (Fig. 6.2-4)

When the point P_{mBt} is in the region of two pair mesh Y_{BC}, deviations of tangent plane Δp_{Bt} and Δp_{Ct} at P_{mBt} and P_{mCt} respectively are equal because of Eq. (6.2-4), so that F_{q2B} and F_{q2C} are expressed as follows.

$$F_{q2B} = K(\theta_2)\{\Delta p_{YBC}-\Delta p_e(\theta_2)\}$$
$$F_{q2C} = K(\theta_2-2\theta_{2p})\{\Delta p_{YBC}-\Delta p_e(\theta_2-2\theta_{2p})\}$$

where $\Delta p_{YBC} = \Delta p_{Bt} = \Delta p_{Ct}$, which are shown by
$\Delta q_{2Bt} = \Delta q_{2Ct} = \Delta p_{Bt}/(\tan\phi_{b20}\tan\eta_{b20}+1)$ in Fig. 6.2-4.

Substituting F_{q2B} and F_{q2C} for Eq. (6.2-6) yields Δp_{YBC} as follows.

$$\left.\begin{aligned}\Delta p_{YBC} &= \{K(\theta_2)\Delta p_e(\theta_2)+K(\theta_2-2\theta_{2p})\Delta p_e(\theta_2-2\theta_{2p})+F_{q2}\}/\{K(\theta_2)+K(\theta_2-2\theta_{2p})\}\\ F_{q2} &= F_{q2B}+F_{q2C}\end{aligned}\right\} \qquad (6.2\text{-}9)$$

Δp_{Bt} and Δp_{Ct} corresponding to F_{q2B} and F_{q2C} incline to the axis u_2 by the inclination angles χ_{2mB} and χ_{2mC} of the planes G_{2mBt} and G_{2mCt} respectively.

When the point P_{mBt} is in the other region of two pair mesh Y_{AB}, Δp_{YAB} is obtained in the same way.

$$\Delta p_{YAB} = \{K(\theta_2+2\theta_{2p})\Delta p_e(\theta_2+2\theta_{2p})+K(\theta_2)\Delta p_e(\theta_2)+F_{q2}\}/\{K(\theta_2+2\theta_{2p})+K(\theta_2)\}$$

Δp_{YAB} is obtained from Eq. (6.2-9) by substituting $\theta_2+2\theta_{2p}$ for θ_2, so that $\Delta p_{Bt}(\theta_2)$ in the region of two pair mesh is expressed as follows by substituting $\Delta p_Y(\theta_2)$ for Δp_{YBC}.

$$\left.\begin{aligned}\Delta p_{Bt}(\theta_2) &= \Delta p_{YAB} = \Delta p_Y(\theta_2+2\theta_{2p}) && \text{(in the region of } Y_{AB})\\ &= \Delta p_{YBC} = \Delta p_Y(\theta_2) && \text{(in the region of } Y_{BC})\end{aligned}\right\} \qquad (6.2\text{-}10)$$

(c) $\Delta p_{Bt}(\theta_2)$ in the region of three pair mesh Z_{KAB}, Z_{ABC} or Z_{BCD}

When the point P_{mBt} is in the region of three pair mesh Z_{ABC}, deviations of tangent plane $\Delta p_{At} = \Delta p_{Bt} = \Delta p_{Ct} = \Delta p_{ZABC}$ at the simultaneous points of contact P_{mAt}, P_{mBt}

and P_{mCt} are obtained in the same way as in the region of two pair mesh.

$$\Delta p_{ZABC} = \{K(\theta_2+2\theta_{2p})\Delta p_e(\theta_2+2\theta_{2p})+K(\theta_2)\Delta p_e(\theta_2)+K(\theta_2-2\theta_{2p})\Delta p_e(\theta_2-2\theta_{2p})+F_{q2}\} / \{K(\theta_2+2\theta_{2p})+K(\theta_2)+K(\theta_2-2\theta_{2p})\} \quad (6.2\text{-}11)$$

where $F_{q2} = F_{q2A} + F_{q2B} + F_{q2C}$

$$F_{q2A} = K(\theta_2+2\theta_{2p})\{\Delta p_{ZABC}-\Delta p_e(\theta_2+2\theta_{2p})\}$$
$$F_{q2B} = K(\theta_2)\{\Delta p_{ZABC}-\Delta p_e(\theta_2)\}$$
$$F_{q2C} = K(\theta_2-2\theta_{2p})\{\Delta p_{ZABC}-\Delta p_e(\theta_2-2\theta_{2p})\}$$

Δp_{At}, Δp_{Bt} and Δp_{Ct} corresponding to F_{q2A}, F_{q2B} and F_{q2C} incline to the axis u_2 by the inclination angles χ_{2mA}, χ_{2mB} and χ_{2mC} of the planes G_{2mAt}, G_{2mBt} and G_{2mCt} at P_{mAt}, P_{mBt} and P_{mCt} respectively.

Δp_{ZKAB} and Δp_{ZBCD} in the regions Z_{KAB} and Z_{BCD} are obtained from Eq. (6.2-11) by substituting $\theta_2+2\theta_{2p}$ and $\theta_2-2\theta_{2p}$ for θ_2 respectively, so that $\Delta p_{Bt}(\theta_2)$ in the region of three pair mesh is expressed as follows by substituting $\Delta p_Z(\theta_2)$ for Δp_{ZABC}.

$$\left.\begin{aligned} \Delta p_{Bt}(\theta_2) &= \Delta p_{ZKAB} = \Delta p_Z(\theta_2+2\theta_{2p}) && \text{(in the region of } Z_{KAB}) \\ &= \Delta p_{ZABC} = \Delta p_Z(\theta_2) && \text{(in the region of } Z_{ABC}) \\ &= \Delta p_{ZBCD} = \Delta p_Z(\theta_2-2\theta_{2p}) && \text{(in the region of } Z_{BCD}) \end{aligned}\right\} \quad (6.2\text{-}12)$$

In regions of more than four pair mesh, they are obtained in the same way.

(5) Relation between regions of mesh and equivalent tooth load F_{q2}

$\Delta p_{Bt}(\theta_2)$ is a function of F_{q2} only when $K(\theta_2)$ and $\Delta p_e(\theta_2)$ are given, so that the region of mesh varies according to F_{q2} as follows.

(a) Equivalent tooth load F_{q2} which realizes the region of one pair mesh X_B or three pair mesh Z_{ABC}.

In Fig. 6.2-3, the following relation is always valid in the region of X_B.

$$\Delta p_X(\theta_2) \leqq \Delta p_e(\theta_2)$$
$$F_{q2} \leqq 0$$

The angle of rotation θ_{2AC} at the intersection A_c of the modified tooth profiles II, $A_{2e}(\Delta p_{Ae})$ and $C_{2e}(\Delta p_{Ce})$, is obtained as follows.

$$\Delta p_e(\theta_{2AC}+2\theta_{2p}) - \Delta p_e(\theta_{2AC}-2\theta_{2p}) = 0$$

The region of one pair mesh X_B is above the point A_c in the direction of the axis Δp_e, therefore,

$$\Delta p_e(\theta_{2AC}-2\theta_{2p}) < \Delta p_X(\theta_{2AC})$$
$$F_{q2AC} < F_{q2}$$

where $F_{q2AC} = K(\theta_{2AC})\{\Delta p_e(\theta_{2AC}-2\theta_{2p})-\Delta p_e(\theta_{2AC})\}$

Therefore, the equivalent tooth load F_{q2} which realizes the region of one pair mesh X_B is obtained as follows.

$$F_{q2AC} < F_{q2} \leqq 0 \qquad (6.2\text{-}13)$$

The equivalent tooth load F_{q2} which realizes the region of three pair mesh Z_{ABC} is obtained as follows because the region of Z_{ABC} is below the point A_c and above K_D in the direction of the axis Δp_e.

$$F_{q2KD} < F_{q2} \leqq F_{q2AC} \qquad (6.2\text{-}14)$$

where F_{q2KD} : F_{q2} when $\Delta p_Z(\theta_2)$ passes the point K_D which is the lowest limit of the region of three pair mesh Z_{ABC} in Fig. 6.2-3.

In other regions of three pair mesh Z_{KAB} and Z_{BCD}, F_{q2} is obtained in the same way.

The region below the point K_D is that of five pair mesh and the equivalent tooth load F_{q2} which realizes it is obtained in the same way.

(b) Equivalent tooth load F_{q2} which realizes the region of two pair mesh Y_{BC}

In Fig. 6.2-3, the angle of rotation θ_{2BC} at the intersection B_C of the modified tooth profiles II, $B_{2e}(\Delta p_{Be})$ and $C_{2e}(\Delta p_{Ce})$, is obtained as follows.

$$\Delta p_e(\theta_{2BC})-\Delta p_e(\theta_{2BC}-2\theta_{2p}) = 0$$

Because the region of two pair mesh Y_{BC} is below the point B_C in the direction of the axis Δp_e,

$$\Delta p_Y(\theta_{2BC}) < \Delta p_e(\theta_{2BC})$$
$$F_{q2} < 0$$

The angle of rotation θ_{2AD} at the intersection A_D of the modified tooth profiles II, $A_{2e}(\Delta p_{Ae})$ and $D_{2e}(\Delta p_{De})$, in Fig. 6.2-3 is obtained as follows.

$$\Delta p_e(\theta_{2AD}+2\theta_{2p})-\Delta p_e(\theta_{2AD}-4\theta_{2p}) = 0$$

Because the region of two pair mesh Y_{BC} is above the point A_D in the direction of the axis Δp_e,

$$\Delta p_e(\theta_{2AD}-4\theta_{2p}) < \Delta p_Y(\theta_{2AD})$$
$$F_{q2AD} < F_{q2}$$

where $F_{q2AD} = \{K(\theta_{2AD})+K(\theta_{2AD}-2\theta_{2p})\}\Delta p_e(\theta_{2AD}-4\theta_{2p})-K(\theta_{2AD})\Delta p_e(\theta_{2AD})$
$-K(\theta_{2AD}-2\theta_{2p})\Delta p_e(\theta_{2AD}-2\theta_{2p})$

Therefore, the equivalent tooth load F_{q2} which realizes the region of two pair mesh Y_{BC} is obtained as follows.

$$F_{q2AD} < F_{q2} < 0 \tag{6.2-15}$$

The equivalent tooth load F_{q2} which realizes the other region of two pair mesh Y_{AB} is obtained in the same way.

The region below the point A_D is that of four pair mesh and the equivalent tooth load F_{q2} which realizes it is obtained in the same way.

(c) Relation between regions of mesh and an equivalent tooth load F_{q2}

The regions of mesh under a certain equivalent tooth load F_{q2} are expressed as follows by Eqs. (6.2-13), (6.2-14) and (6.2-15).

$$\left.\begin{array}{ll} F_{q2AC} < F_{q2} \leqq 0 & : Y_{AB},\ X_B \text{ and } Y_{BC} \\ F_{q2AD} < F_{q2} \leqq F_{q2AC} & : Z_{KAB},\ Y_{AB},\ Z_{ABC},\ Y_{BC} \text{ and } Z_{BCD} \\ F_{q2KD} < F_{q2} \leqq F_{q2AD} & : Z_{KAB},\ Z_{ABC},\ Z_{BCD} \text{ and regions of four pair mesh} \end{array}\right\} \tag{6.2-16}$$

6. 2. 3 Path of contact h_{mt} of the gear pair under load

Figure 6.2-5(a) shows two kinds of path of contact under load, h_{mXY} and h_{mYZ}, which are expressed by the deviation of tangent plane $\triangle p_{Bt}(\theta_2)$ at P_{mBt}, and Fig. 6.2-5(b) shows the radius of base cylinder R_{b2mBt} corresponding to h_{mYZ}. The regions of mesh are determined through Eq. (6.2-16) by giving a certain load F_{q2}, so that the path of contact h_{mt} under load is obtained as follows.

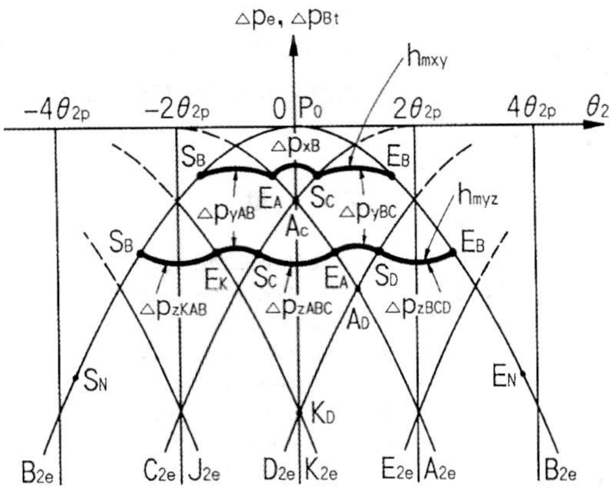

(a) Δp_{mBt} of paths of contact h_{mxy} and h_{myz}

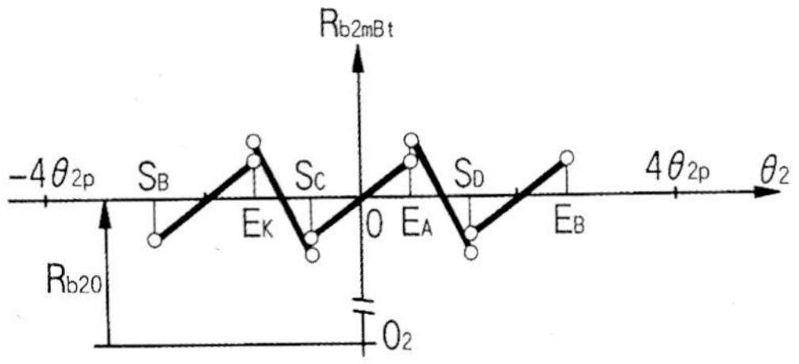

(b) R_{b2mBt} of path of contact h_{myz}

Fig.6.2-5 Examples of $\triangle p_{Bt}$ and R_{b2mBt} of path of contact under load

(1) Path of contact h_{mXY} composed of the regions of one and two pair mesh ($F_{q2AC} < F_{q2} \leqq 0$)

The deviation of tangent plane $\triangle p_{Bt}(\theta_2)$ can be expressed as follows by Eqs. (6.2-8), (6.2-9) and (6.2-10).

$$\left.\begin{array}{llll} \triangle p_{Bt}(\theta_2) & = \triangle p_{YAB} & = \triangle p_Y(\theta_2 + 2\theta_{2p}) & (Y_{AB} : S_B \text{ to } E_A : \theta_{2SB} \leqq \theta_2 \leqq \theta_{2EA}) \\ & = \triangle p_{XB} & = \triangle p_X(\theta_2) & (X_B : E_A \text{ to } S_C : \theta_{2EA} < \theta_2 < \theta_{2SC}) \\ & = \triangle p_{YBC} & = \triangle p_Y(\theta_2) & (Y_{BC} : S_C \text{ to } E_B : \theta_{2SC} \leqq \theta_2 \leqq \theta_{2EB}) \end{array}\right\} \tag{6.2-17}$$

In this example, Δp_{Bt} is continuous but not differentiable at points S_B, E_A, S_C and E_B at which the region of mesh changes, while the radius of base cylinder R_{b2mBt} is discontinuous at the points.

Substituting Eq. (6.2-17) for Eqs. (6.2-1) and (6.2-2) yields the path of contact h_{mXY} which is discontinuous at points S_B, E_A, S_C and E_B because of the discontinuous radius of base cylinder.

The angles of rotation θ_{2SB} and θ_{2EB} at the starting point of mesh S_B and the ending one E_B are obtained as follows.

$$\text{Starting point of mesh } S_B: \Delta p_Y(\theta_{2SB}+2\theta_{2p}) = \Delta p_e(\theta_{2SB})$$
$$\text{Ending point of mesh } E_B: \Delta p_Y(\theta_{2EB}) = \Delta p_e(\theta_{2EB})$$

The starting point E_A and the ending one S_C of one pair mesh on h_{mXY} mean the ending and the starting points of two pair mesh respectively, so that the angles of rotation θ_{2EA} and θ_{2SC} at E_A and S_C and the actual contact ratio m_p are obtained as follows.

$$\theta_{2EA} = \theta_{2EB}-2\theta_{2p}, \quad \theta_{2SC} = \theta_{2SB}+2\theta_{2p}$$
$$m_p = (\theta_{2EB}-\theta_{2SB})/2\theta_{2p} = (\theta_{2EB}-\theta_{2SC})/2\theta_{2p}+1 \qquad (6.2\text{-}18)$$

(2) Path of contact h_{mYZ} composed of the regions of two and three pair mesh ($F_{q2AD} < F_{q2} \leqq F_{q2AC}$)

The deviation of tangent plane $\Delta p_{Bt}(\theta_2)$ can be expressed as follows by Eqs. (6.2-9), (6.2-10), (6.2-11) and (6.2-12).

$$\left.\begin{aligned}
\Delta p_{Bt}(\theta_2) &= \Delta p_{ZKAB} = \Delta p_Z(\theta_2+2\theta_{2p}) & (Z_{KAB}: S_B \text{ to } E_K: \theta_{2SB} \leqq \theta_2 \leqq \theta_{2EK}) \\
&= \Delta p_{YAB} = \Delta p_Y(\theta_2+2\theta_{2p}) & (Y_{AB}: E_K \text{ to } S_C: \theta_{2EK} < \theta_2 < \theta_{2SC}) \\
&= \Delta p_{ZABC} = \Delta p_Z(\theta_2) & (Z_{ABC}: S_C \text{ to } E_A: \theta_{2SC} \leqq \theta_2 \leqq \theta_{2EA}) \\
&= \Delta p_{YBC} = \Delta p_Y(\theta_2) & (Y_{BC}: E_A \text{ to } S_D: \theta_{2EA} < \theta_2 < \theta_{2SD}) \\
&= \Delta p_{ZBCD} = \Delta p_Z(\theta_2-2\theta_{2p}) & (Z_{BCD}: S_D \text{ to } E_B: \theta_{2SD} \leqq \theta_2 \leqq \theta_{2EB})
\end{aligned}\right\} \quad (6.2\text{-}19)$$

In this example, Δp_{Bt} is continuous but not differentiable at points S_B, E_K, S_C, E_A, S_D and E_B at which the region of mesh changes, while the radius of base cylinder R_{b2mBt} is discontinuous at the points as shown in Fig. 6.2-5(b).

Substituting Eq. (6.2-19) for Eqs. (6.2-1) and (6.2-2) yields the path of contact h_{mYZ} which is discontinuous at points S_B, E_K, S_C, E_A, S_D and E_B because of the discontinuous radius of base cylinder R_{b2mBt}.

The angles of rotation θ_{2SB} and θ_{2EB} at the starting point of mesh S_B and the ending one E_B are obtained as follows.

$$\text{Starting point of mesh } S_B: \Delta p_Z(\theta_{2SB}+2\theta_{2p}) = \Delta p_e(\theta_{2SB})$$
$$\text{Ending point of mesh } E_B: \Delta p_Z(\theta_{2EB}-2\theta_{2p}) = \Delta p_e(\theta_{2EB})$$

The angles of rotation θ_{2EK}, θ_{2SC}, θ_{2EA} and θ_{2SD} at E_K, S_C, E_A and S_D respectively on h_{mYZ} and the actual contact ratio m_p are obtained as follows.

$$\theta_{2EK}=\theta_{2EB}-4\theta_{2p},\quad \theta_{2SC}=\theta_{2SB}+2\theta_{2p},\quad \theta_{2EA}=\theta_{2EB}-2\theta_{2p},\quad \theta_{2SD}=\theta_{2SB}+4\theta_{2p}$$
$$m_p=(\theta_{2EB}-\theta_{2SB})/2\theta_{2p}=(\theta_{2EB}-\theta_{2SD})/2\theta_{2p}+2 \qquad (6.2\text{-}20)$$

The paths of contact h_{mt} composed of other regions of mesh are obtained in the same way.

The load deflects the modified tooth surfaces II and realizes the multiple tooth pair mesh, in the region of which $\triangle p_{Bt}$ becomes a periodic function of $2\theta_{2p}$ as shown in Eqs. (6.2-17) and (6.2-19).

In this theory, the region from S_B to E_B is assumed to exist within that from S_N to E_N, where S_N and E_N correspond to the starting and ending points of mesh calculated from the dimensions respectively as shown in Fig. 6.2-5(a). When the point S_B or E_B exceeds the region from S_N to E_N because of excess load, $\triangle p_{Bt}$ of h_{mt} becomes discontinuous, so that another procedure becomes necessary.

6．2．4　Summary

When a gear pair which has involute helicoids for tooth surfaces II (elastic) and their conjugate ones for tooth surfaces I (rigid) is given, the path of contact with its common contact normal at each point between the modified tooth surfaces II which are modified slightly and convexly from the tooth surfaces II and the rigid tooth surfaces I under load is analyzed, which is summarized as follows.

(1) A point of contact with its common contact normal, the fundamental requirements for multiple tooth pair contact, the regions of multiple tooth pair mesh and the equivalent tooth load are obtained under load, where the tooth-pair spring constant is given as a function of the angle of rotation.

(2) A path of contact with its contact normal at each point between the modified tooth surfaces II and the rigid tooth surfaces I under load is obtained, which is a periodic function of angular pitch $2\theta_{2p}$ and is generally discontinuous at the points where the region of mesh changes.

(3) The actual contact ratio which depends on load is also calculated.

6．3　Equations of motion of gears (62), (64)

The gear pair is supposed to have both rotational motion and vibration caused by the springs of the tooth pair and the gear shafts at the same time, therefore the motion of the gear pair is analyzed under the assumption that it is superimposed by the following two motions.

(1) Rotational motion which is isolated from the vibration of the gear system and

(2) Forced vibration of the gear system caused by the periodic variation of load generated by item (1).

In this section, the equations of motion are obtained based on assumption (1). The other item (2) is discussed in 6.4.

6．3．1　Path of contact and one pitch of rotational motion

Figure 6.3-1 shows schematically the path of contact $h_{mYZ}(\theta_2)$ corresponding to Fig. 6.2-5 which is composed of the deviation of tangent plane and the variation of the radius of base cylinder in the regions of three and two pair mesh given by Eq. (6.2-19). Fig. 6.3-1 shows an example of helical gears to make it simple, but it is almost the same for other kinds

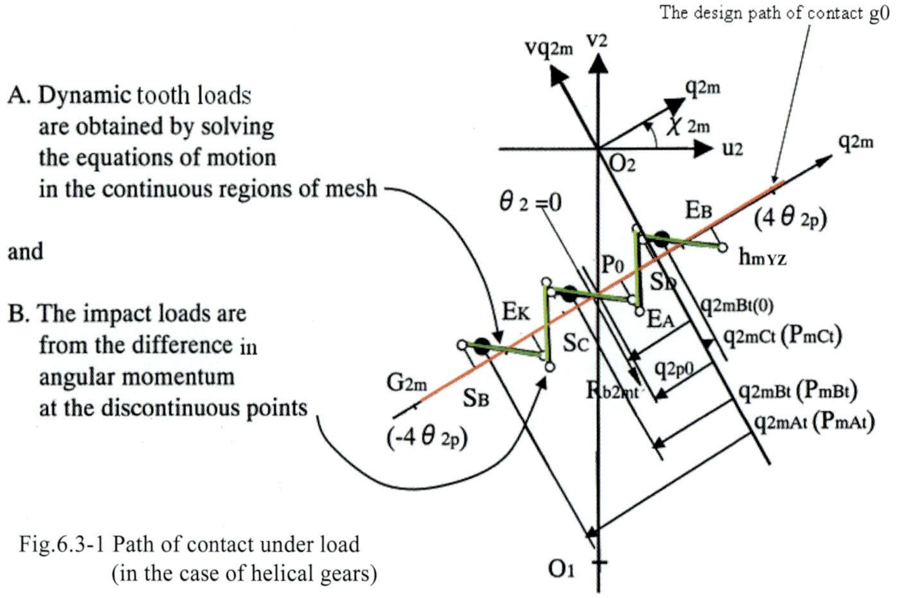

Fig.6.3-1 Path of contact under load
(in the case of helical gears)

of gears except the locations of P_0 and $\theta_2 = 0$. The path of contact $h_{mYZ}(\theta_2)$ is a polygonal line in the neighborhood of the path of contact $\mathbf{g}_0$ under no loads which is discontinuous at the points S_B, E_K, S_C, E_A and S_D because of the discontinuous radius of base cylinder.

The path of contact $h_{mYZ}(\theta_2)$ is obtained as follows. When the deviations of tangent plane $\triangle p_Z(\theta_2)$ and $\triangle p_Y(\theta_2)$ in the regions of three and two pair mesh are given by Eq. (6. 2-19), the paths of contact $h_{mZ}(\theta_2)$ and $h_{mY}(\theta_2)$ are obtained by substituting them for Eqs. (6. 2-1) and (6. 2-2) as follows.

$$h_{mZ}(\theta_2) = h_{mZ}\{u_{2mt}(\theta_2),\ v_{2mt}(\theta_2),\ z_{2mt}(\theta_2);\ 0_2\} \quad (S_C \text{ to } E_A:\ \theta_{2SC} < \theta_2 < \theta_{2EA})$$
$$h_{mY}(\theta_2) = h_{mY}\{u_{2mt}(\theta_2),\ v_{2mt}(\theta_2),\ z_{2mt}(\theta_2);\ 0_2\} \quad (E_A \text{ to } S_D:\ \theta_{2EA} < \theta_2 < \theta_{2SD})$$

Therefore, the path of contact $h_{mYZ}(\theta_2)$ is expressed as follows.

$$\left.\begin{aligned}
h_{mYZ}(\theta_2) &= h_{mZ}(\theta_2 + 2\theta_{2p}) && (Z_{KAB}: S_B \text{ to } E_K:\ \theta_{2SB} < \theta_2 < \theta_{2EK})\\
&= h_{mY}(\theta_2 + 2\theta_{2p}) && (Y_{AB}: E_K \text{ to } S_C:\ \theta_{2EK} < \theta_2 < \theta_{2SC})\\
&= h_{mZ}(\theta_2) && (Z_{ABC}: S_C \text{ to } E_A:\ \theta_{2SC} < \theta_2 < \theta_{2EA})\\
&= h_{mY}(\theta_2) && (Y_{BC}: E_A \text{ to } S_D:\ \theta_{2EA} < \theta_2 < \theta_{2SD})\\
&= h_{mZ}(\theta_2 - 2\theta_{2p}) && (Z_{BCD}: S_D \text{ to } E_B:\ \theta_{2SD} < \theta_2 < \theta_{2EB})
\end{aligned}\right\} \quad (6.\ 3\text{-}1)$$

$h_{mYZ}(\theta_2)$ is a periodic function of $2\theta_{2p}$, so that it is enough that the rotational motion of the gear pair is examined during one pitch from S_C to S_D via E_A.

The rotational motion of the gear pair whose path of contact is discontinuous as shown in Fig. 6. 3-1 is considered as follows, where the modified tooth surfaces are smooth and have no pitch errors.

(1) From S_C to E_A (region of three pair mesh) : they mesh smoothly along $h_{mZ}(\theta_2)$.
(2) At E_A (region of impact) : a point of contact changes during an infinitesimal time τ from $h_{mZ}(\theta_2)$ to $h_{mY}(\theta_2)$ discontinuously.
(3) From E_A to S_D (region of two pair mesh) : they mesh smoothly along $h_{mY}(\theta_2)$.

(4) At S_D (region of impact) : a point of contact changes during an infinitesimal time τ from $h_{mY}(\theta_2)$ to $h_{mZ}(\theta_2 - 2\theta_{2p})$ discontinuously.

The gear pair is supposed to repeat the motions from (1) to (4) mentioned above in the steady state, therefore the equations of motion in the region of mesh are obtained as follows, where the times and locations of each point of contact are shown in Table 6.3-1.

Table 6.3-1 Times and locations of points

Location of point	Time	θ_1	θ_2
S_C	$t_{SC}+\tau$	θ_{1SC}	θ_{2SC}
$h_{mZ}(0)$	0	0	0
E_A	$t_{EA}+\tau$	θ_{1EA}	θ_{2EA}
S_D	$t_{SD}+\tau$	θ_{1SD}	θ_{2SD}

6．3．2 Equations of motion in the region of three pair mesh from S_C to E_A

(1) Equations of motion

The equations of motion are obtained by Eq. (2.5-3) as follows.

$$J_1(d^2\theta_1/dt^2) = F_{q1A}R_{b1mAt} + F_{q1B}R_{b1mBt} + F_{q1C}R_{b1mCt} + T_1 \quad (6.3\text{-}2)$$
$$J_2(d^2\theta_2/dt^2) = F_{q2A}R_{b2mAt} + F_{q2B}R_{b2mBt} + F_{q2C}R_{b2mCt} + T_2 \quad (6.3\text{-}3)$$

where J_1, J_2 : moments of inertia of gears I and II,
θ_1, θ_2 : angles of rotation of gears I and II,
T_1, T_2 : input and output torques of gears I and II,
F_{q1A}, F_{q1B}, F_{q1C} : tooth loads in the direction of axis q_1 shared by gear I at points P_{mAt}, P_{mBt} and P_{mCt},
F_{q2A}, F_{q2B}, F_{q2C} : tooth loads in the direction of axis q_2 shared by gear II at points P_{mAt}, P_{mBt} and P_{mCt},
R_{b1mAt}, R_{b1mBt}, R_{b1mCt} : radii of base cylinder of gear I at points P_{mAt}, P_{mBt} and P_{mCt} and
R_{b2mAt}, R_{b2mBt}, R_{b2mCt} : radii of base cylinder of gear II at points P_{mAt}, P_{mBt} and P_{mCt} respectively.

Because the tooth surface I is the rigid conjugate one of the involute helicoid having the radius of base cylinder R_{b20} and the helix angle ψ_{b20}, the fundamental requirement for contact (Eq. (6.1-11)) at a point of contact P_{mAt} between the tooth surface I and the involute helicoid and the law of action and reaction yield the following equations.

$$\left.\begin{aligned} R_{b1mAt}\cos\psi_{b1mAt} &= (R_{b20}/i_0)\cos\psi_{b20} \\ F_{q1A}/\cos\psi_{b1mAt} &= -F_{q2A}/\cos\psi_{b20} \end{aligned}\right\} \quad (6.3\text{-}4)$$

where ψ_{b1mAt} : inclination angle of the common contact normal n_{mAt} on the plane of action of gear I at P_{mAt}.

The same relationships as Eq. (6.3-4) are valid at points of contact P_{mBt} and P_{mCt}, therefore the following equations are obtained.

$$\left.\begin{aligned} F_{q1A}R_{b1mAt} &= -F_{q2A}(R_{b20}/i_0) \\ F_{q1B}R_{b1mBt} &= -F_{q2B}(R_{b20}/i_0) \\ F_{q1C}R_{b1mCt} &= -F_{q2C}(R_{b20}/i_0) \end{aligned}\right\} \quad (6.3\text{-}5)$$

Because P_{mAt}, P_{mBt} and P_{mCt} are simultaneous points of contact, the radii of base cylinder of gear II must be equal due to the fundamental requirements for multiple tooth pair contact (Eq. (6.2-5)). Therefore Eq. (6.3-6) is obtained by defining the common radius of base cylinder as $R_{b2mZ}(\theta_2)$.

$$R_{b2mAt} = R_{b2mBt} = R_{b2mCt} = R_{b2mZ}(\theta_2) \quad (6.3\text{-}6)$$

By substituting Eqs. (6.3-5) and (6.3-6) for Eqs. (6.3-2) and (6.3-3) and simplifying, Eq. (6.3-7) is obtained.

$$\left.\begin{aligned} J_1(d^2\theta_1/dt^2) &= -F_{q2}(R_{b20}/i_0)+T_1 \\ J_2(d^2\theta_2/dt^2) &= F_{q2}R_{b2mZ}(\theta_2)+T_2 \\ F_{q2} &= F_{q2A}+F_{q2B}+F_{q2C} \end{aligned}\right\} \quad (6.3\text{-}7)$$

(2) Increment of equivalent tooth load F_{q2dZ} and radius of base cylinder $R_{b2mZ}(\theta_2)$

Equivalent tooth load F_{q2} is supposed to be expressed in the region of three pair mesh as follows.

$$\left.\begin{aligned} F_{q2} &= F_{q2S}+F_{q2dZ} \quad (|F_{q2dZ}| \ll |F_{q2S}|) \\ F_{q2S} &= F_{q2SA}+F_{q2SB}+F_{q2SC} = -T_2/R_{b20} \end{aligned}\right\} \quad (6.3\text{-}8)$$

where F_{q2S} : static equivalent tooth load and

F_{q2dZ} : increment of equivalent tooth load in the region of three pair mesh.

Using Eq. (6.3-8), the deviation of tangent plane $\triangle p_Z(\theta_2)$ in the region of three pair mesh from S_C to E_A can be expressed by the sum of the static deviation $\triangle p_{ZS}(\theta_2)$ due to F_{q2S} and the dynamic deflection $z_d(\theta_2)$ due to F_{q2dZ} as follows.

$$\triangle p_Z(\theta_2) = \triangle p_{ZS}(\theta_2)+z_d(\theta_2) \quad (6.3\text{-}9)$$

where

$$\begin{aligned} \triangle p_{ZS}(\theta_2) = \{&K(\theta_2+2\theta_{2p})\triangle p_e(\theta_2+2\theta_{2p})+K(\theta_2)\triangle p_e(\theta_2)+K(\theta_2-2\theta_{2p})\triangle p_e(\theta_2-2\theta_{2p}) \\ &+F_{q2S}\}/\{K(\theta_2+2\theta_{2p})+K(\theta_2)+K(\theta_2-2\theta_{2p})\} \end{aligned} \quad (6.3\text{-}10)$$

$$z_d(\theta_2) = F_{q2dZ}/\{K(\theta_2+2\theta_{2p})+K(\theta_2)+K(\theta_2-2\theta_{2p})\} \quad (6.3\text{-}11)$$

By substituting Eq. (6.3-9) for Eq. (6.2-1), $R_{b2mZ}(\theta_2)$ is obtained as follows.

$$\left.\begin{aligned} R_{b2mZ}(\theta_2) &= R_{b20} + d\Delta p_Z(\theta_2)/d\theta_2 \\ &= R_{b20} + d\Delta p_{ZS}(\theta_2)/d\theta_2 + dz_d(\theta_2)/d\theta_2 \end{aligned}\right\} \quad (6.3\text{-}12)$$

(3) Fundamental requirement for contact

Using Eq. (6. 3-4), following equations at points of contact P_{mAt}, P_{mBt} and P_{mCt} are obtained.

$$\begin{aligned} R_{b1mAt}\cos\phi_{b1mAt} &= R_{b1mBt}\cos\phi_{b1mBt} \\ &= R_{b1mCt}\cos\phi_{b1mCt} = (R_{b20}/i_0)\cos\phi_{b20} \end{aligned}$$

where ϕ_{b1mBt}、ϕ_{b1mCt} : inclination angles of common contact normals $\mathbf{n}_{mBt}$ and $\mathbf{n}_{mCt}$ on planes of action of gear I at points P_{mBt} and P_{mCt} respectively.

Therefore, using Eqs. (6. 1-10) and (6. 3-12), the fundamental requirement for contact is obtained as follows.

$$\begin{aligned} (R_{b20}/i_0)(d\theta_1/dt) &= R_{b2mZ}(\theta_2)(d\theta_2/dt) \\ &= (R_{b20} + d\Delta p_{ZS}/d\theta_2 + dz_d/d\theta_2)(d\theta_2/dt) \end{aligned} \quad (6.3\text{-}13)$$

Finally, the rotational motion of the gear pair in the region of three pair mesh from S_C to E_A is expressed by Eqs. (6. 3-7), (6. 3-8), (6. 3-11), (6. 3-12) and (6. 3-13) as follows.

$$\left.\begin{aligned} J_1(d^2\theta_1/dt^2) &= -F_{q2}(R_{b20}/i_0) + T_1 \\ J_2(d^2\theta_2/dt^2) &= F_{q2}R_{b2mZ}(\theta_2) + T_2 \\ (R_{b20}/i_0)(d\theta_1/dt) &= R_{b2mZ}(\theta_2)(d\theta_2/dt) \\ F_{q2} &= F_{q2S} + F_{q2dZ} \\ F_{q2dZ} &= \{K(\theta_2-2\theta_{2p})+K(\theta_2)+K(\theta_2+2\theta_{2p})\}z_d(\theta_2) \\ R_{b2mZ}(\theta_2) &= R_{b20} + d\Delta p_{ZS}/d\theta_2 + dz_d/d\theta_2 \\ \theta_{2SC} < \theta_2 &< \theta_{2EA} \end{aligned}\right\} \quad (6.3\text{-}14)$$

Equation (6. 3-14) represents simultaneous equations with three unknown quantities θ_1, θ_2 and z_d.

(4) Assumptions of dynamic deflection z_d in the rotational motion of the gear pair

(a) The dynamic deflection z_d is sufficiently small when compared with the deviation of tangent plane Δp_{ZS} caused by the tooth surface modification Δp_e and the static equivalent tooth load F_{q2S}. Therefore, the dynamic variation of the radius of base cylinder $dz_d/d\theta_2$ is neglected to R_{b20} because it is infinitesimal of the second order.

$$R_{b2mZ}(\theta_2) = R_{b20} + d\Delta p_{ZS}/d\theta_2 + dz_d/d\theta_2 \fallingdotseq R_{b20} + d\Delta p_{ZS}/d\theta_2 \quad (6.3\text{-}15)$$

The rotational motion is assumed to be that of the rigid gear pair having a pair of tooth surfaces with the deviation of tangent plane Δp_{ZS}.

(b) The pair of tooth surfaces has no separation, so that the deviation of tangent plane is always continuous but not differentiable at the points where the region of mesh changes.

Under the assumptions above, substituting Eq. (6.3-15) for Eq. (6.3-13), differentiating Eq. (6.3-13) by time and substituting the differentiated Eq. (6.3-13) for Eq. (6.3-14), Eq. (6.3-16) is obtained as follows, which has the increment of equivalent tooth load F_{q2dZ} in the region of three pair mesh.

$$\left.\begin{aligned} f_Z(\theta_2) &= T_2(d\Delta p_{ZS}/d\theta_2)/J_2-(d^2\Delta p_{ZS}/d\theta_2{}^2)(d\theta_2/dt)^2 \\ F_{q2dZ} &= Mf_Z(\theta_2) \\ J_1(d^2\theta_1/dt^2) &= -F_{q2dZ}(R_{b20}/i_0) \\ J_2(d^2\theta_2/dt^2) &= F_{q2dZ}R_{b20}+F_{q2S}(d\Delta p_{ZS}/d\theta_2) \\ \theta_{2SC} &< \theta_2 < \theta_{2EA} \\ \text{where} \quad M &= J_1J_2/\{R_{b20}{}^2(J_1+J_2/i_0{}^2)\} \text{ (equivalent mass)} \end{aligned}\right\} \quad (6.3\text{-}16)$$

6.3.3 Equations of motion in the region of two pair mesh from E_A to S_D

In the same way, equations of motion in the region of two pair mesh from E_A to S_D is obtained by substituting subscript y for z in Eq. (6.3-16) as follows.

$$\left.\begin{aligned} f_Y(\theta_2) &= T_2(d\Delta p_{YS}/d\theta_2)/J_2-(d^2\Delta p_{YS}/d\theta_2{}^2)(d\theta_2/dt)^2 \\ F_{q2dY} &= Mf_Y(\theta_2) \\ J_1(d^2\theta_1/dt^2) &= -F_{q2dY}(R_{b20}/i_0) \\ J_2(d^2\theta_2/dt^2) &= F_{q2dY}R_{b20}+F_{q2S}(d\Delta p_{YS}/d\theta_2) \\ \theta_{2EA} &< \theta_2 < \theta_{2SD} \end{aligned}\right\} \quad (6.3\text{-}17)$$

In another case where the path of contact h_{mt} is another curve which is composed of the regions of one and two pair mesh or three and four pair mesh for examples, the equations of motion are obtained in the same way.

In addition, Eqs. (6.3-16) and (6.3-17) represent that the equations of motion of all the gear pairs having involute helicoids for one member can be replaced by those of the involute spur gears whose radii of base cylinder are R_{b20}/i_0 and R_{b20}. In other words, the rotational motions of the involute helical and involute face gears are represented by the same equations of motion.

6.3.4 Approximation accuracy of the dynamic model (equations of motion) in this theory

In this theory, the gear pair is assumed to have the tooth surface modification and the elastic deflection only for one member, so that the path of contact with its contact normal at each point and the equations of motion of this model are different from those of a stricter model which has the modification and deflection for both members. Therefore, whether the dynamic model has sufficient accuracy or not is discussed compared with the stricter model, where the gear pair and their symbols are the same except that gear I also has modification and deflection.

When modified tooth surfaces I and II are in contact at a point P_{mt} at angles of rotation θ_1 and θ_2 under load, the deviation of the tangent plane W_{mt} at the point P_{mt} is defined as follows. The deviation of the tangent plane W_{mt} in the direction of axis q_2

from the tangent plane W_{mp2} at the point P_{m2} which locates at θ_2 on the true path of contact h_m (Eq. (6.1-3)) is defined by $\triangle p_{2t}$, while the deviation of W_{mt} in the direction of axis q_1 from the tangent plane W_{mp1} at the point P_{m1} which locates at θ_1 on h_m is defined by $\triangle p_{1t}$. When $\theta_1 = i_0\theta_2$, the point P_{m1} superimposes the point P_{m2}. The normals $\mathbf{n}_{mt}$, $\mathbf{n}_{m1}$ and $\mathbf{n}_{m2}$ at the points P_{mt}, P_{m1} and P_{m2} respectively are assumed to be parallel and simultaneous points of contact are indicated by subscripts A, B and C. In this theory, because the modified tooth surfaces I had no modification and no deflection, $P_{m2} = P_m$ and $P_{m1} = P_{mt}$ were obtained.

The equations of motion of the gear pair are given by Eqs. (6.3-2) and (6.3-3) as follows.

$$J_1(d^2\theta_1/dt^2) = F_{q1A}R_{b1mAt} + F_{q1B}R_{b1mBt} + F_{q1C}R_{b1mCt} + T_1 \quad \text{(a)}$$
$$J_2(d^2\theta_2/dt^2) = F_{q2A}R_{b2mAt} + F_{q2B}R_{b2mBt} + F_{q2C}R_{b2mCt} + T_2 \quad \text{(b)}$$

When the deviation of tangent plane of the modified tooth surface I at a point P_{mAt} is given by $\triangle p_{1At}$, the radius of base cylinder R_{b1mAt} and the tooth load F_{q1A} are expressed as follows.

$$\left.\begin{aligned} R_{b1mAt} &= R_{b1mA} + d\triangle p_{1At}/d\theta_1 \\ F_{q1A} &= F_{q1SA} + F_{q1dA} \end{aligned}\right\} \quad \text{(c)}$$

where R_{b1mA} is the radius of base cylinder at the point of contact P_{m1A} on h_m and $d\triangle p_{1At}/d\theta_1$ and F_{q1dA} are infinitesimal of the first order.

When the deviation of tangent plane of the modified tooth surface II is given by $\triangle p_{2At}$, the following equations are obtained in the same way.

$$\left.\begin{aligned} R_{b2mAt} &= R_{b20} + d\triangle p_{2At}/d\theta_2 \\ F_{q2A} &= F_{q2SA} + F_{q2dA} \end{aligned}\right\} \quad \text{(d)}$$

where $d\triangle p_{2At}/d\theta_2$ and F_{q2dA} are infinitesimal of the first order.

The fundamental requirement for contact between an involute helicoid and its conjugate surface and the law of action and reaction (Eq. (6.3-4)) at the point P_{m1A} yield the following equations.

$$\left.\begin{aligned} R_{b1mA}\cos\phi_{b1mA} &= (R_{b20}/i_0)\cos\phi_{b20} \\ F_{q1A}/\cos\phi_{b1mA} &= -F_{q2A}/\cos\phi_{b20} \end{aligned}\right\} \quad \text{(e)}$$

where ϕ_{b1mA} is the inclination angle of the normal $\mathbf{n}_{m1A}$ on the plane of action of gear I at the point P_{m1A} and equals ϕ_{b1mAt} because the normal $\mathbf{n}_{m1A}$ is parallel to $\mathbf{n}_{mAt}$.

The same relationships are there at the simultaneous points of contact P_{mBt} and P_{mCt}, so that the following equations are obtained.

6. Rotational motion of gears and dynamic increment of tooth load

$$\left.\begin{aligned} F_{q1A}R_{b1mA} &= -F_{q2A}(R_{b20}/i_0) \\ F_{q1B}R_{b1mB} &= -F_{q2B}(R_{b20}/i_0) \\ F_{q1C}R_{b1mC} &= -F_{q2C}(R_{b20}/i_0) \end{aligned}\right\} \qquad \text{(f)}$$

At the simultaneous points of contact P_{mAt}, P_{mBt} and P_{mCt}, the normal components $\triangle p_{1nt}$ of the deviation of tangent plane are equal, so that $\triangle p_{1nt}$ can be defined as follows.

$$\left.\begin{aligned} \triangle p_{1nt} &= \int_0^{\theta 1} (d\triangle p_{1At}/d\theta_1)\cos\phi_{b1mAt}\,d\theta_1 + \triangle p_{1nt}(0) \\ &= \int_0^{\theta 1} (d\triangle p_{1Bt}/d\theta_1)\cos\phi_{b1mBt}\,d\theta_1 + \triangle p_{1nt}(0) \\ &= \int_0^{\theta 1} (d\triangle p_{1Ct}/d\theta_1)\cos\phi_{b1mCt}\,d\theta_1 + \triangle p_{1nt}(0) \end{aligned}\right\} \qquad \text{(g)}$$

where $\triangle p_{1Bt}$, $\triangle p_{1Ct}$: deviations of tangent plane of the modified tooth surfaces I at the points P_{mBt} and P_{mCt} respectively and when $\theta_1 = 0$, P_{mAt}, P_{mBt} and P_{mCt} are simultaneous points of contact.

By substituting Eqs. (c), (d), (f) and (g) for Eq. (a) and by simplifying, the following equations are obtained.

$$\left.\begin{aligned} J_1(d^2\theta_1/dt^2) &= -F_{q2d}(R_{b20}/i_0) - F_{q2S}(d\triangle p_{12t}/d\theta_1) \\ F_{q2S} &= F_{q2SA} + F_{q2SB} + F_{q2SC} = -T_2/R_{b20} \\ F_{q2d} &= F_{q2dA} + F_{q2dB} + F_{q2dC} \\ d\triangle p_{12t}/d\theta_1 &= (d\triangle p_{1nt}/d\theta_1)/\cos\phi_{b20} \end{aligned}\right\} \qquad \text{(h)}$$

where F_{q2S} : static equivalent tooth load,
F_{q2d} : increment of equivalent tooth load and
$\triangle p_{12t}$: displacement of $\triangle p_{1t}$ in the direction of axis q_2.

Equation (b) is transformed in the same way.

$$\left.\begin{aligned} J_2(d^2\theta_2/dt^2) &= F_{q2d}R_{b20} + F_{q2S}(d\triangle p_{2t}/d\theta_2) \\ \triangle p_{2t} &= \triangle p_{2At} = \triangle p_{2Bt} = \triangle p_{2Ct} \end{aligned}\right\} \qquad \text{(i)}$$

where $\triangle p_{2Bt}$, $\triangle p_{2Ct}$: deviations of tangent plane of the modified tooth surfaces II at the points P_{mBt} and P_{mCt} respectively.

The normal velocities at the point of contact P_{mAt} are obtained as follows.

$$\begin{aligned} R_{b1mAt}(d\theta_1/dt)\cos\phi_{b1mAt} &= (R_{b1mA} + d\triangle p_{1At}/d\theta_1)(d\theta_1/dt)\cos\phi_{b1mA} \\ &= (R_{b20}/i_0 + d\triangle p_{12t}/d\theta_1)(d\theta_1/dt)\cos\phi_{b20} \\ R_{b2mAt}(d\theta_2/dt)\cos\phi_{b20} &= (R_{b20} + d\triangle p_{2At}/d\theta_2)(d\theta_2/dt)\cos\phi_{b20} \\ &= (R_{b20} + d\triangle p_{2t}/d\theta_2)(d\theta_2/dt)\cos\phi_{b20} \end{aligned}$$

Because the same relationships are there at the simultaneous points of contact P_{mBt} and P_{mCt}, the fundamental requirement for contact is obtained as follows.

$$(R_{b20}/i_0 + d\Delta p_{12t}/d\theta_1)(d\theta_1/dt) = (R_{b20} + d\Delta p_{2t}/d\theta_2)(d\theta_2/dt) \qquad (j)$$

By eliminating $(d^2\theta_1/dt^2)$ and $(d^2\theta_2/dt^2)$ from Eqs. (h), (i) and (j), the increment of equivalent tooth load F_{q2d} is obtained as follows.

$$\begin{aligned} F_{q2d} = & -MF_{q2S}R_{b20}\{(d\Delta p_{12t}/d\theta_1)/(i_0 J_1)+(d\Delta p_{2t}/d\theta_2)/J_2\} \\ & -M\{d^2(\Delta p_{2t}-\Delta p_{12t})/d\theta_2^2\}(d\theta_2/dt)^2 \end{aligned} \qquad (k)$$

Equations (h), (i) and (k) represent the equations of motion of the stricter model which has the surface modification and deflection for both members.

Equations (h), (i) and (k) are represented by Eq. (6.3-16) or (6.3-17) in this theory, so that the approximation accuracy of the dynamic model can be discussed by comparing those equations.

Considering $\Delta p_{2t} - \Delta p_{12t} = \Delta p_{ZS}$, the difference between F_{q2d} of Eq. (k) and F_{q2dZ} of Eq. (6.3-16) is obtained as follows,.

$$F_{q2d} - F_{q2dZ} = -F_{q2S}(d\Delta p_{12t}/d\theta_1)i_0/R_{b20} = -F_{q2S}(d\Delta p_{12t}/d\theta_2)/R_{b20} \qquad (l)$$

By substituting Eq. (l) for Eqs. (h) and (i), Eq. (m) is obtained as follows.

$$\left.\begin{aligned} J_1(d^2\theta_1/dt^2) &= -F_{q2dZ}(R_{b20}/i_0) \\ J_2(d^2\theta_2/dt^2) &= F_{q2dZ}R_{b20} + F_{q2S}\{d(\Delta p_{2t}-\Delta p_{12t})/d\theta_2\} \end{aligned}\right\} \qquad (m)$$

Equation (m) is equal to those of Eq. (6.3-16), which means that the equations of motion of the two different models are the same because the operating torques are the same although the tooth loads differ a little according to each base cylinder. Therefore their solutions (rotational motion) are the same.

The difference between the two models is $F_{q2d} - F_{q2dZ}$ only shown by Eq. (l). Therefore, it is enough to actualize $F_{q2d} \fallingdotseq F_{q2dZ}$ in order to assure almost the same accuracy of this model as that of the stricter one, the conditions of which are as follows.

(1) Normal displacements $\Delta p_{12t} \ll \Delta p_{2t}$

When tooth surface I can be assumed to be nearly rigid compared with tooth surface II, for example, a gear pair having gear I made of steel and gear II of plastics almost satisfies $F_{q2d} \fallingdotseq F_{q2dZ}$.

(2) $|F_{q2S}(d\Delta p_{12t}/d\theta_2)/R_{b20}| \ll |F_{q2dZ}|$

When a certain load F_{q2S} is given, taking a sufficiently large $d\theta_2/dt$ can realize $F_{q2d} \fallingdotseq F_{q2dZ}$ because F_{q2dZ} includes $(d\theta_2/dt)^2$. For example, most of the gear noise in vehicles occurs under conditions of high velocity and light load, so that the approximation accuracy is sufficient to deal with them. The experimental results mentioned in 6.6 show that the term which does not include $(d\theta_2/dt)^2$ has little influence on the amplitude of the dynamic increment of tooth load F_{q2dn} under the operating condition beyond 2000r/min.

Under the conditions of high velocity and light load, the two different models can be considered to be actually the same one as follows.

$$\left.\begin{aligned} F_{q2d} &= F_{q2dZ} \fallingdotseq -M\{d^2(\triangle p_{2t}-\triangle p_{12t})/d\theta_2{}^2\}(d\theta_2/dt)^2 \\ J_1(d^2\theta_1/dt^2) &= M(R_{b20}/i_0)\{d^2(\triangle p_{2t}-\triangle p_{12t})/d\theta_2{}^2\}(d\theta_2/dt)^2 \\ J_2(d^2\theta_2/dt^2) &= -MR_{b20}\{d^2(\triangle p_{2t}-\triangle p_{12t})/d\theta_2{}^2\}(d\theta_2/dt)^2 \end{aligned}\right\} \quad (n)$$

Equation (n) shows that the single flank error under load which means $\triangle p_{2t}-\triangle p_{12t}$ can represent the dynamic performance of the gear pair in some degree if the shapes of the error resemble each other. However, because the equations of motion are not valid at the discontinuous points of the path of contact where the region of mesh changes, the solution $F_{q2d} \fallingdotseq F_{q2dZ}$ is insufficient for the load acting on the gear system, so that the dynamic impact loads which occur at the discontinuous points must be obtained.

6. 3. 5 Summary

When a path of contact with its contact normal at each point of a gear pair which has symmetrical convex tooth surfaces (elastic) modified slightly from true involute helicoids for tooth surfaces II and the conjugate tooth surfaces (rigid) of the true involute helicoids for tooth surfaces I is given under load as functions of the angle of rotation, the rotational motion of the gear pair is analyzed, which is summarized as follows.

(1) The equations of motion of the gear pair mentioned above are obtained.

(2) The equations of motion of all kinds of gear pairs from helical to hypoid gears having involute helicoids for one member can be expressed by those of the involute spur gears whose radii of base cylinder are R_{b20}/i_0 and R_{b20}.

(3) The approximation accuracy of the dynamic model of this theory is examined, compared with a stricter one which has the surface modification and deflection on both members.

6. 4 Dynamic increment of tooth load and variation of bearing loads [62], [64]

6. 4. 1 Solutions of equations of motion

The solutions of θ_1 and θ_2 are expected to be given by linear combinations of the angular displacements $\omega_{10}t$ and $\omega_{20}t$ and their deviations $\triangle\theta_1$ and $\triangle\theta_2$ respectively as follows.

$$\left.\begin{aligned} \theta_1 &= \omega_{10}t + \triangle\theta_1 \\ \theta_2 &= \omega_{20}t + \triangle\theta_2 \end{aligned}\right\} \quad (6.4\text{-}1)$$

where ω_{10} and ω_{20} : mean angular velocities of gears I and II

$\triangle\theta_1$ and $\triangle\theta_2$: deviations of angular displacement which are assumed to be infinitesimal of the first order.

Using Eq. (6.4-1), the right-hand terms $f_Z(\theta_2)$ and $f_Y(\theta_2)$ of Eqs. (6.3-16) and (6.3-17) can be approximated by $f_Z(\omega_{20}t)$ and $f_Y(\omega_{20}t)$, so that the right-hand terms of the equations of motion are expressed by functions of time t.

Solving Eq. (6.3-16) regarding gear Ⅰ, the angular acceleration, the angular velocity and the angular displacement are obtained as follows, where subscript z means that a point of contact is on the path of contact $h_{mZ}(\theta_2)$ of three pair mesh and $(d^2\triangle p_{ZS}/d\theta_2^2)\times(d\triangle\theta_2/dt)$ in $f_Z(\theta_2)$ is neglected from the assumption of Eq. (6.4-1) because it is infinitesimal of the second order.

$$\left.\begin{aligned} f_Z(\theta_2) &= T_2(d\triangle p_{ZS}/d\theta_2)/J_2-(d^2\triangle p_{ZS}/d\theta_2^2)\,\omega_{20}^2 \\ d^2\theta_{1Z}/dt^2 &= -F_{q2dZ}(R_{b20}/i_0)/J_1 = -M(R_{b20}/i_0)f_Z(\theta_2)/J_1 \\ d\theta_{1Z}/dt &= \omega_{10}+\triangle\omega_{1Z} \qquad (\triangle\omega_{1Z} \ll \omega_{10}) \\ \triangle\omega_{1Z} &= -(MR_{b20}/i_0/J_1)\int^t f_Z(\theta_2)dt + C_{1Z} \\ \theta_{1Z} &= \omega_{10}t+\int^t \triangle\omega_{1Z}dt + C_{2Z} \end{aligned}\right\} \qquad (6.4\text{-}2)$$

In the same way, the solution of Eq. (6.3-17) is obtained as follows, where subscript y means that a point of contact is on the path of contact $h_{mY}(\theta_2)$ of two pair mesh.

$$\left.\begin{aligned} f_Y(\theta_2) &= T_2(d\triangle p_{YS}/d\theta_2)/J_2-(d^2\triangle p_{YS}/d\theta_2^2)\,\omega_{20}^2 \\ d^2\theta_{1Y}/dt^2 &= -F_{q2dY}(R_{b20}/i_0)/J_1 = -M(R_{b20}/i_0)f_Y(\theta_2)/J_1 \\ d\theta_{1Y}/dt &= \omega_{10}+\triangle\omega_{1Y} \qquad (\triangle\omega_{1Y} \ll \omega_{10}) \\ \triangle\omega_{1Y} &= -(M\,R_{b20}/i_0/J_1)\int^t f_Y(\theta_2)dt + C_{1Y} \\ \theta_{1Y} &= \omega_{10}t+\int^t \triangle\omega_{1Y}dt + C_{2Y} \end{aligned}\right\} \qquad (6.4\text{-}3)$$

The angular acceleration, the angular velocity and the angular displacement of gear Ⅱ are obtained in the same way.

6．4．2 Boundary conditions and integration constants

(1) Relation between angle of rotation θ_2 and time t in the region of mesh

The relations between the angle of rotation θ_2 and time t in the region of mesh are obtained from Table 6.3-1 as follows, where the variation in angular velocity is supposed to be negligibly small to ω_{20} and $2t_p$ is a pitch of time.

$$\begin{aligned} t_{EA} &= t_{SC}+(\theta_{2EA}-\theta_{2SC})/\omega_{20} \\ t_{SD} &= t_{EA}+(\theta_{2SD}-\theta_{2EA})/\omega_{20} = t_{SC}+2t_p \quad (\theta_{2p} = \omega_{20}t_p) \end{aligned}$$

(2) Integration constants C_{1Z} and C_{1Y}

Neglecting the dynamic deflection, the angle of rotation θ_{1EA} at time t_{EA} is obtained from Eq. (6.4-2) as follows.

$$\theta_{1EA}-\theta_{1SC} = (\omega_{10}+C_{1Z})(t_{EA}-t_{SC})-(M\,R_{b20}/i_0/J_1)\int_{tSC}^{tEA}\{\int^t f_Z(\theta_2)dt\}dt$$

Substituting Eq. (6.3-15) for Eq. (6.3-13) and integrating it from t_{SC} to t_{EA},

$$(R_{b20}/i_0)(d\theta_1/dt) = (R_{b20}+d\Delta p_{ZS}/d\theta_2)(d\theta_2/dt)$$

$$(R_{b20}/i_0)(\theta_{1EA}-\theta_{1SC}) = R_{b20}(\theta_{2EA}-\theta_{2SC})+\int_{\theta 2SC}^{\theta 2EA}(d\Delta p_{ZS}/d\theta_2)d\theta_2$$

Considering $\omega_{10}(R_{b20}/i_0)(t_{EA}-t_{SC}) = R_{b20}(\theta_{2EA}-\theta_{2SC})$, C_{1Z} is obtained by eliminating $(R_{b20}/i_0)(\theta_{1EA}-\theta_{1SC})$ from the equation above as follows.

$$C_{1Z} = \left[(M\,R_{b20}/i_0/J_1)\int_{tSC}^{tEA}\{\int^{t} f_Z(\theta_2)dt\}dt + \int_{\theta 2SC}^{\theta 2EA}(d\Delta p_{ZS}/d\theta_2)d\theta_2/(R_{b20}/i_0)\right]/(t_{EA}-t_{SC}) \qquad (6.4\text{-}4)$$

C_{1Y} is obtained in the same way as follows.

$$C_{1Y} = \left[(M\,R_{b20}/i_0/J_1)\int_{tEA}^{tSD}\{\int^{t} f_Y(\theta_2)dt\}dt + \int_{\theta 2EA}^{\theta 2SD}(d\Delta p_{YS}/d\theta_2)d\theta_2/(R_{b20}/i_0)\right]/(t_{SD}-t_{EA}) \qquad (6.4\text{-}5)$$

(3) Integration constants C_{2Z} and C_{2Y}

When $t = 0$, $\theta_{1Z} = 0$, therefore $C_{2Z} = 0$.

Angular displacement is continuous at $t = t_{EA}$, which means $\theta_{1Z}(t_{EA}) = \theta_{1Y}(t_{EA})$, so that C_{2Y} is obtained as follows.

$$C_{2Y} = \int^{tEA}\Delta\omega_{1Z}dt - \int^{tEA}\Delta\omega_{1Y}dt$$

6. 4. 3 Impact loads and rotational motion in a steady state

(1) Impact loads

At points S_C and E_B, where the path of contact is discontinuous, the angular velocities also become discontinuous, so that impact loads are generated. Since the angular velocities before and after the time of impact are already given by Eqs. (6.4-2) and (6.4-3), the impact loads are obtained as follows from the difference in angular momentum.

$$\left.\begin{aligned} S_C:\ & F_{dSC} = J_1\{d\theta_{1Z}(t_{SC})/dt - d\theta_{1Y}(t_{SC}+2t_p)/dt\}/\{(R_{b20}/i_0)\cdot\tau\} \\ E_A:\ & F_{dEA} = J_1\{d\theta_{1Y}(t_{EA})/dt - d\theta_{1Z}(t_{EA})/dt\}/\{(R_{b20}/i_0)\cdot\tau\} \end{aligned}\right\} \qquad (6.4\text{-}6)$$

$d\theta_{1Y}(t)/dt$ and $\Delta\omega_{1Y}(t)$ are periodic functions of $2t_p$ based on Eq. (6.3-1), therefore they are expressed by $d\theta_{1Y}(t_{SC}+2t_p)/dt$ and $\Delta\omega_{1Y}(t_{SC}+2t_p)$ at time t_{SC}.

(2) Rotational motion during one pitch in a steady state

The increment of kinetic energy $\triangle E_{1SC}$ of gear I during the region of impact at Sc is obtained as follows, where the variations $\triangle\omega_{1Z}$ and $\triangle\omega_{1Y}$ are infinitesimal of the first order to ω_{10}.

$$\begin{aligned}\text{Region of impact at Sc}: \triangle E_{1SC} &= J_1[\{d\theta_{1Z}(t_{SC})/dt\}^2-\{d\theta_{1Y}(t_{SC}+2t_p)/dt\}^2]/2\\ &= J_1[\{\omega_{10}+\triangle\omega_{1Z}(t_{SC})\}^2-\{\omega_{10}+\triangle\omega_{1Y}(t_{SC}+2t_p)\}^2]/2\\ &= J_1\omega_{10}\{\triangle\omega_{1Z}(t_{SC})-\triangle\omega_{1Y}(t_{SC}+2t_p)\}\end{aligned}$$

The increments of kinetic energy in the other regions are obtained in the same way.

$$\begin{aligned}\text{Region of three pair mesh from Sc to EA}: \triangle E_{1Z} &= J_1\omega_{10}\{\triangle\omega_{1Z}(t_{EA})-\triangle\omega_{1Z}(t_{SC})\}\\ \text{Region of impact at EA}: \triangle E_{1EA} &= J_1\omega_{10}\{\triangle\omega_{1Y}(t_{EA})-\triangle\omega_{1Z}(t_{EA})\}\\ \text{Region of two pair mesh from EA to SD}: \triangle E_{1Y} &= J_1\omega_{10}\{\triangle\omega_{1Y}(t_{SD})-\triangle\omega_{1Y}(t_{EA})\}\\ &= J_1\omega_{10}\{\triangle\omega_{1Y}(t_{SC}+2t_p)-\triangle\omega_{1Y}(t_{EA})\}\end{aligned}$$

Therefore, the increment of kinetic energy $\triangle E_1$ of gear I during one pitch is as follows.

$$\triangle E_1 = \triangle E_{1SC}+\triangle E_{1Z}+\triangle E_{1EA}+\triangle E_{1Y} = 0 \qquad (6.4\text{-}7)$$

Equation (6.4-7) indicates that the energies $\triangle E_{1Z}$ and $\triangle E_{1Y}$ supplied (or lost) in the regions of mesh are returned back through the energies $\triangle E_{1SC}$ and $\triangle E_{1EA}$ in the regions of impact, so that gear I continues to rotate steadily.

The situation is exactly the same for gear II.

6．4．4 Dynamic increment of tooth load expressed by Fourier series

Since the gear pair rotates steadily as indicated by Eq. (6.4-7), the dynamic increment of tooth load F_{q2d} during one pitch is given by Eqs. (6.3-16), (6.3-17) and (6.4-6) as follows.

$$\left.\begin{aligned}F_{q2d} &= F_{dSC} && (t_{SC} \leqq t \leqq t_{SC}+\tau)\\ &= F_{q2dZ} && (t_{SC} < t < t_{EA})\\ &= F_{dEA} && (t_{EA} \leqq t \leqq t_{EA}+\tau)\\ &= F_{q2dY} && (t_{EA} < t < t_{SD})\end{aligned}\right\} \qquad (6.4\text{-}8)$$

Substituting $t-(t_{SC}+t_{SD})/2$ for t, F_{q2d} is expressed by the following Fourier series, because it is a periodic function defined during $[-t_p, t_p]$, which is fragmentary but smooth.

$$F_{q2d}(t) = \sum_{n=1}^{\infty}\{A_n\cos(n\pi t/t_p) + B_n\sin(n\pi t/t_p)\} \qquad (6.4\text{-}9)$$

where $A_n = \int_{-tp}^{tp} F_{q2d}(\xi)\cos(n\pi\xi/t_p)d\xi/t_p$, $B_n = \int_{-tp}^{tp} F_{q2d}(\xi)\sin(n\pi\xi/t_p)d\xi/t_p$

$n = 1, 2, \cdots, \infty$.

When F_{q2dn} is the amplitude of the dynamic increment of tooth load of the n-th order,

Eq. (6.4-9) is expressed as follows.

$$F_{q2d}(t) = \sum_{n=1}^{\infty} F_{q2dn} \sin(n\omega t + \alpha_n) \qquad (6.4\text{-}10)$$

where $F_{q2dn} = \sqrt{(A_n^{\,2} + B_n^{\,2})}$, $\omega = \pi / t_p = \pi \omega_{20} / \theta_{2p}$, $\alpha_n = \tan^{-1}(A_n / B_n)$.

$F_{q2d}(t)$ given by Eq. (6.4-10) denotes the external load of the forced vibration system of the gear pair.

6.4.5 Equation of vibration and frequency characteristics of amplified tooth load (66)

When a point of contact is in the region of three pair mesh, Eq. (6.3-13) is expressed as follows.

$$(R_{b20} / i_0)(d\theta_1/dt) = \{R_{b20} + (d\Delta p_{ZS}/d\theta_2)\}(d\theta_2/dt) + dz_d/dt$$

Differentiating both sides of the equation by time t,

$$(R_{b20}/i_0)(d^2\theta_1/dt^2) = \{R_{b20} + (d\Delta p_{ZS}/d\theta_2)\}(d^2\theta_2/dt^2) + (d^2\Delta p_{ZS}/d\theta_2^{\,2})(d\theta_2/dt)^2 + d^2 z_d/d^2t \qquad (6.4\text{-}11)$$

Using Eq. (6.3-16), eliminating $d^2\theta_1/dt^2$ and $d^2\theta_2/dt^2$ and simplifying through neglecting the term which is infinitesimal of the second order, Eq. (6.4-12) is obtained.

$$\left.\begin{aligned} &M(d^2 z_d/dt^2) + K_Z z_d = M\{T_2(d\Delta p_{ZS}/d\theta_2)/J_2 - (d^2\Delta p_{ZS}/d\theta_2^{\,2})\,\omega_{20}^{\,2}\} = Mf_Z(\theta_2) \\ &K_Z = K(\theta_2 - 2\theta_{2p}) + K(\theta_2) + K(\theta_2 + 2\theta_{2p}) \end{aligned}\right\} \qquad (6.4\text{-}12)$$

Equation (6.4-12) is the equation of vibration of the gear pair in the region of three pair mesh.

In the same way, that in the region of two pair mesh is obtained as follows.

$$\left.\begin{aligned} &M(d^2 y_d/dt^2) + K_Y y_d = M\{T_2(d\Delta p_{YS}/d\theta_2)/J_2 - (d^2\Delta p_{YS}/d\theta_2^{\,2})\,\omega_{20}^{\,2}\} = Mf_Y(\theta_2) \\ &K_Y = K(\theta_2 - 2\theta_{2p}) + K(\theta_2) \end{aligned}\right\} \qquad (6.4\text{-}13)$$

Equations (6.4-12) and (6.4-13) mean that the gear pair system with the equivalent mass M and the spring constant K_Z or K_Y is vibrated by the external load $Mf_Z(\theta_2)$ or $Mf_Y(\theta_2)$ in the region of three or two pair mesh respectively. Therefore during one pitch, it is supposed to be vibrated by the load given by Eq. (6.4-10) which includes the impact loads F_{dSC} and F_{dEA} at the points where the region of mesh changes. In the result, the equation of vibration of the gear pair is expressed approximately by adding the damping term as follows, where X is the dynamic deflection corresponding to the dynamic increment of load F_{q2d}.

$$M(d^2X/dt^2)+2\gamma\sqrt{(MK_m)}(dX/dt)+K_m X=\sum_{n=1}^{\infty} F_{q2dn}\sin(n\omega t+\alpha_n) \qquad (6.4\text{-}14)$$

where K_m : mean tooth-pair spring constant during one pitch,

$$K_m = \left(\int_{\theta 2SC}^{\theta 2EA} K_Z\, d\theta_2 + \int_{\theta 2EA}^{\theta 2SD} K_Y\, d\theta_2\right)/(2\theta_{2p}) \text{ and}$$

$2\gamma\sqrt{(MK_m)}(dX/dt)$: damping force (γ is damping coefficient).

The amplitude of n-th order F_{vn} of the load amplified by the vibration of the gear pair is obtained from the particular solution of Eq. (6.4-15) as follows.

$$\left.\begin{aligned} F_{vn} &= G_n(\omega)F_{q2dn} \\ G_n(\omega) &= 1/\sqrt{(\{1-(n\omega/\omega_m)^2\}^2+4\gamma^2(n\omega/\omega_m)^2)} \end{aligned}\right\} \qquad (6.4\text{-}15)$$

where $G_n(\omega)$:transfer ratio of n-th order (transfer function) and $\omega_m^2 = K_m/M$.

The amplitude of the load amplified by the vibration of the gear pair has the frequency characteristics shown by Eq. (6.4-15).

6．4．6 Variation of bearing loads

When a path of contact is given by Eq. (6.3-1) and a point of contact P_{mBt} on the modified tooth surface B_{2e} under load moves from S_C to S_D via E_A in Fig.6.3-1, the variation of bearing loads is calculated, where the gear vibration system is assumed to have little influence.

(1) Points of contact

When a point of contact P_{mBt} is in the region of three pair mesh from S_C to E_A ($\theta_{2SC} \leqq \theta_2 \leqq \theta_{2EA}$), the point P_{mBt} and its common contact normal $\mathbf{n}_{mBt}$ are expressed by Eqs. (6.2-1) and (6.2-19) as follows.

$$\left.\begin{aligned} &P_{mBt}(q_{2mBt}, -R_{b2mBt}, z_{2mBt};\ O_{q2}) \\ &n_{mBt}(\pi/2-\chi_{2mBt}, \phi_{b20};\ O_2) \\ &q_{2mBt} = q_{2m}(\theta_2)+\triangle q_{2Z} \\ &z_{2mBt} = (q_{2mBt}-q_{2p0})\tan\eta_{b20} \\ &R_{b2mBt} = R_{b2mZ} = R_{b20}+\triangle R_{b2Z} \\ &\chi_{2mBt} = \chi_{20}-\xi_{2mB} \\ &\text{where } \triangle q_{2Z} = \triangle p_Z(\theta_2)/(\tan\eta_{b20}\tan\phi_{b20}+1) \\ &\triangle R_{b2Z} = d\triangle p_Z(\theta_2)/d\theta_2 \text{ and} \\ &q_{2p0} : \text{location of the design point } P_0 \text{ on the axis } q_2. \end{aligned}\right\} \qquad (6.4\text{-}16)$$

Deviations of tangent plane and variations of the radius of base cylinder of the simultaneous points of contact P_{mAt} and P_{mCt} are the same as those of the point P_{mBt}, so that they are expressed as follows.

$$\begin{aligned}
&P_{mAt}(q_{2mAt},\ -R_{b2mAt},\ z_{2mAt};\ O_{q2}) && P_{mCt}(q_{2mCt},\ -R_{b2mCt},\ z_{2mCt};\ O_{q2})\\
&n_{mAt}(\pi/2-\chi_{2mAt},\ \phi_{b20};\ O_2) && n_{mCt}(\pi/2-\chi_{2mCt},\ \phi_{b20};\ O_2)\\
&q_{2mAt} = q_{2m}(\theta_2+2\theta_{2p})+\triangle q_{2Z} && q_{2mCt} = q_{2m}(\theta_2-2\theta_{2p})+\triangle q_{2Z}\\
&z_{2mAt} = (q_{2mAt}-q_{2p0})\tan\eta_{b20} && z_{2mCt} = (q_{2mCt}-q_{2p0})\tan\eta_{b20}\\
&R_{b2mAt} = R_{b2mZ} = R_{b20}+\triangle R_{b2Z} && R_{b2mCt} = R_{b2mZ} = R_{b20}+\triangle R_{b2Z}\\
&\chi_{2mAt} = \chi_{20}-\xi_{2mA} && \chi_{2mCt} = \chi_{20}-\xi_{2mC}
\end{aligned}$$

When a point of contact P_{mBt} is in the region of two pair mesh from E_A to S_D, they are obtained in the same way.

(2) Load shared on each point of contact

When the dynamic increment of tooth load F_{q2d} can be given by Eq. (6.4-10) and the point of contact P_{mBt} is in the region of three pair mesh from S_C to E_A, the tooth load F_{q2B} shared at the point P_{mBt} is expressed as follows.

$$\left.\begin{aligned}
F_{q2B} &= F_{q2SB}+F_{q2dB}\\
F_{q2SB} &= K(\theta_2)\{\triangle p_{ZS}(\theta_2)-\triangle p_e(\theta_2)\}\\
F_{q2dB} &= K(\theta_2)F_{q2d}/\{K(\theta_2-2\theta_{2p})+K(\theta_2)+K(\theta_2+2\theta_{2p})\}
\end{aligned}\right\} \quad (6.4\text{-}17)$$

where F_{q2SB} : static tooth load shared at P_{mBt} and
F_{q2dB} : dynamic increment of tooth load shared at P_{mBt}.

Loads F_{q2A} and F_{q2C} at the points P_{mAt} and P_{mCt} are obtained in the same way as follows.

$$\begin{aligned}
F_{q2A} &= F_{q2SA}+F_{q2dA}\\
F_{q2SA} &= K(\theta_2+2\theta_{2p})\ \{\triangle p_{ZS}(\theta_2)-\triangle p_e(\theta_2+2\theta_{2p})\}\\
F_{q2dA} &= K(\theta_2+2\theta_{2p})F_{q2d}/\{K(\theta_2-2\theta_{2p})+K(\theta_2)+K(\theta_2+2\theta_{2p})\}
\end{aligned}$$

$$\begin{aligned}
F_{q2C} &= F_{q2SC}+F_{q2dC}\\
F_{q2SC} &= K(\theta_2-2\theta_{2p})\{\triangle p_{ZS}(\theta_2)-\triangle p_e(\theta_2-2\theta_{2p})\}\\
F_{q2dC} &= K(\theta_2-2\theta_{2p})F_{q2d}/\{K(\theta_2-2\theta_{2p})+K(\theta_2)+K(\theta_2+2\theta_{2p})\}
\end{aligned}$$

Equivalent tooth load F_{q2} is obtained as follows.

$$\begin{aligned}
F_{q2} &= F_{q2S}+F_{q2d}\\
F_{q2S} &= F_{q2SA}+F_{q2SB}+F_{q2SC}\\
F_{q2d} &= F_{q2dA}+F_{q2dB}+F_{q2dC}
\end{aligned}$$

When a point of contact P_{mBt} is in the region of two pair mesh, they are obtained in the same way.

(3) Bearing loads

Using Eqs. (2.5-1), (6.4-16) and (6.4-17), bearing loads caused by F_{q2B} are obtained in the coordinate system O_{q2} as follows.

$$\left.\begin{aligned}
B_{z2B} &= -F_{q2B}\tan\psi_{b20} \\
B_{q2fB}+B_{q2rB} &= -F_{q2B} \\
B_{q2fB}b_{20} &= -F_{q2B}\{z_{2mBt}-q_{2mBt}\tan\psi_{b20}-z_{2r}\} \\
&= -F_{q2B}\{(\tan\eta_{b20}-\tan\psi_{b20})q_{2mBt}-q_{2p0}\tan\eta_{b20}-z_{2r}\} \\
B_{vq2fB}+B_{vq2rB} &= 0 \\
B_{vq2rB}b_{20} &= F_{q2B}R_{b2mBt}\tan\psi_{b20}
\end{aligned}\right\} \quad (6.4\text{-}18)$$

where B_{z2B} : load on bearing b_{2a} in the direction of axis z_2,
B_{q2fB} and B_{q2rB} : loads on bearings b_{2f} and b_{2r} in the direction of axis q_2,
B_{vq2fB} and B_{vq2rB} : loads on bearings b_{2f} and b_{2r} in the direction of axis v_{q2},
z_{2f} and z_{2r} : locations of bearings b_{2f} and b_{2r} on the axis z_2 and
b_{20} : distance between b_{2f} and b_{2r} $(z_{2f}-z_{2r}>0)$.
The direction of each load is positive when it is in the positive direction of each axis.

The bearing loads caused by F_{q2B} are expressed by Eq. (3.1-1) in the coordinate system O_2 as follows.

$$\left.\begin{aligned}
B_{z2B} &= -F_{q2B}\tan\psi_{b20} && \text{(load on bearing } b_{2a} \text{ in the direction of axis } z_2) \\
B_{u2fB} &= B_{q2fB}\cos\chi_{2mBt}-B_{vq2fB}\sin\chi_{2mBt} && \text{(load on bearing } b_{2f} \text{ in the direction of axis } u_2) \\
B_{v2fB} &= B_{q2fB}\sin\chi_{2mBt}+B_{vq2fB}\cos\chi_{2mBt} && \text{(load on bearing } b_{2f} \text{ in the direction of axis } v_2) \\
B_{u2rB} &= B_{q2rB}\cos\chi_{2mBt}-B_{vq2rB}\sin\chi_{2mBt} && \text{(load on bearing } b_{2r} \text{ in the direction of axis } u_2) \\
B_{v2rB} &= B_{q2rB}\sin\chi_{2mBt}+B_{vq2rB}\cos\chi_{2mBt} && \text{(load on bearing } b_{2r} \text{ in the direction of axis } v_2)
\end{aligned}\right\} \quad (6.4\text{-}19)$$

The other bearing loads caused by tooth loads F_{q2A} and F_{q2C} are obtained by substituting subscript A or C for B in Eqs. (6.4-18) and (6.4-19).

The bearing loads in the region of three pair mesh from S_C to E_A are obtained by summing those caused by F_{q2A}, F_{q2B} and F_{q2C} as follows.

$$\left.\begin{aligned}
B_{z2} &= B_{z2A}+B_{z2B}+B_{z2C} \\
B_{u2f} &= B_{u2fA}+B_{u2fB}+B_{u2fC} \\
B_{v2f} &= B_{v2fA}+B_{v2fB}+B_{v2fC} \\
B_{u2r} &= B_{u2rA}+B_{u2rB}+B_{u2rC} \\
B_{v2r} &= B_{v2rA}+B_{v2rB}+B_{v2rC}
\end{aligned}\right\} \quad (6.4\text{-}20)$$

The bearing loads in the region of two pair mesh are obtained in the same way.

Therefore the amplitude of the variation of bearing loads during one pitch is obtained by calculating the maximum and the minimum values of Eq. (6.4-20) during the period from S_C to S_D via E_A $(\theta_{2SC}\leqq\theta_2<\theta_{2SD})$.

6.4.7 Example of the variation of bearing loads

To show the simplest example, the variation of bearing loads in the case of involute helical gears is calculated as follows.

The conditions for no variation of bearing loads for an involute helical gear pair are obtained by Eq. (3.1-2) as follows.

6. Rotational motion of gears and dynamic increment of tooth load

$$\left.\begin{array}{l}\Delta q_{2Z} = \Delta R_{b2Z} = 0,\ \ \xi_{2mA} = \xi_{2mB} = \xi_{2mC} = 0,\ \ \eta_{b20} = \phi_{b20} \\ F_{q2d} = 0\end{array}\right\} \quad (6.4\text{-}21)$$

Substituting Eq. (6.4-21) for Eqs. (6.4-16), (6.4-17), (6.4-18), (6.4-19) and (6.4-20) and simplifying them, the bearing loads B_{z20} and B_{u2f0} under the conditions for no variation of bearing loads are obtained as follows.

$B_{z20} = -F_{q2S}\tan\phi_{b20}$ (load on bearing b_{2a} in the direction of axis z_2)

$B_{u2f0} = B_{q2f0}\cos\chi_{20} - B_{vq2f0}\sin\chi_{20}$ (load on bearing b_{2f} in the direction of axis u_2)

where $B_{q2f0} = F_{q2S}(q_{2p0}\tan\phi_{b20} + z_{2r})/b_{20}$

$B_{vq2f0} = -F_{q2S}R_{b20}\tan\phi_{b20}/b_{20}$

Other radial loads of bearing are obtained in the same way. Therefore, the variations of bearing load are obtained as follows.

(1) Variation of load ΔB_{z2} on bearing b_{2a} in the direction of axis z_2

$$\Delta B_{z2} = B_{z2} - B_{z20} = -F_{q2d}\tan\phi_{b20} \quad (6.4\text{-}22)$$

where $B_{z2} = -(F_{q2A} + F_{q2B} + F_{q2C})\tan\phi_{b20}$

Variation of bearing load ΔB_{z2} is determined by F_{q2d} only and it has no relation with the choice of the inclination angle η_{b20} of the path of contact.

(2) Variation of load ΔB_{u2f} on bearing b_{2f} in the direction of axis u_2

Because of $\xi_{2m} = 0$ in involute helical gears, bearing load B_{u2f} is obtained as follows.

$$B_{u2f} = \sum_{B}^{A,B,C} B_{q2fB}\cos\chi_{20} - \sum_{B}^{A,B,C} B_{vq2fB}\sin\chi_{20} \quad (6.4\text{-}23)$$

$$\sum_{B}^{A,B,C} B_{q2fB} = B_{q2fA} + B_{q2fB} + B_{q2fC}$$

$$= \{(F_{q2S} + F_{q2d})(q_{2p0}\tan\eta_{b20} + z_{2r}) - (\tan\eta_{b20} - \tan\phi_{b20})\sum_{B}^{A,B,C}(F_{q2B}q_{2mBt})\}/b_{20}$$

$$\sum_{B}^{A,B,C} B_{vq2fB} = B_{vq2fA} + B_{vq2fB} + B_{vq2fC}$$

$$= -\{(F_{q2S} + F_{q2d})R_{b20} - F_{q2S}\Delta R_{b2Z}\}\tan\phi_{b20}/b_{20}$$

$$\sum_{B}^{A,B,C}(F_{q2B}q_{2mBt}) = F_{q2A}q_{2mAt} + F_{q2B}q_{2mBt} + F_{q2C}q_{2mCt}$$

Therefore, the variation of bearing load ΔB_{u2f} is obtained as follows.

$$\Delta B_{u2f} = B_{u2f} - B_{u2f0}$$

$$= [F_{q2d}(q_{2p0}\tan\eta_{b20} + z_{2r}) + (\tan\eta_{b20} - \tan\phi_{b20})\{F_{q2S}q_{2p0} - \sum_{B}^{A,B,C}(F_{q2B}q_{2mBt})\}]\cos\chi_{20}/b_{20}$$

$$+ (F_{q2d}R_{b20} - F_{q2S}\Delta R_{b2Z})\tan\phi_{b20}\sin\chi_{20}/b_{20} \quad (6.4\text{-}24)$$

ΔB_{u2f} depends on ΔR_{b2Z}, F_{q2d} and η_{b20}.

When $\Delta R_{b2Z} = 0$, $F_{q2d} = 0$, which means $F_{q2B} = F_{q2SB}$, so that Eq. (6.4-25) is obtained as follows.

$$\Delta B_{u2f} = (\tan\eta_{b20} - \tan\phi_{b20}) \sum_{B}^{A,B,C} F_{q2SB}(q_{2p0} - q_{2mBt})\cos\chi_{20}/b_{20} \quad (6.4\text{-}25)$$

Equation (6.4-25) means that the variation of bearing load occurs by the wrong selection of the inclination angle η_{b20} of the path of contact even if $F_{q2d} = 0$ in involute helical gears.

Other variations of radial loads of bearings are obtained in the same way.

6．4．8 Summary

The rotational motion of a gear pair is analyzed by solving the equations of motion and the variation of bearing loads is obtained, the results of which are summarized as follows.

(1) The equations of motion are solved in the regions of mesh where the paths of contact are smooth (differentiable) to obtain algebraically the dynamic increments of tooth load.

(2) The impact loads generated at the points where the path of contact is discontinuous are obtained from the difference in angular momentum.

(3) In the rotational motion of the gear pair, the energies supplied (or lost) in the regions of mesh are returned back through the impact loads at the points where the region of mesh changes, so that they continue to rotate steadily.

(4) The dynamic increment of tooth load during one pitch is a periodic function composed of the dynamic increments of tooth load in the regions of mesh and the impact loads and is expressed by a Fourier series which is the external load acting on the gear vibration system.

(5) The variation of bearing loads is calculated, which consists of both the dynamic increment of tooth load and the fluctuation of the static load (or torque) that is generated mainly by the wrong selection of the inclination angle η_{b20} of the path of contact.

6．5 Dynamic increment of tooth load of a gear pair having a single flank error composed of symmetrical convex curves [64]

In this section, the newly developed method mentioned from 6.1 to 6.4 is applied to a gear pair having a single flank error composed of symmetrical convex curves which is the most important for practical applications and the algebraic solution of the dynamic increment of tooth load is presented.

6．5．1 Deviation of tangent plane and radius of base cylinder under no loads

The deviation of tangent plane $\Delta p_e(\theta_2)$ and the radius of base cylinder $R_{b2me}(\theta_2)$

under no loads are given by Eq. (6.1-14) as follows, when a symmetrical convex curve of the single flank error is given.

$$\left.\begin{aligned} \Delta p_e(\theta_2) &= a_1 \theta_2^{\,2}/2 \\ R_{b2me}(\theta_2) &= R_{b20} + a_1 \theta_2 \end{aligned}\right\} \qquad (6.5\text{-}1)$$

where a_1 : base circle variation coefficient ($a_1 < 0$) under no loads.

Figure 6.5-1 shows $\Delta p_e(\theta_2)$ and $R_{b2mc}(\theta_2)$.

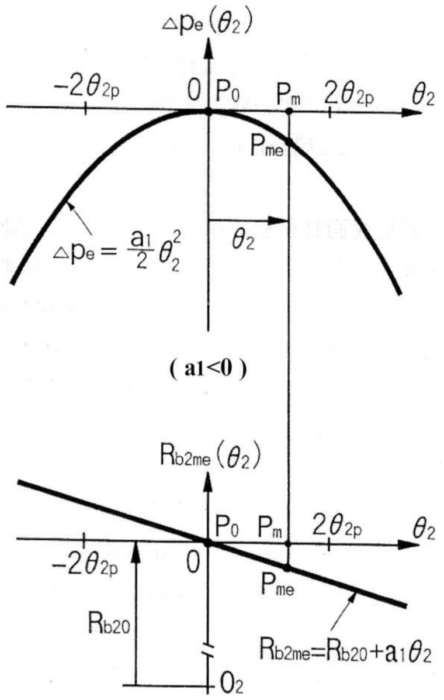

Fig.6.5-1 Assumption of $\triangle p_e$ and R_{b2me}

6. 5. 2 Deviation of tangent plane and radius of base cylinder under static tooth load

(1) Tooth pair spring constant $K(\theta_2)$ [67]

The tooth-pair spring constant corresponding to F_{q2} is assumed as follows, which is given by the approximate equation of Umezawa et al.

$$K(\theta_2) = K_p \cdot \exp(-C_p \,|\, \theta_2^{\,3} \,|) \qquad (6.5\text{-}2)$$

where K_p and C_p : constants determined by the gear dimensions.

(2) Static deviation of tangent plane $\Delta p_{XS}(\theta_2)$ and static radius of base cylinder $R_{b2mXS}(\theta_2)$ in the region of one pair mesh

Substituting $\Delta p_{XS}(\theta_2)$ for Δp_{XB} given by Eq. (6.2-8) corresponding to the static equivalent tooth load F_{q2S} and approximating $\Delta p_{XS}(\theta_2)$ by a parabola with axis $\theta_2 = 0$, $\Delta p_{XS}(\theta_2)$ and $R_{b2mXS}(\theta_2)$ are obtained as follows.

$$\left.\begin{aligned} \Delta p_{XS}(\theta_2) &= a_{1X} \theta_2^{\,2}/2 + F_{q2S}/K(0) \\ R_{b2mXS}(\theta_2) &= R_{b20} + d\Delta p_{XS}/d\theta_2 = R_{b20} + a_{1X}\theta_2 \end{aligned}\right\} \qquad (6.5\text{-}3)$$

where a_{1X} : base circle variation coefficient in the region of one pair mesh.

(3) Static deviation of tangent plane $\Delta p_{YS}(\theta_2)$ and static radius of base cylinder $R_{b2mYS}(\theta_2)$ in the region of two pair mesh

In the same way, substituting $\Delta p_{YS}(\theta_2)$ for Δp_{YBC} given by Eq. (6.2-9) corresponding to the static equivalent tooth load F_{q2S} and approximating $\Delta p_{YS}(\theta_2)$ by a parabola with axis $\theta_2 = \theta_{2p}$, $\Delta p_{YS}(\theta_2)$ and $R_{b2mYS}(\theta_2)$ are obtained as follows.

$$\left.\begin{aligned} \Delta p_{YS}(\theta_2) &= a_{1Y}(\theta_2 - \theta_{2p})^2/2 + \Delta p_e(\theta_{2p}) + F_{q2S}/(2K(\theta_{2p})) \\ R_{b2mYS}(\theta_2) &= R_{b20} + d\Delta p_{YS}/d\theta_2 = R_{b20} + a_{1Y}(\theta_2 - \theta_{2p}) \end{aligned}\right\} \qquad (6.5\text{-}4)$$

where a_{1Y} : base circle variation coefficient in the region of two pair mesh.

(4) Static deviation of tangent plane $\Delta p_{ZS}(\theta_2)$ and static radius of base cylinder $R_{b2mZS}(\theta_2)$ in the region of three pair mesh

In the same way, substituting $\Delta p_{ZS}(\theta_2)$ for Δp_{ZABC} given by Eq. (6.2-11) corresponding to the static equivalent tooth load F_{q2S} and approximating $\Delta p_{ZS}(\theta_2)$ by a parabola with axis $\theta_2 = 0$, $\Delta p_{ZS}(\theta_2)$ and $R_{b2mZS}(\theta_2)$ are obtained as follows.

$$\left.\begin{aligned} \Delta p_{ZS}(\theta_2) &= a_{1Z}\theta_2^{\,2}/2 + \{2K(2\theta_{2p})\Delta p_e(2\theta_{2p}) + K(0)\Delta p_e(0) + F_{q2S}\}/\{2K(2\theta_{2p}) + K(0)\} \\ R_{b2mZS}(\theta_2) &= R_{b20} + d\Delta p_{ZS}/d\theta_2 = R_{b20} + a_{1Z}\theta_2 \end{aligned}\right\} \qquad (6.5\text{-}5)$$

where a_{1Z} : base circle variation coefficient in the region of three pair mesh.

Those in the regions of four pair mesh and more are obtained in the same way.

Substituting equations from (6.5-1) to (6.5-5) for Eqs. (6.2-17) and (6.2-19), the deviation of tangent plane and the radius of base cylinder under the static equivalent tooth load F_{q2S} are obtained, which are already shown in Fig. 6.2-5. In addition, using Eq. (6.3-1), the path of contact h_{mYZ} is obtained, which is already shown in Fig. 6.3-1.

6. 5. 3 Times and locations (angles of rotation) of points of contact during one pitch

When the gear pair is in the region (h_{mYZ}) of three or two pair mesh shown in Fig. 6.2-5, the path of contact is a polygonal line discontinuous at points E_K, S_C, E_A and E_D as shown in Fig. 6.3-1. Since it is a periodic function of $2\theta_{2p}$ and symmetrical about $\theta_2 = 0$, the times and the locations during one pitch can be determined as shown in Table 6.5-1, where $h_{mZ}(0)$ is away from P_0 by the static deflection. When the path of contact is h_{mXY}, $h_{mX}(0)$ is used only instead of $h_{mZ}(0)$.

Table 6.5-1 Times and locations of points

Location	Time	θ_1	θ_2
$h_{mY}(-\theta_{2p})$	$-t_p$	$-\theta_{1p}$	$-\theta_{2p}$
E_A	$-t_q$	$-\theta_{1q}$	$-\theta_{2q}$
$h_{mZ}(0)$	0	0	0
S_C	t_q	θ_{1q}	θ_{2q}
$h_{mY}(\theta_{2p})$	t_p	θ_{1p}	θ_{2p}

where $t_q = \theta_{2q}/\omega_{20}$, $t_p = \theta_{2p}/\omega_{20}$

6. 5. 4 Solutions of equations of motion

(1) Dynamic increments of tooth load in the regions of mesh

The dynamic increments of tooth loads F_{q2dYS}, F_{q2dZ} and F_{q2dYE} are obtained as follows from Eqs. (6.3-16) and (6.3-17), where ω_{20} is the mean angular velocity of gear II.

$$f_Z(t) = a_{1Z}\{T_2\omega_{20}t/J_2 - \omega_{20}^{\,2}\}$$
$$f_Y(t) = a_{1Y}\{T_2(\omega_{20}t - \theta_{2p})/J_2 - \omega_{20}^{\,2}\}$$
$$f_Y(t+2t_p) = a_{1Y}\{T_2(\omega_{20}(t+2t_p) - \theta_{2p})/J_2 - \omega_{20}^{\,2}\}$$

$$\left.\begin{array}{lll} F_{q2dYS} = Mf_Y(t+2t_p) & (-2t_p+t_q < t < -t_q) \\ F_{q2dZ} = M\,f_Z(t) & (-t_q < t < t_q) \\ F_{q2dYE} = M\,f_Y(t) & (t_q < t < 2t_p-t_q) \end{array}\right\} \quad (6.5\text{-}6)$$

When the path of contact is h_{mXY}, a_{1X} is used instead of a_{1Z}.

(2) Angular velocities

Angular velocities are obtained from Eqs. (6.4-2) and (6.4-3) as follows.

$$\left.\begin{array}{l}
d\theta_{1YS}(t)/dt = d\theta_{1YE}(t+2t_p)/dt \\
\quad = \omega_{10}+\Delta\omega_{1Y}(t+2t_p) \quad (\omega_{10} \ll \Delta\omega_{1Y}) \quad (-2t_p+t_q < t < -t_q) \\
\Delta\omega_{1Y}(t+2t_p) = -M(R_{b20}/i_0/J_1)\int^t f_Y(t+2t_p)dt + C_{1y} \\
\int^t f_Y(t+2t_p)dt = a_{1Y}(t+2t_p)\{T_2\omega_{20}(t+2t_p)/(2J_2)-\omega_{20}{}^2 - T_2\theta_{2p}/J_2\} \\
d\theta_{1Z}(t)/dt = \omega_{10}+\Delta\omega_{1Z}(t) \quad (\omega_{10} \ll \Delta\omega_{1Z}) \quad (-t_q < t < t_q) \\
\Delta\omega_{1Z}(t) = -(MR_{b20}/i_0/J_1)\int^t f_Z(t)dt + C_{1Z} \\
\int^t f_Z(t)dt = a_{1Z}t(T_2\omega_{20}t/(2J_2)-\omega_{20}{}^2) \\
d\theta_{1YE}(t)/dt = \omega_{10}+\Delta\omega_{1Y}(t) \quad (\omega_{10} \ll \Delta\omega_{1Y}) \quad (t_q < t < 2t_p-t_q) \\
\Delta\omega_{1Y}(t) = -(M\,R_{b20}/i_0/J_1)\int^t f_Y(t)dt + C_{1Y} \\
\int^t f_Y(t)dt = a_{1Y}t(T_2\omega_{20}t/(2J_2)-\omega_{20}{}^2 - T_2\theta_{2p}/J_2)
\end{array}\right\} \quad (6.5\text{-}7)$$

Considering that the second term in the right-hand side of Eq. (6.4-4) is 0 because of an odd function, the integration constant C_{1Z} is obtained as follows.

$$\begin{aligned} C_{1Z} &= \left[(M\,R_{b20}/i_0/J_1)\int_{-tq}^{tq} a_{1Z}t(T_2\omega_{20}t/2J_2-\omega_{20}{}^2)\,dt \right.\\ &\quad \left. +\int_{-\theta 2q}^{\theta 2q} a_{1Z}\theta_2 d\theta_2/(R_{b20}/i_0)\right]/(2t_q) \\ &= M(R_{b20}/i_0)\,a_{1Z}T_2\omega_{20}t_q{}^2/(6J_1J_2) \end{aligned}$$

C_{1Y} is obtained from Eq. (6.4-5) in the same way.

$$\begin{aligned} C_{1Y} &= \left[(M\,R_{b20}/i_0/J_1)\int_{tq}^{2tp-tq} a_{1Y}t(T_2\omega_{20}t/2J_2-\omega_{20}{}^2- T_2\theta_{2p}/J_2)\,dt \right.\\ &\quad \left. +\int_{\theta 2q}^{2\theta 2p-\theta 2q} a_{1Y}(\theta_2-\theta_{2p})d\theta_2/(R_{b20}/i_0)\right]/(2(t_p-t_q)) \\ &= -M(R_{b20}/i_0)\,a_{1Y}\{T_2\omega_{20}(2t_p{}^2+2t_pt_q-t_q{}^2)/(6J_2)+\omega_{20}{}^2t_p\}/J_1 \end{aligned}$$

6. 5. 5 Impact loads and dynamic increment of tooth load

The impact loads are obtained from Eq. (6.4-6) as follows, where τ is the time of impact (infinitesimal).

$$\left.\begin{aligned}
F_{dSC} &= J_1\{d\theta_{1Z}(-t_q)/dt - d\theta_{1YS}(-t_q)/dt\}/(R_{b20}\cdot\tau/i_0)\\
&= J_1\{\triangle\omega_{1Z}(-t_q) - \triangle\omega_{1Y}(-t_q+2t_p)\}/(R_{b20}\cdot\tau/i_0)\\
&= -M\,[\omega_{20}{}^2\{a_{1Y}t_p+(a_{1Z}-a_{1Y})t_q\}\\
&\qquad -T_2\omega_{20}\{t_q{}^2(a_{1Y}-a_{1Z})+a_{1Y}t_p(t_p-2t_q)\}/(3J_2)]\,/\tau\\
F_{dEA} &= J_1\{d\theta_{1YE}(t_q)/dt - d\theta_{1Z}(t_q)/dt\}/(R_{b20}\cdot\tau/i_0)\\
&= J_1\{\triangle\omega_{1Y}(t_q) - \triangle\omega_{1Z}(t_q)\}/(R_{b20}\cdot\tau/i_0)\\
&= -M\,[\omega_{20}{}^2\{a_{1Y}t_p+(a_{1Z}-a_{1Y})t_q\}\\
&\qquad +T_2\omega_{20}\{t_q{}^2(a_{1Y}-a_{1Z})+a_{1Y}t_p(t_p-2t_q)\}/(3J_2)]\,/\tau
\end{aligned}\right\}\quad (6.5\text{-}8)$$

Dynamic increment of tooth load F_{q2d} during one pitch is obtained from Eq. (6.4-8) as follows.

$$\left.\begin{aligned}
F_{q2d} &= F_{q2dYS} && (-t_p < t < -t_q)\\
&= F_{dSC} && (-t_q \leqq t \leqq -t_q+\tau)\\
&= F_{q2dZ} && (-t_q < t < t_q)\\
&= F_{dEA} && (t_q \leqq t \leqq t_q+\tau)\\
&= F_{q2dYE} && (t_q < t < t_p)
\end{aligned}\right\}\quad (6.5\text{-}9)$$

Dynamic increment of tooth load F_{q2d} is a periodic function given by Eq. (6.5-9), which is shown schematically in Fig. 6.5-2.

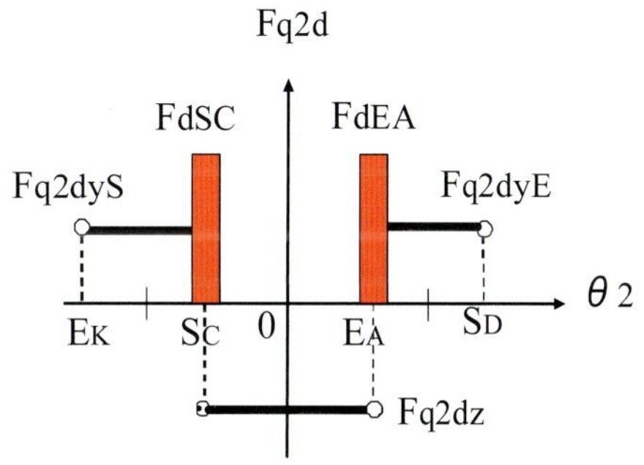

Fig.6.5-2 Solutions of the equations of motion and the impact loads

6．5．6 Dynamic increment of tooth load expressed by Fourier series

F_{q2d} is expressed by the following Fourier series because of a periodic function defined during $[-t_p,\ t_p]$, which is fragmentary but smooth.

$$F_{q2d}(t) = \sum_{n=1}^{\infty}\{A_n\cos(n\pi t/t_p) + B_n\sin(n\pi t/t_p)\} \quad (6.5\text{-}10)$$

$$\begin{aligned}
A_n &= \int_{-tp}^{tp} F_{q2d}(\xi)\cos(n\pi\xi/t_p)\,d\xi/t_p\\
&= -2M\omega_{20}{}^2\,[\{(a_{1Z}-a_{1Y})t_q/t_p+a_{1Y}\}\cos(n\pi t_q/t_p)-\{(a_{1Z}-a_{1Y})/(n\pi)\}\sin(n\pi t_q/t_p)]
\end{aligned}$$

$$\begin{aligned}
B_n &= \int_{-tp}^{tp} F_{q2d}(\xi)\sin(n\pi\xi/t_p)\,d\xi/t_p\\
&= -2M(T_2\omega_{20}/J_2)\,[\{(a_{1Y}-a_{1Z})t_q-a_{1Y}t_p\}(1/(n\pi))\cos(n\pi t_q/t_p)\\
&\quad +\{(a_{1Y}-a_{1Z})(t_q{}^2/(3t_p)-t_p/(n\pi)^2)+a_{1Y}(t_p-2t_q)/3\}\sin(n\pi t_q/t_p)]
\end{aligned}$$

Simplifying A_n and B_n, Eq. (6.5-11) is obtained as follows, where $\zeta = t_q/t_p$ $(0 \leqq \zeta \leqq 1)$.

$$\left.\begin{aligned}
A_n &= -2M\omega_{20}^2 U_n \\
U_n &= \{(a_{1Z}-a_{1Y})\zeta + a_{1Y}\}\cos(n\pi\zeta) - \{(a_{1Z}-a_{1Y})/(n\pi)\}\sin(n\pi\zeta) \\
B_n &= -2M(T_2\theta_{2p}/J_2)V_n \\
V_n &= (1/(n\pi))\{(a_{1Y}-a_{1Z})\zeta - a_{1Y}\}\cos(n\pi\zeta) \\
&\quad +\{(a_{1Y}-a_{1Z})(\zeta^2/3-1/(n\pi)^2) + a_{1Y}(1-2\zeta)/3\}\sin(n\pi\zeta)
\end{aligned}\right\} \tag{6.5-11}$$

U_n and V_n are coefficients which depend on the base circle variation coefficients (a_{1Z} and a_{1Y}) and the position (ζ) of the point where the region of mesh changes, which means that they depend on the amount of tooth surface modification and the load.

The amplitude F_{q2dn} of the n-th order is given by Eq. (6.4-10) as follows, where a_{1X} is substituted for a_{1Z} when the path of contact is in the region of one pair mesh and a_{1W} is for a_{1Y} when it passes in the region of four pair mesh.

$$F_{q2dn} = 2M\sqrt{[(\omega_{20}^2 U_n)^2 + \{(T_2\theta_{2p}/J_2)V_n\}^2]} \tag{6.5-12}$$

where $M = J_1J_2/\{R_{b20}^2(J_1+J_2/i_0^2)\}$,

$U_n = \{(a_{1X}-a_{1Y})\zeta + a_{1Y}\}\cos(n\pi\zeta) - \{(a_{1X}-a_{1Y})/(n\pi)\}\sin(n\pi\zeta) \quad (1\leqq m_p\leqq 2)$

$\quad = \{(a_{1Z}-a_{1Y})\zeta + a_{1Y}\}\cos(n\pi\zeta) - \{(a_{1Z}-a_{1Y})/(n\pi)\}\sin(n\pi\zeta) \quad (2\leqq m_p\leqq 3)$

$\quad = \{(a_{1Z}-a_{1W})\zeta + a_{1W}\}\cos(n\pi\zeta) - \{(a_{1Z}-a_{1W})/(n\pi)\}\sin(n\pi\zeta) \quad (3\leqq m_p\leqq 4)$,

$V_n = (1/(n\pi))\{(a_{1Y}-a_{1X})\zeta - a_{1Y}\}\cos(n\pi\zeta)$

$\quad +\{(a_{1Y}-a_{1X})(\zeta^2/3-1/(n\pi)^2) + a_{1Y}(1-2\zeta)/3\}\sin(n\pi\zeta) \quad (1\leqq m_p\leqq 2)$

$\quad = (1/(n\pi))\{(a_{1Y}-a_{1Z})\zeta - a_{1Y}\}\cos(n\pi\zeta)$

$\quad +\{(a_{1Y}-a_{1Z})(\zeta^2/3-1/(n\pi)^2) + a_{1Y}(1-2\zeta)/3\}\sin(n\pi\zeta) \quad (2\leqq m_p\leqq 3)$

$\quad = (1/(n\pi))\{(a_{1W}-a_{1Z})\zeta - a_{1W}\}\cos(n\pi\zeta)$

$\quad +\{(a_{1W}-a_{1Z})(\zeta^2/3-1/(n\pi)^2) + a_{1W}(1-2\zeta)/3\}\sin(n\pi\zeta) \quad (3\leqq m_p\leqq 4)$,

$n = 1, 2, 3, \cdots$,

$\zeta = 2 - m_p \quad (1\leqq m_p\leqq 2)$

$\quad = m_p - 2 \quad (2\leqq m_p\leqq 3)$

$\quad = 4 - m_p \quad (3\leqq m_p\leqq 4)$,

when $m_p \geqq m_{p0}$, $m_p = m_{p0}$,

ω_{20} : mean angular velocity of gear II,

m_p and m_{p0} : actual and theoretical contact ratios,

a_{1X}, a_{1Y}, a_{1Z} and a_{1W} : base circle variation coefficients of one, two, three and four pair mesh.

Under the conditions of high velocity and light load, F_{q2dn} can be approximated effectively by the term including ω_{20}^2 only in Eq. (6.5-12) because of $\omega_{20}^2 \gg (T_2\theta_{2p}/J_2)$, so that Eq. (6.5-13) is obtained as follows.

$$F_{q2dn} = 2M\omega_{20}^2 |U_n| \tag{6.5-13}$$

Therefore, the dynamic increment of tooth load F_{q2d} caused by the rotational motion of the gear pair is expressed by Eqs. (6.4-10) and (6.5-12) as follows, where $\alpha_n = \tan^{-1}(A_n/B_n)$.

$$F_{q2d}(t) = \sum_{n=1}^{\infty} F_{q2dn}\sin(n\pi\omega_{20}t/\theta_{2p}+\alpha_{n}) \qquad (6.5\text{-}14)$$

6．5．7 Amplification of F_{q2dn} through the gear vibration system

When the gear pair is rigidly supported with bearings, the amplitude F_{In} of n-th order of the load amplified by the gear vibration system composed of the tooth pair and the shaft is expressed by Eq.(6.4-15) as follows.

$$F_{In} = I_{bn}(\omega)I_{tn}(\omega)G_{n}(\omega)F_{q2dn} \qquad (6.5\text{-}15)$$

where $G_{n}(\omega)$: transfer function of n-th order of the gear vibration system composed of the tooth-pair spring and the equivalent mass and
$I_{bn}(\omega)$ and $I_{tn}(\omega)$: transfer functions of n-th order of the gear vibration system composed of the bending and torsional springs of the shaft and the equivalent mass respectively.

Since the main dimensions of tooth pairs and shafts and the disposition of bearings are determined approximately through the requirements for their performances, their strength and their resonance frequencies, when a gear apparatus is designed, $I_{bn}\cdot I_{tn}\cdot G_{n}$ are almost determined automatically at the same time. Therefore, in order to assure an allowable level of F_{In} in the broad range of frequency of the gear apparatus, it is important that

(1) The dimensions of gears and bearings are designed so as to make the transfer functions $I_{bn}\cdot I_{tn}\cdot G_{n}$ smaller and
(2) The tooth surfaces are modified to reduce the amplitude F_{q2dn} of the dynamic increment of tooth load acting on the gear vibration system, which is effective as an index of the dynamic performance of gears.

6．5．8 Summary

The newly developed method is applied to a gear pair with a single flank error composed of symmetrical convex curves which is the most practical and an algebraic equation for the amplitude of the dynamic increment of tooth load is derived, which is given by a comparatively simple expression relating the following factors.

(1) Dimensions of gears (moment of inertia, radius of base cylinder and number of teeth)
(2) Amount of tooth surface modification and the tooth-pair spring constant
(3) Operating conditions (rotational velocity and torque)

6．6 Verification by experiment through helical gear pairs having a single flank error composed of symmetrical convex curves (65)

In this section, the dynamic increment of tooth load obtained by Eq.(6.5-12) is verified by experiment through involute helical gears and it is shown that it can be used as an index for dynamic performances.

6. 6. 1 Experiment

(1) Test gear pairs

Three pairs of helical gears with different tooth surface modifications are manufactured for test gear pairs. Table 6.6-1 shows their dimensions, the tooth-pair spring constant given by the approximate equation of Umezawa et al.(67) They are ground after heat treatment into three different convex tooth surfaces, accuracies of which are shown in Fig. 6.6-1.

Table 6.6-1 Dimensions of a pair of test samples

	Gear I		Gear II
Number of teeth	37		35
Center distance (mm)		122.00	
Normal module		2.895	
Normal pressure angle (deg.)		16.779	
Helix angle (deg.)		31.306	
Pitch diameter (mm)	125.39		118.61
Base diameter (mm)	118.24		111.85
Outside diameter (mm)	133.05		125.70
Face width (mm)		20.00	
Transverse contact ratio		1.740	
Overlap contact ratio		1.085	
Spring rate (N/μm)	$K(\theta_2)=103\exp(-185\|\theta_2\|^3)$		

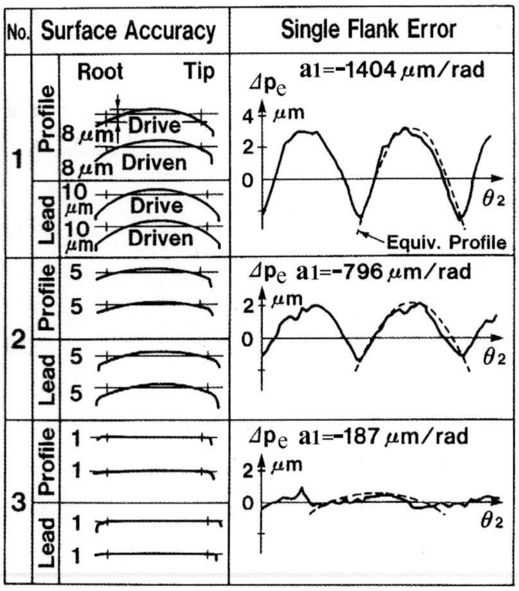

Fig.6.6-1 Single flank errors of test samples

Surface variations and pitch errors in the same gear pair are negligibly small. Broken parabolas indicate the approximate curves corresponding to the single flank errors of the gear pairs and a_1 denotes the base-circle variation coefficient under no loads. The errors of profile and lead are measured and compared in the same effective range respectively.

(2) Calculated results of the amplitude F_{q2dn} of the dynamic increment of tooth load

The amplitude F_{q2dn} of the dynamic increment of tooth load for a gear pair with symmetrical convex tooth surfaces is given by Eq. (6.5-12), in which U_n, V_n and ζ are determined by transmitted torque T_2, the tooth-pair spring constant $K(\theta_2)$ and the base circle variation coefficient a_1 under no loads which represents a tooth surface modification. Therefore, F_{q2dn} is a function composed of five variables, M (equivalent mass of the gear pair), K, a_1, T_2 and ω_{20}. In this experiment, the gear pairs have the same dimensions (the same M and K) and three different surface modifications, which means that F_{q2dn} becomes a function of two variables, T_2 and ω_{20}, with a parameter a_1.

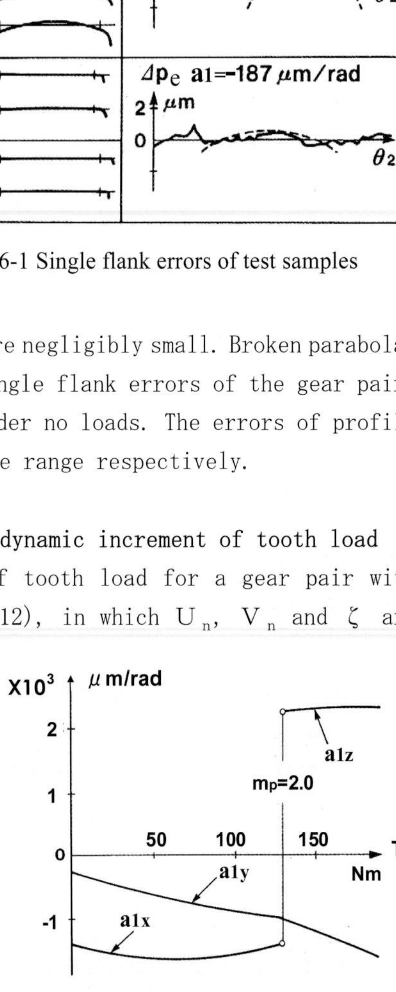

Fig.6.6-2(a) Base circle coefficients a_{1x}, a_{1y} and a_{1z}

Figures 6.6-2(a), (b) and (c) show the calculated

results of the base circle variation coefficients (a_{1x}, a_{1y} and a_{1z}), actual contact ratio m_p, ζ, U_n and V_n of the test gear pair 1. The increase of load varies each base-circle variation coefficient monotonously as shown in Fig. 6. 6-2(a), while the variation of m_p gives a sort of periodic change to ζ in Fig. 6. 6-2(b) and results in the complicated load characteristics of U_n and V_n as shown in Fig. 6. 6-2(c). In this example, m_p reaches barely 2. 0 at T_2 = 130 Nm on account of its large amount of modification. When m_p reaches the theoretical contact ratio according to the increase of T_2, ζ becomes constant and U_n and V_n become monotonous functions corresponding to the variations of base circle variation coefficient. (Note 6. 6-1).

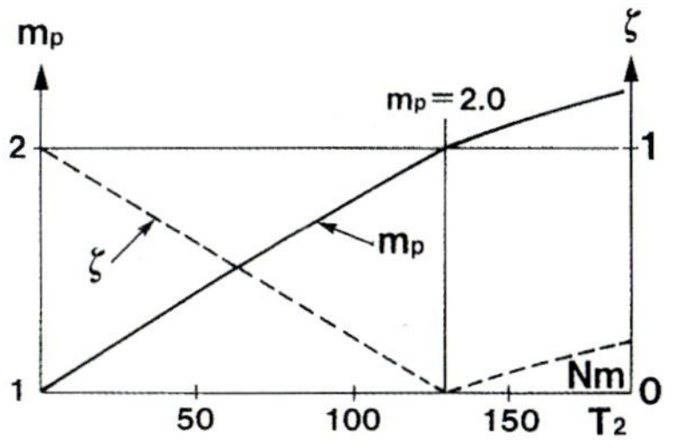

Fig.6.6-2(b) Actual contact ratio mp and ζ

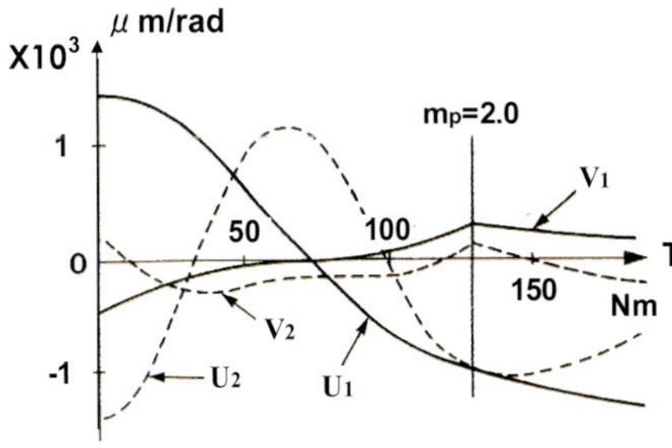

Fig.6.6-2(c) U_n and V_n

Figures 6. 6-3(a) and (b) show the calculated F_{q2dn} based on Fig. 6. 6-2, which is a function of T_2 and ω_{20} with the parameter a_1. F_{q2dn} is a monotonous function with regard to ω_{20}, while it fluctuates complicatedly with the increase of T_2 until the actual contact ratio reaches the theoretical one. (Note 6. 6-2).

Figure 6. 6-4 shows the influence of the tooth surface modification which is represented by a_1. Bigger modification (algebraically smaller a_1) gives influence to the load characteristics of F_{q2dn} in such a manner that it increases the value of F_{q2dn} itself and the load T_2 giving the minimum F_{q2dn}. In the case of a_1 = −187 μ m/rad, m_{po} = 2. 825 (theoretical) is realized around T_2 = 50 Nm, therefore F_{q2dn} becomes monotonous in the region of more than T_2 = 50 Nm.

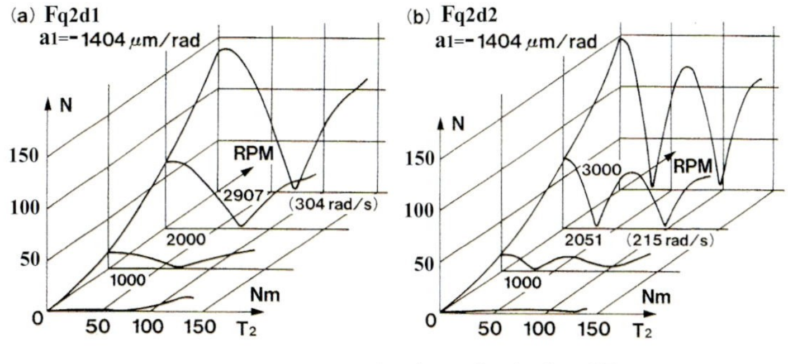

Fig.6.6-3(a) Schematic view of calculated F_{q2dn}

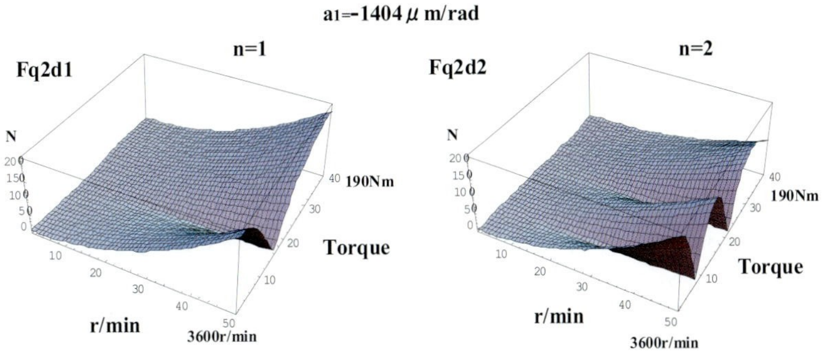

Fig.6.6-3(b) Schematic view of calculated F_{q2dn}

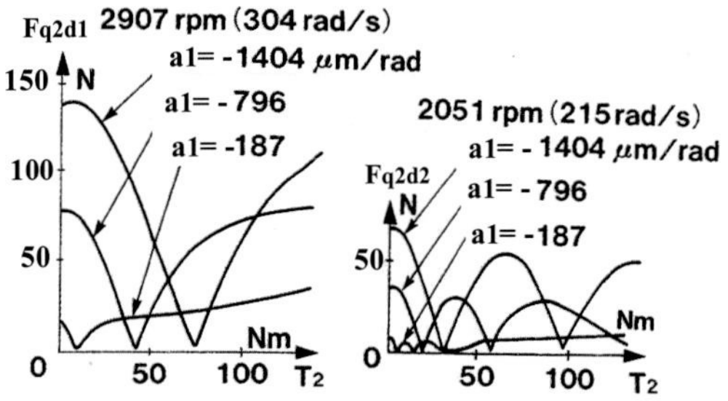

Fig.6.6-4 Variation of calculated F_{q2dn} due to a_1 and T_2

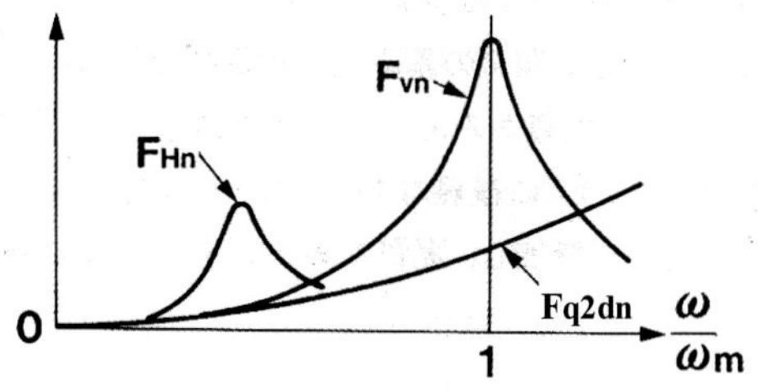

Fig.6.6-5 Relationship among F_{q2dn}, F_{vn} and F_{Hn}

(3) Method of verifying F_{q2dn}

The dynamic increment of tooth load which is the external load acting on the tooth surfaces forces the gear pair itself and the other components (the shaft, the housing and so on) to vibrate and causes gear noise and vibration.

Figure 6.6-5 shows schematically the relationships among F_{q2dn}, F_{vn} amplified through the gear pair and F_{Hn} amplified through the other components (the shaft and the housing), where ω_m denotes the resonance velocity of the gear pair.

Using Eq. (6.5-15), F_{vn} and F_{Hn} can be expressed as follows.

$$\left.\begin{aligned} F_{vn} &= G_n(\omega)F_{q2dn} \\ F_{Hn} &= H_n(\omega)F_{vn} = H_n(\omega)G_n(\omega)F_{q2dn} \\ &= \lambda_n(\omega)F_{q2dn} \\ \lambda_n(\omega) &= H_n(\omega)G_n(\omega) \end{aligned}\right\} \qquad (6.6\text{-}1)$$

where $\omega = \pi/t_p = \pi\,\omega_{20}/\theta_{2p}$ and

$G_n(\omega)$ and $H_n(\omega)$: transfer functions of the gear pair and the shaft and housing.

Based on Eq. (6.6-1), F_{q2dn} is verified by measuring F_{Hn} as follows.

(a) Verification of the load characteristic $F_{q2dn}(T_2)$

When angular velocity is fixed to a certain value ω_{20} chosen arbitrarily, $\lambda_n(\omega)$ becomes constant, so that the measured F_{Hn} must always be proportional to the calculated F_{q2dn} under any load at a certain ω_{20} and a_1.

(b) Verification of the velocity characteristic $F_{q2dn}(\omega_{20})$

The result of item (a) gives a constant of proportionality $\lambda_n(\omega)$ between F_{Hn} and $F_{q2dn}(\omega_{20})$ under various loads at a certain ω_{20} and a_1. By varying ω_{20}, a corresponding $\lambda_n(\omega)$ can be obtained in the same way, so that $F_{Hn}/\lambda_n(\omega)$ must always be equal to the calculated $F_{q2dn}(\omega_{20})$ under a certain load T_2 and a_1.

(c) Verification of the surface modification characteristic $F_{q2dn}(a_1)$

According to Eq. (6.5-12), a_1 varies U_n and V_n only which have no relation with ω_{20}, so that it is obtained by comparing the load characteristics $F_{q2dn}(T_2)$ under different a_1s at a certain ω_{20} as shown in Fig, 6.6-4.

(4) Method of measuring F_{Hn}

Figure 6.6-6 shows the method of measuring F_{Hn} schematically. Test gear pairs are put in the same gear box and the sound pressure is measured through a fixed microphone at different loads by changing the velocity continuously.

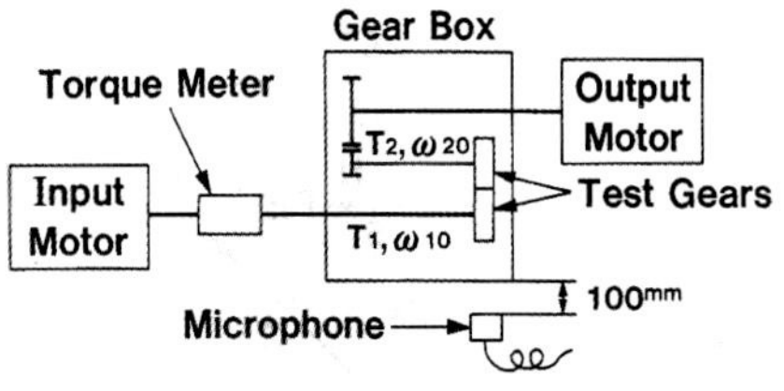

Fig.6.6-6 Schematic view of test apparatus

Figure 6.6-7 is an example of the results which shows the first component of the tooth mesh frequency under a certain load. One of the most remarkable resonance peaks which always appears at any load regardless of the test gear pairs is selected (the peak around 2900r/min, shown by ＊ in this example) and the F_{Hn} is measured by changing the load at this peak.

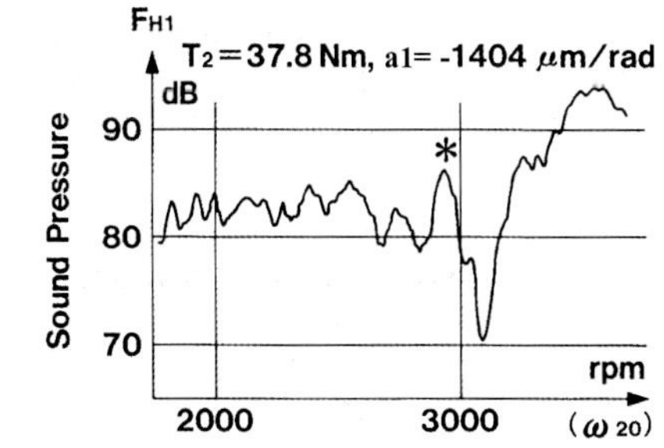

Fig.6.6-7 Example of measured sound pressure F_{H1}

6．6．2 Results of the experiment

Figure 6.6-8(1) shows the measured F_{H1} (the first component of the tooth mesh frequency) with the calculated F_{q2d1} in the broken curve, while Fig.6.6-8(2) shows the measured F_{H2} (the second component) with the calculated F_{q2d2}. In the region where the load is too small, for example from 0 to 15 Nm in the case of (a), the measured values are omitted from the discussion because of their lack of reliability. Resistance torque due to lubricants and bearings are neglected.

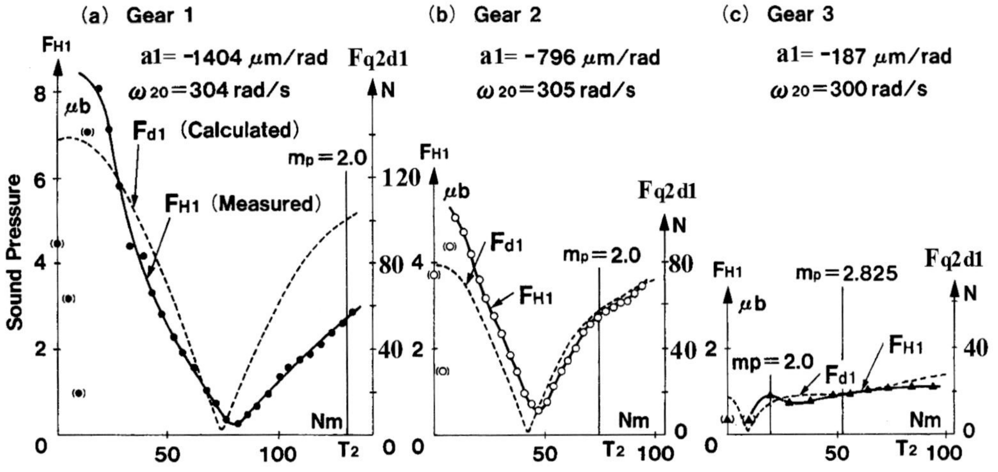

Fig.6.6-8(1) Measured F_{H1} and calculated F_{q2d1}

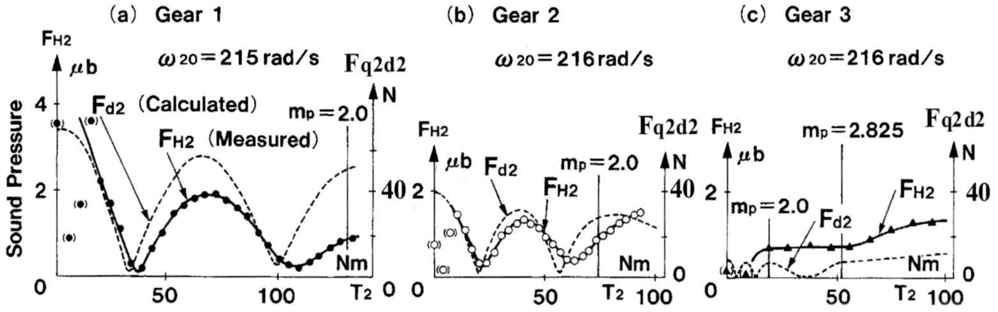

Fig.6.6-8(2) Measured F_{H2} and calculated F_{q2d2}

(1) Load characteristics

Figures 6.6-8(1)(a) and 6.6-8(2)(a) show that the measured curves, F_{H1} and F_{H2} have a good correlation with the calculated ones, F_{q2d1} and F_{q2d2} respectively. The same good correlations are also shown in Figs. 6.6-8(1)(b) and (2)(b) and Figs. 6.6-8(1)(c) and (2)(c) for other gear pairs with different amounts of tooth surface modification.

(2) Surface modification characteristics

Comparison among Figs. 6.6-8(1)(a), (b) and (c) shows that the relationships between F_{q2d1} and a_1 shown on Fig. 6.6-4 are verified by those between the measured F_{H1} and a_1. Figures 6.6-8(2)(a), (b) and (c) show the same for the second component.

(3) Angular velocity characteristics

Figure 6.6-9 shows the angular velocity characteristics of F_{q2d1} for the test gear 1. Six resonance points are chosen in the region from 2000 to 3500r/min, where each constant of proportionality λ_1 between F_{H1} and F_{q2d1} is obtained as shown in Fig. 6.6-9 (a) and F_{H1}/λ_1 is plotted with the calculated F_{q2d1} (broken curve) under three different loads in Fig. 6.6-9(b). F_{H1}/λ_1 has a good correlation with the F_{q2d1} except at T_2 = 75.7 Nm, where F_{q2d1} has the minimum value which could not be caught by the corresponding F_{H1} because it was too small compared with the surrounding noise.

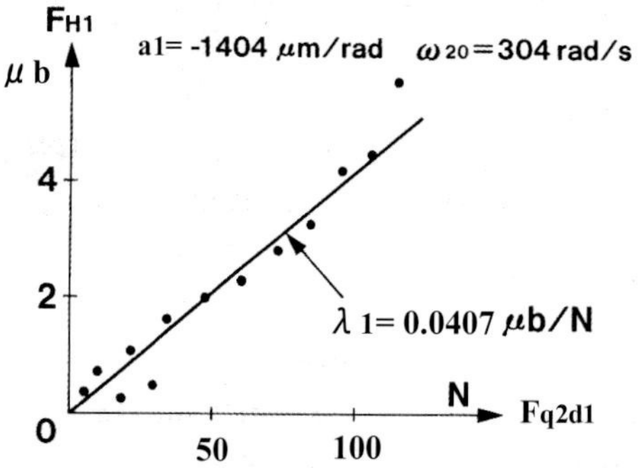

Fig.6.6-9(a) Ratio λ_1 between F_{H1} and F_{q2d1}

(4) Correlation between F_{Hn} and F_{q2dn}

Figure 6.6-10 shows the correlation between the calculated F_{q2dn} and the measured F_{Hn}

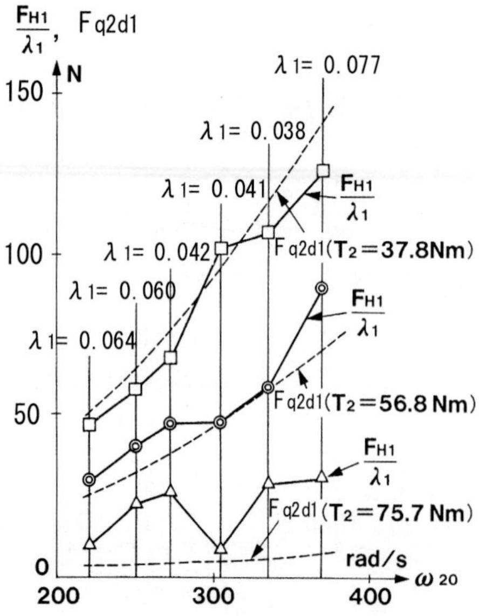

Fig.6.6-9(b) Correlation between F_{H1}/λ_1 and F_{q2d1}

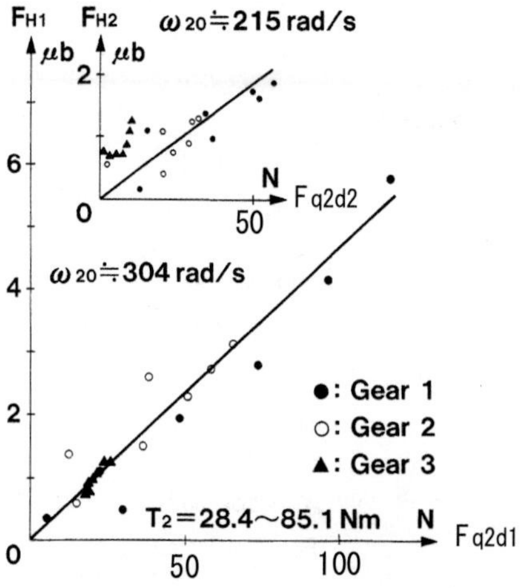

Fig.6.6-10 Relations between F_{Hn} and F_{q2dn}

in Figs. 6.6-8(1) and 6.6-8(2), where seven points are plotted every 9.5 Nm in the load from 28 Nm to 85 Nm. It shows that there are a good correlation between F_{H1} and F_{q2d1} and also a fair correlation between F_{H2} and F_{q2d2}.

6．6．3 Index of dynamic performance of gear pairs

The amplitude F_{vn} of the tooth load amplified by vibration of the gear pair forced by F_{q2dn} is given by Eq. (6.4-15) as follows.

$$F_{vn} = G_n(\omega)F_{q2dn} \leqq F_{q2S}$$
$$G_n(\omega) = 1/\sqrt{[\{1-(n\omega/\omega_m)^2\}^2 + 4\gamma^2(n\omega/\omega_m)^2]}$$

Therefore, when the maximum value of F_{vn} is F_{Rn}, it is given as follows.

$$F_{Rn} = F_{q2dn}/(2\gamma) \leqq F_{q2S} \qquad (6.6\text{-}2)$$

These F_{vn} and F_{Rn} are considered to be the objects of the present dynamic load theories. F_{q2dn} is obtained by the modification of tooth surface (a_1) and the operating conditions (ω_{20} and T_2), so that the amplitude of dynamic increment of tooth load is calculated by Eq. (6.6-2), when the damping coefficient (γ) of the gear pair can be given. For example, if $\gamma = 0.07$, the maximum amplitude F_{Rn} of dynamic increment of tooth load at the resonance point is obtained simply as follows.

$$F_{Rn} = 7F_{q2dn} \leqq F_{q2S}$$

The results mentioned above indicate that F_{q2dn} proposed in this theory can be an index of the dynamic performances like the vibration and noise and the dynamic load of gear pairs.

In designing the dynamic strength of gears, the dimensions of gears, the modification of tooth surface, the operating conditions and the damping coefficient must be chosen for F_{vn} and F_{Rn} not to be beyond F_{q2S}. And a gear pair which has good dynamic performances and rotates smoothly means that it has F_{q2dn} as small as possible.

6．6．4 Variation of bearing loads caused by static load

The variation of bearing loads mentioned in 6.4.6 is discussed here in the example of involute helical gears, which is caused by the variation of points of action or inclination angles of the static load F_{q2S}.

In involute helical gears, the variations $\Delta\chi_2$, $\Delta\phi_{b2}$ and ΔR_{b2} caused by tooth modification and deflection are negligibly small, so that the calculated variations of bearing loads caused by them are sufficiently small compared with those by the dynamic increment of tooth load F_{q2d}. Therefore, here, the variation of bearing load ΔB_{u2f} by Eq. (6.4-25) based on the wrong selection (η_{b20}) of tooth bearing is calculated, where $\eta_{b20} = 40.0°$ is given against $\eta_{b20} = \phi_{b20} = 29.8°$ which realizes no variation of bearing loads.

Figure 6.6-11 shows the case of tooth modification $a_1 = -1404\,\mu$m/rad, where (a) shows F_{q2d1} (same as Fig. 6.6-3(b)) and (b) shows ΔB_{u2f} which is shown by dividing the torque from 4.73 to 170 Nm into four equal parts and one pitch into 16 equal parts. Fig. 6.6-11 shows that ΔB_{u2f} is negligibly small compared with F_{q2d1} in the region of high rotational velocity such as around 3000r/min. Compared with Fig.6.6-8(1), there may be some influence in the region where F_{q2d1} becomes the minimum but it is thought to be negligible generally.

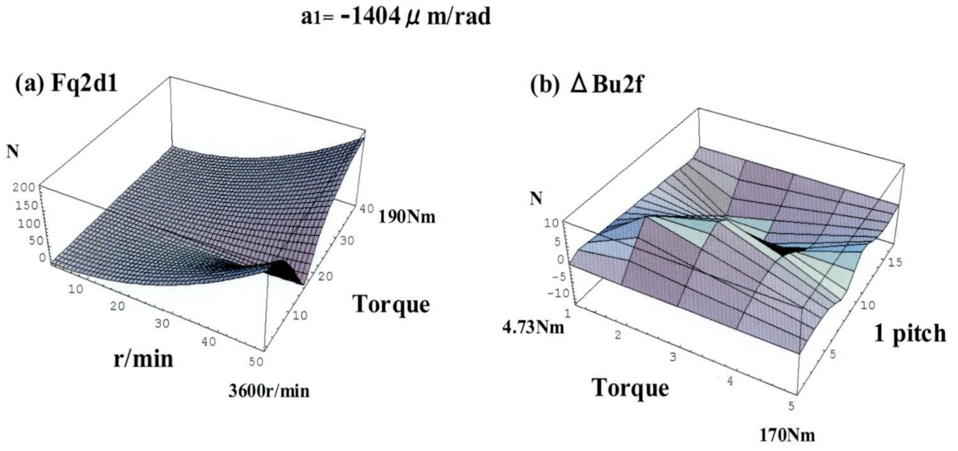

Fig.6.6-11 Variation $\triangle B_{u2f}$ of bearing load due to static load

Figure 6.6-12 shows the case of a_1 = $-187\ \mu$m/rad, where F_{q2d1} becomes smaller, but $\triangle B_{u2f}$ becomes much smaller than F_{q2d1}, therefore the influence is thought to be negligible.

The problem remains in the case of gear pairs with tooth surfaces except involute helicoids.

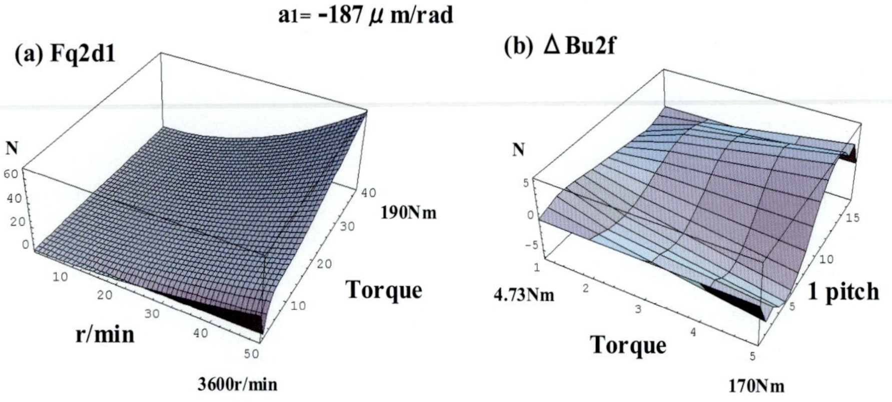

Fig.6.6-12 Variation $\triangle B_{u2f}$ of bearing load due to static load

6. 6. 5 Summary

The validity of the amplitude F_{q2dn} of dynamic increment of tooth load obtained by Eq.(6.5-12) in 6.5 is verified by experiment through involute helical gears.

(1) F_{q2dn} is a function of two variables, velocity and operating load, with the parameter of tooth surface modification, when the dimensions and material of a gear pair are given. The characteristics of load, velocity and surface modification are clarified and verified by experiment.

(2) F_{q2dn} is an index of dynamic performances, such as dynamic load and vibration.

(3) The variation of bearing loads caused by the variation of points of action or inclination angles of the static load F_{q2s} is negligibly small compared with F_{q2dn} in involute helical gears.

(Note 6.6-1) : Figures 6.6-2(a) and (c) are modified to reference (65) because of (Note 6.1-1).
(Note 6.6-2) : F_{q2dn} in the figures after Fig. 6.6-3 are multiplied by 9.8 to F_{q2dn} of reference (65), because the author mistook the variable M in reference (65).

6．7　Summary and references

When a gear pair which has involute helicoids for tooth surfaces II (elastic) and their conjugate ones for tooth surfaces I (rigid) is given, the path of contact with its common contact normal at each point between the modified surfaces II which are modified slightly and convexly from tooth surfaces II and tooth surfaces I is obtained, the equations of motion are solved and the dynamic increment of tooth load is obtained and verified by experiment, which is summarized as follows.

(1) A path of contact with its common contact normal at each point between the modified tooth surfaces II and tooth surfaces I under no loads is obtained as functions of the angle of rotation expressed by the deviation of tangent plane and the inclination angle of the path of contact caused by the surface modification, which are determined by measuring the single flank error and the helix form deviation. And the fundamental requirements for multiple tooth pair contact of the gear pair are clarified, which says that the deviations of tangent plane and the radii of base circle at the simultaneous points of contact must be equal.

(2) The equivalent tooth load is introduced, under which the path of contact with its common contact normal at each point between the modified tooth surfaces II and tooth surfaces I, the regions of multiple tooth pair mesh and the actual contact ratio are obtained. The path of contact under load is generally discontinuous at the points where the region of mesh changes.

(3) The equations of motion of the gear pair are clarified. The equations of motion of all kinds of gear pairs from helical to hypoid gears having involute helicoids for one member can be expressed by those of the involute spur gears whose radii of base cylinder are R_{b20}/i_0 and R_{b20}. So that the solution, namely the dynamic increment of tooth load is also expressed the same from cylindrical to hypoid gears. In addition, the approximation accuracy of the dynamic model of this theory is examined, compared with a stricter one which has surface modifications and deflections on both members.

(4) The equations of motion are solved algebraically to obtain the dynamic increment of tooth load. The dynamic increment of tooth load during one pitch is a periodic function composed of the dynamic increments of tooth load in the regions of mesh and the impact loads at the discontinuous points where the region of mesh changes and is expressed by a Fourier series, which is the external load acting on the gear vibration system. In addition, the variation of bearing loads is calculated,

(5) The new method mentioned above is applied to a gear pair with a single flank error composed of symmetrical convex curves which are the most practical and an algebraic equation for the amplitude of the dynamic increment of tooth load is derived, which is given by a comparatively simple expression relating (a) dimensions of gears (moment of inertia, radius of base cylinder and number of teeth), (b) amount of tooth surface modification and the tooth-pair spring constant and (c) operating conditions (rotational velocity and torque).

(6) The validity of the amplitude F_{q2dn} of the dynamic increment of tooth load obtained by item (5) is verified by experiment through involute helical gears. F_{q2dn} is a function

of two variables, velocity and operating load, with the parameter of tooth surface modification, when the dimensions and material of a gear pair are given. The characteristics of load, velocity and surface modification are clarified and verified by experiment. The variation of bearing loads caused by that of the static load is negligibly small compared with F_{q2dn} in involute helical gears, judging from the results of the experiment and the calculated bearing loads.

References

(61) Honda, S., "Basic theory on tooth contact and dynamic loads of gears (A path of contact of a gear pair having modified involute helicoids for one member)", JSME Int.J, Series C, Vol.42, No.4 (1999), 991-1002.

(62) Honda, S., "Basic theory on tooth contact and dynamic loads of gears (Dynamic incremental loads and variation of bearing loads of a gear pair having modified involute helicoids for one member)", JSME Int.J, Series C, Vol.43, No.2 (2000), 445-454.

(63) Honda, S., "Rotational vibration of a helical gear pair with modified tooth surfaces (Modified tooth surface and its equivalent tooth profile)", JSME Int.J, Series C, Vol.36, No.1 (1993), 125-134.

(64) Honda, S., "Rotational vibration of a helical gear pair with modified tooth surfaces (Rotational motion of a gear pair and its dynamic increment of tooth load)", JSME Int.J, Series C, Vol.36, No.3 (1993), 375-385.

(65) Honda, S., "Rotational vibration of a helical gear pair with modified tooth surfaces (Verification of the new theory by experiment and a new design method for dynamic performance)", JSME Int.J, Series C, Vol.38, No.1 (1995), 112-121.

(66) 会田俊夫，佐藤進，由井雄二郎，福間洋，歯車の振動，騒音に関する基礎的研究，機論，34-268(1968-12), 2226, 2237, 2246, 2254.

(67) Umezawa, K., Suzuki, T. and Sato, T., "Vibration of Power Transmission Helical Gears (Approximate Equation of Tooth Stiffness)", Bulletin of JSME, Vol.29, No.251, May 1986, 1605-1611.

7. Design of a pair of tooth surfaces transmitting inconstant rotational motion

With regard to gear pairs with an inconstant gear ratio, methods of design and manufacture have been discussed in theories of non-circular gears [71], but the surface of action and the tooth surfaces have not been calculated concretely as far as the author knows. In this chapter, the calculation methods of tooth surfaces of gear pairs with parallel axes ($\Sigma = 0$) are discussed concerning the two items mentioned below and it is indicated that the new tooth geometry can be applied to gear pairs with an inconstant gear ratio.

(1) In 7.1, how to design tooth surfaces I and II and a tooth surface of a rack transmitting inconstant rotational motion under a given center distance.

(2) In 7.2, how to design a tooth surface of a rack driven by a screw with constant lead and tooth surface I (pinion) which transmit inconstant rotational motion. They are used practically for automobile steering gears.

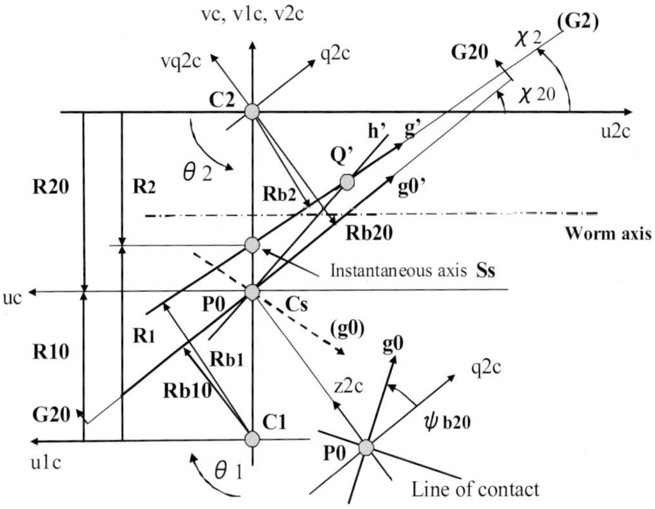

Fig.7.1-1(a) Path of contact **h** in the coordinate systems C_2 and Cq_2

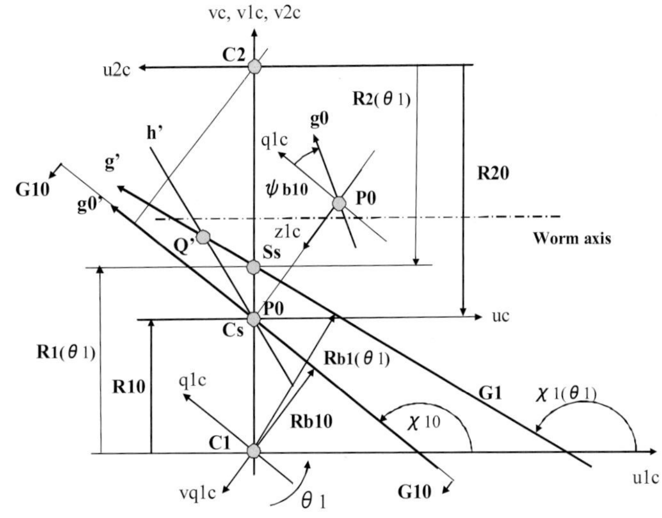

Fig.7.1-1(b) Path of contact **h** in the coordinate systems C_1 and Cq_1

7.1 Tooth surfaces I and II and a tooth surface of a rack transmitting inconstant rotational motion under a given center distance

Figure 7.1-1(a) shows a design point P_0, common contact normals $\mathbf{g}_0$ and $\mathbf{g}$ and a path of contact $\mathbf{h}$ in the coordinate systems C_1, C_S and C_2 when $\Sigma = 0$, where the sign (') means the orthographic projection. Figure 7.1-1(a) is shown from the positive direction of the axis z_{2c} according to the basic theory in Chapter 2 and the common contact normal $\mathbf{g}_0$ is represented by a solid or broken line.

Figure 7.1-1(b) is Fig. 7.1-1(a) seen from the positive direction of the axis z_{1c}. In this chapter, Fig. 7.1-1(b) is mainly used because gear I (pinion) is dealt with mostly.

7.1.1 Coordinate systems C_1, C_2 and C_S and design point P_0

When the center distance E, the shaft angle $\Sigma = 0$ and the reference gear ratio $i(0) = i_{S0}$ at angle of rotation $\theta_1 = 0$ are given, the design point P_0 and the coordinate systems C_1, C_2 and C_S are determined as shown in Figs. 7.1-1(a) and (b), where R_{10} and R_{20} are obtained as follows.

$$R_{10} = E/(i_{S0} + 1)$$
$$R_{20} = i_{S0} R_{10}$$

Therefore, the design point P_0 is expressed in the coordinate systems C_1, C_2 and C_S as follows.

$$P_0(0,\ R_{10},\ 0\ ;\ C_1),\quad P_0(0,\ -R_{20},\ 0\ ;\ C_2),\quad P_0(0,\ 0,\ 0\ ;\ C_S)$$

7.1.2 Common contact normal $\mathbf{g}_0$ through P_0 and coordinate systems C_{q1} and C_{q2}

The inclination angle of the common contact normal $\mathbf{g}_0$ through P_0 is given in the coordinate system C_S as follows.

$$\mathbf{g}_0(\phi_0,\ \phi_{n0}\ ;\ C_S)$$

In this theory, the direction of $\mathbf{g}_0$ coincides with that of the normal velocity when the angular velocity is positive, so that the solid line $(0 \leqq \phi_{n0} \leqq \pi/2)$ and the broken one $(\pi/2 \leqq \phi_{n0} \leqq \pi)$ in Fig. 7.1-1(a) represent the normals of both tooth surfaces at P_0.

When $\mathbf{g}_0$ is given, the plane of action G_{10} including $\mathbf{g}_0$ and the coordinate system C_{q1} are determined and the inclination angle of $\mathbf{g}_0$ is expressed by Eq. (2.1-26) in the coordinate system C_1 as follows.

$$\mathbf{g}_0(\phi_{10} = \pi/2 - \chi_{10},\ \phi_{b10}\ ;\ C_1)$$

$P_0(q_{1c0},\ -R_{b10},\ 0\ ;\ C_{q1})$ is expressed by Eq. (2.1-3) as follows, where the positive direction of the axis q_{1c} coincides with that of $\mathbf{g}_0$.

$$q_{1c0} = R_{10}\sin\chi_{10},\quad R_{b10} = -R_{10}\cos\chi_{10}$$

The plane of action G_{20} including $\mathbf{g}_0$ and the coordinate system C_{q2} are obtained in the same way.

7.1.3 Path of contact and instantaneous gear ratio $i(\theta_1)$

(1) Path of contact

In Fig. 7.1-1(b), when the common contact normal $\mathbf{g}_0$ moves to $\mathbf{g}$ according to the rotation of gears, a point of contact and a path of contact $\mathbf{h}$ are represented by Q and P_0Q

respectively.

The path of contact **h** is expressed by Eq. (2.3-2), the variables of which satisfy the following conditions.

(a) The inclination angle $\psi_{b1}(\theta_1)$ on the plane of action G_1 including **g** $(\phi_1(\theta_1) = \pi/2 - \chi_1(\theta_1), \psi_{b1}(\theta_1); C_1)$ is always constant, where θ_1 is the angle of rotation of gear I (pinion).

$$\psi_{b1}(\theta_1) = \psi_{b10}$$

(b) The inclination angle of the path of contact on the plane of action G_1 is assumed to be constant as follows.

$$\eta_{b1}(\theta_1) = \psi_{b10}.$$

(c) When $\theta_1 = 0$, the common contact normal **g** coincides with $\mathbf{g}_0$ through P_0.

The path of contact **h** satisfying the conditions mentioned above is expressed by Eq. (2.3-2) in the coordinate system C_{q1} as follows, where $P_0(q_{1c0}, -R_{b10}, 0; C_{q1})$ and $\mathbf{g}_0$ $(\phi_{10} = \pi/2 - \chi_{10}, \psi_{b10}; C_1)$ are given.

$$\left.\begin{aligned}
q_{1c}(\theta_1) &= \int_0^{\theta 1} [R_{b1}(\theta_1)(1-d\chi_1/d\theta_1)\cos^2\psi_{b10}]\, d\theta_1 + q_{1c0} \\
R_{b1}(\theta_1) &= \int_0^{\theta 1} (dR_{b1}/d\theta_1)d\theta_1 + R_{b10} \\
z_{1c}(\theta_1) &= \int_0^{\theta 1} [R_{b1}(\theta_1)(1-d\chi_1/d\theta_1)\cos\psi_{b10}\sin\psi_{b10}]\, d\theta_1 \\
\chi_1(\theta_1) &= \int_0^{\theta 1} (d\chi_1/d\theta_1)d\theta_1 + \chi_{10} = \pi/2 - \phi_1(\theta_1) \\
\mathbf{g}\,&(\phi_1(\theta_1) = \pi/2 - \chi_1(\theta_1),\ \psi_{b10};\ C_1)
\end{aligned}\right\} \quad (7.1\text{-}1)$$

Equation (7.1-1) is determined by giving $(d\chi_1/d\theta_1)$ and $(dR_{b1}/d\theta_1)$. The path of contact **h** is a curve generally, but it is approximated by a straight line in Figs. 7.1-1(a) and (b).

(2) Instantaneous gear ratio $i(\theta_1)$

In Fig. 7.1-1(b), the plane of action G_1 including the common contact normal **g** intersects the plane including the axis z_{1c} and the design point P_0 along an instantaneous axis S_S. Because the common contact normal **g** passes the instantaneous axis S_S, the instantaneous gear ratio $i(\theta_1)$ is obtained as follows, where $R_1(\theta_1)$ is the distance from the axis z_{1c} to the instantaneous axis S_S.

$$i(\theta_1) = \{E - R_1(\theta_1)\}/R_1(\theta_1) = E/R_1(\theta_1) - 1 \quad (7.1\text{-}2)$$
$$\text{or} \quad R_1(\theta_1) = E/(1 + i(\theta_1))$$

Substituting the radius of base circle $R_{b1}(\theta_1)$ for $R_1(\theta_1)$ by Eq. (2.1-3), $i(\theta_1)$ is obtained as follows.

$$i(\theta_1) = -E\cos\chi_1(\theta_1)/R_{b1}(\theta_1) - 1 \qquad (7.1\text{-}3)$$

When the gear ratio $i(\theta_{1E0}) = i_{E0}$ at $\theta_1 = \theta_{1E0}$ is given and a combination of $\chi_1(\theta_1)$ and $R_{b1}(\theta_1)$ which satisfies Eq. (7.1-3) at angle of rotation $\theta_1 = \theta_{1E0}$ is chosen, the path of contact **h** is determined by Eq. (7.1-1).

7.1.4 Surface of action

Since the surface of action is composed of the lines of contact, each of which is the intersection of the plane of action G_1 and the plane of tangent at each point of contact, it is obtained by Eqs. (7.1-1) and (3.5-3) in the coordinate systems C_{q1} and C_1 as follows, where z_{1cq} is a parameter instead of z_{cq} (Fig. 3.5-2).

$$\left.\begin{aligned}
q_{1c}(\theta_1, z_{1cq}) &= \int_0^{\theta 1} [R_{b1}(\theta_1)(1-d\chi_1/d\theta_1)\cos^2\phi_{b10}]\, d\theta_1 + q_{1c0} - z_{1cq}\tan\phi_{b10}\\
R_{b1}(\theta_1, z_{1cq}) &= R_{b1}(\theta_1)\\
z_{1c}(\theta_1, z_{1cq}) &= \int_0^{\theta 1} [R_{b1}(\theta_1)(1-d\chi_1/d\theta_1)\cos\phi_{b10}\sin\phi_{b10}]\, d\theta_1 + z_{1cq}\\
u_{1c}(\theta_1, z_{1cq}) &= q_{1c}(\theta_1, z_{1cq})\cos\chi_1(\theta_1) + R_{b1}(\theta_1)\sin\chi_1(\theta_1)\\
v_{1c}(\theta_1, z_{1cq}) &= q_{1c}(\theta_1, z_{1cq})\sin\chi_1(\theta_1) - R_{b1}(\theta_1)\cos\chi_1(\theta_1)\\
g\,(\phi_1(\theta_1) &= \pi/2 - \chi_1(\theta_1),\ \phi_{b10}\ ;\ C_1)
\end{aligned}\right\} \qquad (7.1\text{-}4)$$

7.1.5 Conjugate tooth surfaces

Transforming Eq. (7.1-4) into the coordinate system C_{r1} through Eq. (3.5-6), the conjugate tooth surface I and the inclination angle of the surface normal **n** of gear I (pinion) are obtained as follows, where the coordinate systems C_{r1} and C_1 coincide when $\theta_1 = 0$.

$$\left.\begin{aligned}
\chi_{r1}(\theta_1) &= \chi_1(\theta_1) - \theta_1 = \pi/2 - \phi_1(\theta_1) - \theta_1\\
\phi_{r1}(\theta_1) &= \phi_1(\theta_1) + \theta_1\\
u_{r1c}(\theta_1, z_{1cq}) &= q_{1c}(\theta_1, z_{1cq})\cos\chi_{r1}(\theta_1) + R_{b1}(\theta_1)\sin\chi_{r1}(\theta_1)\\
v_{r1c}(\theta_1, z_{1cq}) &= q_{1c}(\theta_1, z_{1cq})\sin\chi_{r1}(\theta_1) - R_{b1}(\theta_1)\cos\chi_{r1}(\theta_1)\\
z_{r1c}(\theta_1, z_{1cq}) &= z_{1c}(\theta_1, z_{1cq})
\end{aligned}\right\} \qquad (7.1\text{-}5)$$
$$n\,(\phi_{r1}(\theta_1) = \pi/2 - \chi_{r1}(\theta_1),\ \phi_{b10};\ C_{r1})$$

In the same way, transforming Eq. (7.1-4) into the coordinate systems C_2 and C_{q2} through Eqs. (2.1-1), (2.1-2) and (2.1-7) and the transformed Eq. (7.1-4) into the coordinate system C_{r2} through Eq. (3.5-5), the conjugate tooth surface II and the inclination angle of the surface normal **n** of gear II (gear) are obtained as follows.

$$\left.\begin{aligned}
&\theta_2 = \int_0^{\theta 1} d\theta_1 / i(\theta_1)\\
&\chi_{r2}(\theta_1) = \chi_2(\theta_1) - \theta_2 = \pi/2 - \phi_2(\theta_1) - \theta_2\\
&\phi_{r2}(\theta_1) = \phi_2(\theta_1) + \theta_2\\
&u_{r2c}(\theta_1) = q_{2c}(\theta_1,\ z_{1cq})\cos\chi_{r2}(\theta_1) + R_{b2}(\theta_1)\sin\chi_{r2}(\theta_1)\\
&v_{r2c}(\theta_1) = q_{2c}(\theta_1,\ z_{1cq})\sin\chi_{r2}(\theta_1) - R_{b2}(\theta_1)\cos\chi_{r2}(\theta_1)\\
&z_{r2c}(\theta_1) = z_{2c}(\theta_1,\ z_{1cq})\\
&n(\phi_{r2}(\theta_1) = \pi/2 - \chi_{r2}(\theta_1),\ \phi_{b20};\ C_{r2})
\end{aligned}\right\} \qquad (7.1\text{-}6)$$

7．1．6　Tooth surface of rack

When the coordinate system C_{1crk} is defined, which makes a parallel translation with the velocity $R_1(\theta_1)(d\theta_1/dt)$ of the instantaneous axis in the direction of the axis u_{1c} to the coordinate system C_1, the relations between the coordinate systems C_{1crk} and C_1 are obtained as follows, where the coordinate systems C_{1crk} and C_1 coincide when $\theta_1 = 0$.

$$\left.\begin{aligned}
&u_{1crk}(\theta_1,\ z_{1cq}) = u_{1c}(\theta_1,\ z_{1cq}) - L_{1crk}(\theta_1)\\
&v_{1crk}(\theta_1,\ z_{1cq}) = v_{1c}(\theta_1,\ z_{1cq})\\
&z_{1crk}(\theta_1,\ z_{1cq}) = z_{1c}(\theta_1,\ z_{1cq})\\
&L_{1crk}(\theta_1) = -\int_0^{t} R_1(\theta_1)(d\theta_1/dt)dt = -\int_0^{\theta 1} R_1(\theta_1)d\theta_1\\
&\qquad = \int_0^{\theta 1} \{R_{b1}(\theta_1)/\cos\chi_1(\theta_1)\}d\theta_1
\end{aligned}\right\} \qquad (7.1\text{-}7)$$

In this theory, $R_1(\theta_1)$ is always positive, therefore it is noticeable that $L_{1crk}(\theta_1)$ looks toward the negative direction of the axis u_{1c} when $(d\theta_1/dt)$ is positive.

Transforming Eq. (7.1-4) into the coordinate system C_{1crk} through Eq. (7.1-7), the tooth surface of the rack is obtained,

7．1．7　When the tooth surface of a rack is a plane ($d\chi_1/d\theta_1 = 0$)

Figure 7.1-2 shows the case where the tooth surface of a rack is a plane.

(1) Inclination angle of common contact normal

Because the tooth surface of the rack is a plane, the inclination angle $g(\phi_1(\theta_1) = \pi/2 - \chi_1(\theta_1),\ \phi_{b10};\ C_1)$ must be constant, therefore,

$$\begin{aligned}
&d\chi_1/d\theta_1 = 0\\
&\chi_1(\theta_1) = \chi_{10} = \pi/2 - \phi_{10}
\end{aligned} \qquad (7.1\text{-}8)$$

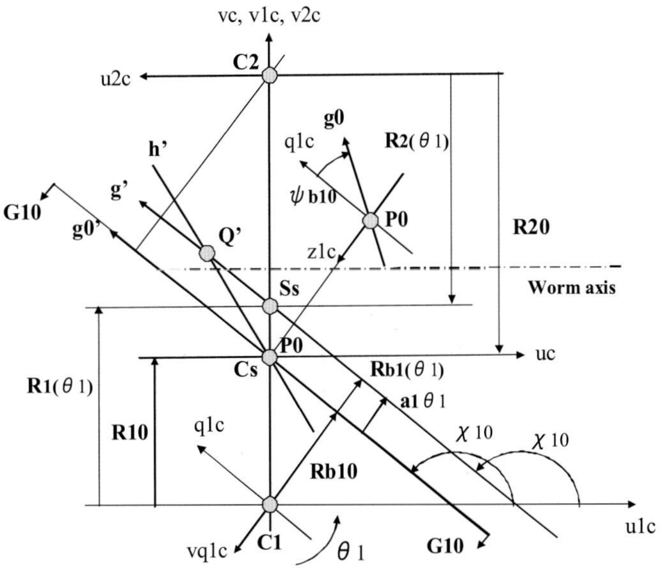

Fig.7.1-2 Path of contact **h** when $\chi(\theta_1) = \chi_{10}$ in the coordinate systems C_1 and C_{q1}

(2) Radius $R_{b1}(\theta_1)$ of base cylinder

It can be chosen at will, but in this theory, $dR_{b1}/d\theta_1 = a_1$ is assumed to simplify handling.

$$\begin{aligned} &dR_{b1}/d\theta_1 = a_1 \\ &R_{b1}(\theta_1) = R_{b10} + a_1\theta_1 \end{aligned} \qquad (7.1\text{-}9)$$

From the given gear ratio i_{E0} at θ_{1E0} and Eq. (7.1-3), a_1 is obtained as follows.

$$\begin{aligned} R_{b10} &= -R_{10}\cos\chi_{10} \\ R_{b1}(\theta_{1E0}) &= R_{b10} + a_1\theta_{1E0} \\ &= -E\cos\chi_{10}/(i_{E0} + 1) \\ a_1 &= -\{E\cos\chi_{10}/(i_{E0} + 1) + R_{b10}\}/\theta_{1E0} \end{aligned} \qquad (7.1\text{-}10)$$

(3) Path of contact and instantaneous gear ratio $i(\theta_1)$

Using Eqs. (7.1-8), (7.1-9) and (7.1-10), the path of contact and the instantaneous gear ratio are obtained as follows.

$$\left.\begin{aligned} q_{1c}(\theta_1) &= (R_{b10}\theta_1 + a_1\theta_1^2/2)\cos^2\phi_{b10} + q_{1c0} \\ R_{b1}(\theta_1) &= R_{b10} + a_1\theta_1 \\ z_{1c}(\theta_1) &= (R_{b10}\theta_1 + a_1\theta_1^2/2)\cos\phi_{b10}\sin\phi_{b10} \\ \chi_1(\theta_1) &= \chi_{10} = \pi/2 - \phi_{10} \\ i(\theta_1) &= -E\cos\chi_{10}/(R_{b10} + a_1\theta_1) - 1 \\ a_1 &= -\{E\cos\chi_{10}/(i_{E0} + 1) + R_{b10}\}/\theta_{1E0} \end{aligned}\right\} \qquad (7.1\text{-}11)$$

(4) Surface of action

Substituting Eq. (7.1-11) for Eq. (7.1-4), the surface of action is obtained as follows.

$$\left.\begin{aligned} q_{1c}(\theta_1, z_{1cq}) &= (R_{b10}\theta_1 + a_1\theta_1^2/2)\cos^2\phi_{b10} + q_{1c0} - z_{1cq}\tan\phi_{b10} \\ R_{b1}(\theta_1, z_{1cq}) &= R_{b1}(\theta_1) = R_{b10} + a_1\theta_1 \\ z_{1c}(\theta_1, z_{1cq}) &= (R_{b10}\theta_1 + a_1\theta_1^2/2)\cos\phi_{b10}\sin\phi_{b10} + z_{1cq} \\ u_{1c}(\theta_1, z_{1cq}) &= q_{1c}(\theta_1, z_{1cq})\cos\chi_{10} + R_{b1}(\theta_1)\sin\chi_{10} \\ v_{1c}(\theta_1, z_{1cq}) &= q_{1c}(\theta_1, z_{1cq})\sin\chi_{10} - R_{b1}(\theta_1)\cos\chi_{10} \\ g\,(\phi_1(\theta_1) &= \phi_{10} = \pi/2 - \chi_{10},\ \phi_{b10}\ ;\ C_1) \end{aligned}\right\} \qquad (7.1\text{-}12)$$

(5) Conjugate tooth surfaces and the tooth surface of a rack

Transforming Eq. (7.1-12) into the coordinate systems C_{r1} and C_{r2} through Eqs. (7.1-5) and (7.1-6), the conjugate tooth surfaces I and II are obtained.

Transforming Eq. (7.1-12) into the coordinate system C_{1crk} through Eq. (7.1-7), the conjugate tooth surface (plane) of the rack is obtained, where $L_{1crk}(\theta_1)$ of Eq. (7.1-7) is obtained as follows.

$$L_{1crk}(\theta_1) = (R_{b10}\theta_1 + a_1\theta_1^2/2)/\cos\chi_{10}$$

7. 1. 8 When tooth surface I is an involute helicoid ($dR_{b1}/d\theta_1 = 0$)

Figure 7.1-3 shows the case where tooth surface I is an involute helicoid and the inclination angle $\chi_1(\theta_1)$ of the common contact normal **g** varies corresponding to Fig. 7.1-1(b).

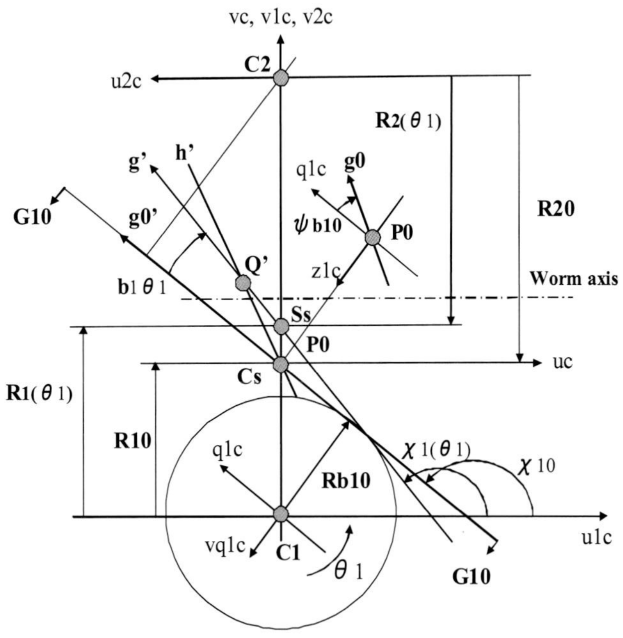

Fig.7.1-3 Path of contact **h** when $R_{b1}(\theta_1) = R_{b10}$ in the coordinate systems C_1 and C_{q1}

(1) Radius $R_{b1}(\theta_1)$ of base cylinder

Because tooth surface I is an involute helicoid, it is obtained.

$$dR_{b1}/d\theta_1 = 0$$
$$R_{b1}(\theta_1) = R_{b10} \qquad (7.1\text{-}13)$$

(2) Inclination angle $\chi_1(\theta_1)$ of common contact normal **g**

$d\chi_1/d\theta_1 = b_1$ (constant) is assumed to simplify handling, therefore,

$$\chi_1(\theta_1) = b_1\theta_1 + \chi_{10} = \pi/2 - \phi_1(\theta_1) \qquad (7.1\text{-}14)$$

Because the distance of instantaneous axis S_s from the axis z_{1c} is expressed by $R_1(\theta_1)$, the instantaneous gear ratio $i(\theta_1)$ is obtained by Eq. (7.1-2) as follows.

$$v_{1c}(\theta_1) = R_1(\theta_1) = -R_{b10}/\cos(b_1\theta_{1E0} + \chi_{10})$$
$$i(\theta_1) = \{E - R_1(\theta_1)\}/R_1(\theta_1) = -(E\cos(b_1\theta_{1E0} + \chi_{10})/R_{b10} + 1)$$

From the given gear ratio $i(\theta_{1E0}) = i_{E0}$, b_1 is obtained as follows.

$$i_{E0} = -E\cos(b_1\theta_{1E0} + \chi_{10})/R_{b10} - 1$$
$$\cos(b_1\theta_{1E0} + \chi_{10}) = -R_{b10}(i_{E0} + 1)/E$$
$$b_1 = [\mathrm{ArcCos}\{-R_{b10}(i_{E0} + 1)/E\} - \chi_{10}]/\theta_{1E0} \qquad (7.1\text{-}15)$$

(3) Path of contact and instantaneous gear ratio $i(\theta_1)$

The path of contact and the instantaneous gear ratio are obtained as follows.

$$\left.\begin{aligned}
q_{1c}(\theta_1) &= R_{b10}(1-b_1)\theta_1\cos^2\phi_{b10} + q_{1c0}\\
R_{b1}(\theta_1) &= R_{b10}\\
z_{1c}(\theta_1) &= R_{b10}(1-b_1)\theta_1\cos\phi_{b10}\sin\phi_{b10}\\
\chi_1(\theta_1) &= b_1\theta_1 + \chi_{10} = \pi/2 - \phi_1(\theta_1)\\
i(\theta_1) &= -E\cos(b_1\theta_1 + \chi_{10})/R_{b10} - 1\\
b_1 &= [\mathrm{ArcCos}\{-R_{b10}(i_{E0} + 1)/E\} - \chi_{10}]/\theta_{1E0}
\end{aligned}\right\} \qquad (7.1\text{-}16)$$

(4) Surface of action

Substituting Eq. (7.1-16) for Eq. (7.1-4), the surface of action is obtained as follows.

$$\left.\begin{aligned}
&q_{1c}(\theta_1, z_{1cq}) = R_{b10}(1-b_1)\theta_1\cos^2\phi_{b10} + q_{1c0} - z_{1cq}\tan\phi_{b10} \\
&R_{b1}(\theta_1, z_{1cq}) = R_{b10} \\
&z_{1c}(\theta_1, z_{1cq}) = R_{b10}(1-b_1)\theta_1\cos\phi_{b10}\sin\phi_{b10} + z_{1cq} \\
&u_{1c}(\theta_1, z_{1cq}) = q_{1c}(\theta_1, z_{1cq})\cos\chi_1(\theta_1) + R_{b10}\sin\chi_1(\theta_1) \\
&v_{1c}(\theta_1, z_{1cq}) = q_{1c}(\theta_1, z_{1cq})\sin\chi_1(\theta_1) - R_{b10}\cos\chi_1(\theta_1) \\
&\quad g\,(\phi_1(\theta_1) = \pi/2 - \chi_1(\theta_1),\ \phi_{b10};\ C_1)
\end{aligned}\right\} \qquad (7.1\text{-}17)$$

(5) Conjugate tooth surfaces and the tooth surface of a rack

Transforming Eq. (7.1-17) into the coordinate systems C_{r1} and C_{r2} through Eqs. (7.1-5) and (7.1-6), the conjugate tooth surfaces I and II are obtained.

Transforming Eq. (7.1-17) into the coordinate system C_{1crk} through Eq. (7.1-7), the conjugate tooth surface of the rack is obtained, where $L_{1crk}(\theta_1)$ of Eq. (7.1-7) is obtained as follows.

$$L_{1crk}(\theta_1) = \int_0^{\theta 1} \{R_{b10}/\cos(b_1\theta_1 + \chi_{10})\}\,d\theta_1$$

7.2 Tooth surface of a rack driven by a screw with constant lead and its mating tooth surface of gear I (pinion) transmitting inconstant rotational motion

In Fig. 7.1-1(b), the coordinate systems C_1, C_S and C_2, the design point P_0, the common contact normals $\mathbf{g}_0$ and $\mathbf{g}$ and the path of contact $\mathbf{h}$ expressed by Eq. (7.1-1) are given. Gear II is replaced by a rack which is driven by a screw with lead L and makes a parallel translation in the direction of the axis u_{1c}.

7.2.1 Instantaneous gear ratio and design point

When the distances of instantaneous axis S_S from the axes of gears I and II are $R_1(\theta_1)$ and $R_2(\theta_1)$, the angular velocities of gears I and II are $\omega_1(\theta_1)$ and $\omega_2(\theta_1)$ and that of the screw is $\omega_m(\theta_1)$ at an angle of rotation θ_1 respectively, the velocity V_{rk} of the rack at this instant in the direction of the axis u_{1c} is obtained as follows.

$$V_{rk} = R_1(\theta_1)\,\omega_1(\theta_1) = R_2(\theta_1)\,\omega_2(\theta_1) = L\omega_m(\theta_1)\,/\,2\pi$$

Therefore, the instantaneous gear ratio $i_m(\theta_1)$ between the screw and gear I is obtained as follows.

$$\left.\begin{aligned}
&i_m(\theta_1) = \omega_m(\theta_1)/\omega_1(\theta_1) = 2\pi R_1(\theta_1)/L \\
\text{or}\quad &R_1(\theta_1) = i_m(\theta_1)L/\,2\pi
\end{aligned}\right\} \qquad (7.2\text{-}1)$$

where L : lead (constant) of the screw driving the rack

From the given gear ratio $i_m(0) = i_{s0}$, design point $P_0(0,\ R_{10},\ 0\ ;\ C_1)$ is obtained as follows.

$$R_{10} = R_1(0) = i_{S0} L/2\pi$$

Using Eq. (2.1-3), the instantaneous gear ratio $i_m(\theta_1)$ is expressed by the radius $R_{b1}(\theta_1)$ of base circle as follows.

$$i_m(\theta_1) = -2\pi R_{b1}(\theta_1)/(L\cos\chi_1(\theta_1)) \qquad (7.2\text{-}2)$$

When the gear ratio $i_m(\theta_{1E0}) = i_{E0}$ at $\theta_1 = \theta_{1E0}$ is given and a combination of $\chi_1(\theta_1)$ and $R_{b1}(\theta_1)$ which satisfies Eq. (7.2-2) at angle of rotation $\theta_1 = \theta_{1E0}$ is chosen, the path of contact **h** is determined by Eq. (7.1-1).

7. 2. 2 When the tooth surface of a rack is a plane ($d\chi_1/d\theta_1 = 0$)

(1) Selection of $\chi_1(\theta_1)$ and $R_{b1}(\theta_1)$

$\chi_1(\theta_1)$ and $R_{b1}(\theta_1)$ are given as same as in Eqs. (7.1-8) and (7.1-9) as follows.

$$\left.\begin{aligned} &d\chi_1/d\theta_1 = 0, && \chi_1(\theta_1) = \chi_{10} = \pi/2 - \phi_{10} \\ &dR_{b1}/d\theta_1 = a_{1m}, && R_{b1}(\theta_1) = R_{b10} + a_{1m}\theta_1 \end{aligned}\right\} \qquad (7.2\text{-}3)$$

From the given gear ratio i_{E0} at θ_{1E0} and Eq. (7.2-2), a_{1m} is obtained as follows.

$$\begin{aligned} R_{b1}(\theta_1) &= R_{b10} + a_{1m}\theta_1 \\ R_1(\theta_{1E0}) &= i_m(\theta_{1E0})L/2\pi = i_{E0} L/2\pi \\ R_{b10} &= -R_1(0)\cos\chi_{10} \\ R_{b1}(\theta_{1E0}) &= -R_1(0)\cos\chi_{10} + a_{1m}\theta_{1E0} = -R_1(\theta_{1E0})\cos\chi_{10} \end{aligned}$$

$$\left.\begin{aligned} a_{1m} &= -\{R_1(\theta_{1E0}) - R_1(0)\}\cos\chi_{10}/\theta_{1E0} \\ &= -(i_{E0} - i_{S0})L\cos\chi_{10}/(2\pi\theta_{1E0}) \end{aligned}\right\} \qquad (7.2\text{-}4)$$

(2) Path of contact and instantaneous gear ratio $im(\theta_1)$

Substituting Eqs. (7.2-3) and (7.2-4) for Eqs. (7.1-1) and (7.2-2), the path of contact and the instantaneous gear ratio are obtained as follows.

$$\left.\begin{aligned} q_{1c}(\theta_1) &= (R_{b10}\theta_1 + a_{1m}\theta_1^{\,2}/2)\cos^2\phi_{b10} + q_{1c0} \\ R_{b1}(\theta_1) &= R_{b10} + a_{1m}\theta_1 \\ z_{1c}(\theta_1) &= (R_{b10}\theta_1 + a_{1m}\theta_1^{\,2}/2)\cos\phi_{b10}\sin\phi_{b10} \\ \chi_1(\theta_1) &= \chi_{10} = \pi/2 - \phi_{10} \\ i_m(\theta_1) &= -2\pi(R_{b10} + a_{1m}\theta_1)/\cos\chi_{10}/L \\ a_{1m} &= -(i_{E0} - i_{S0})L\cos\chi_{10}/(2\pi\theta_{1E0}) \end{aligned}\right\} \qquad (7.2\text{-}5)$$

Equations (7.2-5) and (7.1-11) for the path of contact are the same, where only $i_m(\theta_1)$ and a_{1m} are different.

(3) Surface of action and tooth surfaces of gear I and a rack

Since Eqs. (7.2-5) and (7.1-11) for the path of contact are the same, the surface of action is obtained by substituting a_{1m} for a_1 of Eq. (7.1-12), the tooth surface of gear I is obtained by Eq. (7.1-5) and that of the rack is obtained by Eq. (7.1-7).

7. 2. 3 When tooth surface of gear I is an involute helicoid ($dR_{b1}/d\theta_1 = 0$)

(1) Selection of $\chi_1(\theta_1)$ and $R_{b1}(\theta_1)$

$\chi_1(\theta_1)$ and $R_{b1}(\theta_1)$ are given the same as those in Eqs. (7.1-13) and (7.1-14) as follows.

$$\left.\begin{aligned} &dR_{b1}/d\theta_1 = 0, && R_{b1}(\theta_1) = R_{b10} \\ &d\chi_1/d\theta_1 = b_{1m}, && \chi_1(\theta_1) = b_{1m}\theta_1 + \chi_{10} = \pi/2 - \phi_1(\theta_1) \end{aligned}\right\} \qquad (7.2\text{-}6)$$

From the given gear ratio $i_m(\theta_{1E0}) = i_{E0}$ and Eq. (7.2-2), b_{1m} is obtained as follows.

$$\begin{aligned} i_{E0} &= -2\pi R_{b10}/\cos(b_{1m}\theta_{1E0} + \chi_{10})/L \\ \cos(b_{1m}\theta_{1E0} + \chi_{10}) &= -2\pi R_{b10}/(i_{E0}L) \\ b_{1m} &= [\mathrm{ArcCos}\{-2\pi R_{b10}/(i_{E0}L)\} - \chi_{10}]/\theta_{1E0} \end{aligned} \qquad (7.2\text{-}7)$$

(2) Path of contact and instantaneous gear ratio $i_m(\theta_1)$

Substituting Eqs. (7.2-6) and (7.2-7) for Eqs. (7.1-1) and (7.2-2), the path of contact and the instantaneous gear ratio are obtained as follows.

$$\left.\begin{aligned} q_{1c}(\theta_1) &= R_{b10}(1-b_{1m})\theta_1\cos^2\phi_{b10} + q_{1c0} \\ R_{b1}(\theta_1) &= R_{b10} \\ z_{1c}(\theta_1) &= R_{b10}(1-b_{1m})\theta_1\cos\phi_{b10}\sin\phi_{b10} \\ \chi_1(\theta_1) &= b_{1m}\theta_1 + \chi_{10} = \pi/2 - \phi_1(\theta_1) \\ i_m(\theta_1) &= -2\pi R_{b10}/\cos(b_{1m}\theta_1 + \chi_{10})/L \\ b_{1m} &= [\mathrm{ArcCos}\{-2\pi R_{b10}/(i_{E0}L)\} - \chi_{10}]/\theta_{1E0} \end{aligned}\right\} \qquad (7.2\text{-}8)$$

Eqs. (7.2-8) and (7.1-16) for the path of contact are the same, where only $i_m(\theta_1)$ and b_{1m} are different.

(3) Surface of action and tooth surfaces of gear I and a rack

Since Eqs. (7.2-8) and (7.1-16) for the path of contact are the same, the surface of action is obtained by substituting b_{1m} for b_1 of Eq. (7.1-17), the tooth surface of gear I is obtained by Eq. (7.1-5) and that of the rack is obtained by Eq. (7.1-7).

7. 2. 4 Another design method of a plane tooth surface of a rack driven by a screw with constant lead and its mating tooth surface of gear I transmitting inconstant rotational motion (72)

The design example shown in reference (72) is another one which is composed of the rack driven by a screw with constant lead and the mating gear I. They are designed as follows, where the sector gear in reference (72) is treated as gear I here, so that it is noticeable that subscript 2 in reference (72) is 1 in this subsection to unify the whole theory.

Substituting $\omega_1(t)$ for $d\theta_1/dt$, equations of path of contact at a certain time s are obtained from Eq. (7. 1-1) as follows, where $d\chi_1/d\theta_1 = 0$ because the rack has plane surfaces.

$$\left.\begin{aligned} q_{1c}(s) &= \int_0^s [R_{b1}(t)\,\omega_1(t)\cos^2\phi_{b10}]\,dt + q_{1c0} \\ R_{b1}(s) &= \int_0^s (dR_{b1}/dt)\,dt + R_{b10} \\ z_{1c}(s) &= \int_0^s [R_{b1}(t)\,\omega_1(t)\cos\phi_{b10}\sin\phi_{b10}]\,dt \\ \chi_1(s) &= \chi_{10} = \pi/2 - \phi_{10} \\ \theta_1(s) &= \int_0^s \omega_1(t)\,dt \end{aligned}\right\} \qquad (7.2\text{-}9)$$

where

$R_{b1}(t)$: radius of base circle at t

s : a certain time

(1) Selection of radius $R_{b1}(s)$ of base circle

When the distance of the instantaneous axis from the axis I is $R_1(s)$ and the angular velocity of the screw is $\omega_m(s)$, the velocity V_{rk} of the rack at the instant s is expressed as follows.

$$V_{rk} = R_1(s)\,\omega_1(s) = -R_{b1}(s)\,\omega_1(s)/\cos\chi_{10} = L\omega_m(s)\ /\ 2\pi \qquad (7.2\text{-}10)$$

When the time at θ_{1E0} is s_{E0} and V_{rk} is supposed to be constant, the angular velocities $\omega_1(0)$ and $\omega_1(s_{E0})$ are obtained as follows.

$$\omega_1(0) = V_{rk}/\ R_1(0), \qquad \omega_1(s_{E0}) = V_{rk}/\ R_1(\theta_{1E0})$$
$$R_1(0) = i_{s0}\ L/2\pi, \qquad R_1(\theta_{1E0}) = i_{E0}\ L/2\pi$$

When $\omega_1(s)$ is expressed by a linear equation in the region from $\omega_1(0)$ to $\omega_1(s_{E0})$ where the gear ratio varied, it is obtained as follows.

$$\omega_1(s) = \omega_1(0) + c_1\,s$$
$$c_1 = (\omega_1(s_{E0}) - \omega_1(0))/s_{E0}$$

Therefore, s_{E0} is obtained as follows because of $\theta_{1E0} = \int_0^{sE0} \omega_1(t)\,dt$.

$$s_{E0} = 2\theta_{1E0}/(\omega_1(0) + \omega_1(s_{E0}))$$

$R_{b1}(s)$ is expressed by V_{rk} and $\omega_1(s)$ as follows.

$$R_{b1}(s) = -V_{rk}\cos\chi_{10}/\omega_1(s) = -V_{rk}\cos\chi_{10}/(\omega_1(0) + c_1\,s) \qquad (7.2\text{-}11)$$

(2) Instantaneous gear ratio $i_m(s)$

It is obtained from Eq. (7.2-1) as follows.

$$i_m(s) = 2\pi R_1(s)/L = 2\pi V_{rk}/(L\omega_1(s)) = 2\pi V_{rk}/(L(\omega_1(0)+c_1 s)) \qquad (7.2\text{-}12)$$

(3) Path of contact and instantaneous gear ratio $i_m(s)$

Substituting Eqs. (7.2-10), (7.2-11) and (7.2-12) for Eq. (7.2-9), the path of contact and the instantaneous gear ratio are obtained as follows.

$$\left.\begin{aligned}
q_{1c}(s) &= -V_{rk}\, s\cos\chi_{10}\cos^2\phi_{b10} + q_{1c0}\\
R_{b1}(s) &= -R_1(s)\cos\chi_{10} = -V_{rk}\cos\chi_{10}/\omega_1(s)\\
z_{1c}(s) &= -V_{rk}\, s\cos\chi_{10}\cos\phi_{b10}\sin\phi_{b10}\\
\chi_1(s) &= \chi_{10} = \pi/2 - \phi_{10}\\
\omega_1(s) &= \omega_1(0)+c_1 s\\
c_1 &= (\omega_1(s_{E0})-\omega_1(0))/s_{E0}\\
\theta_1(s) &= \int_0^s \omega_1(t)\,dt = \omega_1(0)s + c_1 s^2/2\\
i_m(s) &= 2\pi V_{rk}/\omega_1(s)/L = 2\pi V_{rk}/(\omega_1(0)+c_1 s)/L
\end{aligned}\right\} \qquad (7.2\text{-}13)$$

In reference (72), Eq. (7.2-13) with $\phi_{b10} = 0$ is analyzed and the numerical example is shown.

Eq. (7.2-13) is expressed in the tangential-polar coordinates $(q_{1c}(s),\ \theta_1(s))$, where $q_{1c}(s)$ and $dq_{1c}(s)/d\theta_1(s) = -V_{rk}\cos\chi_{10}/\omega_1(s) = R_{b1}(s)$ are the same as those in the coordinate system C_{q1}, so that Eq. (4-5) of the pinion profile in reference (72) is the same one which is obtained by transforming Eq. (7.2-13) into the coordinate system C_{r1}. The difference between Eqs. (7.2-13) and (7.2-5) is whether the angular velocity $\omega_1(s)$ is a linear equation of s or the base circle radius $R_{b1}(\theta_1)$ is a linear one of θ_1 only, but Eq. (7.2-5) is simpler in handling because V_{rk} and $\omega_1(s)$ are not necessary to be introduced.

7. 3 Design examples

In design examples, gears I and II are called pinion and gear respectively.

7. 3. 1 Design example (1)

The example in which the center distance is given and the tooth surface of a rack is a plane mentioned in 7.1.7 is calculated here.

(1) Given conditions

(a) Center distance E, shaft angle Σ and gear ratios i_{s0} and i_{E0}

A gear pair whose dimensions are given as follows is calculated.

$$E = 612.492\text{ mm},\ \Sigma = 0.00001°,\ i_{s0} = 20.50(\theta_1 = 0),\ i_{E0} = 17.676(\theta_{1E0} = 35°)$$

(Note 7.3-1) : In order to compare one with the other, $R_{10} = 28.488$ mm and $i_{s0} = 20.50$ are given the same as in reference (72) and $R_{20} = 584.004$ mm and E = 612.492 mm are calculated back.

(Note 7.3-2) : From the gear ratio i_{E072} = 23.60 and the reference circle $R_{1E0} = (i_{E072} / i_{S0}) R_{10}$ = 32.796 mm at θ_{1E0} = 35° in reference (72), E = 612.492 mm and i_{E0} = 17.676 are calculated back.

(b) Design point P_0

The input data are given in the coordinate system C_S and transformed into the coordinate system C_1.

$P_0(0, 0, 0 ; C_S)$,　$P_0(0, 28.488, 0 ; C_1)$

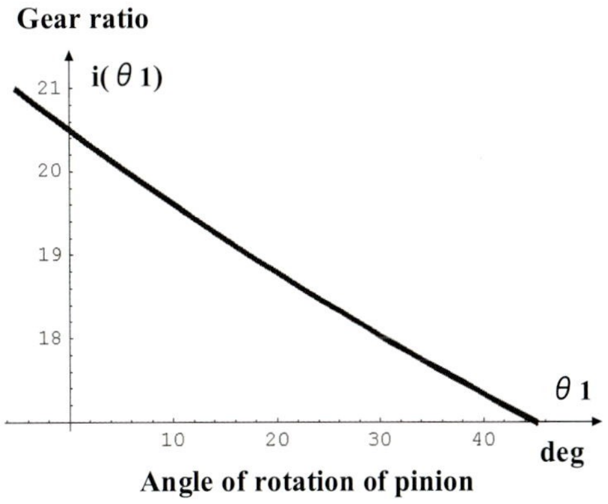

Fig.7.3-1 Variation of gear ratio

(c) Common contact normal $\mathbf{g}_{0D}$ through P_0

The inclination angle of the common contact normal $\mathbf{g}_{0D}$ is given in the coordinate system C_S and transformed into the coordinate system C_1.

$g_{0D}(\phi_0 = -7.5°, \phi_{n0D} = 122.0° ; C_S)$
$g_{0D}(\phi_{10D} = -121.78°, \phi_{b10D} = 3.96° ; C_1)$

(d) Dimensions of the reference rack ($d\chi_1/d\theta_1 = 0$)

They are given as follows, which are almost the same as in reference (72).

Reference circular pitch
$p_{circular}$ = 13.0 mm (from which the number of teeth (not integer) is calculated back)
Whole depth　h_t = 7.2 mm (clearance c_L = 1.0 mm and top land of rack t_{cn} = 1.5 mm)
Addendum　A_d = 2.60 mm
Pinion facewidth　F_p = 20 mm (10 mm from P_0 toward both the large and the small ends), while F_p = 28 mm in reference (72).

(2) Results of calculation

(a) Base circle variation coefficient a_1, radius $R_{b1}(\theta_1)$ of base cylinder and gear ratio $i(\theta_1)$

They are obtained by Eq. (7.1-11) as follows.

a_1 = 5.995mm/rad.
$R_{b1}(\theta_1) = 24.217 + 5.995\theta_1$
$i(\theta_1) = 520.67/(24.217 + 5.995\theta_1) - 1$

Figure 7.3-1 shows the variation of the gear ratio.

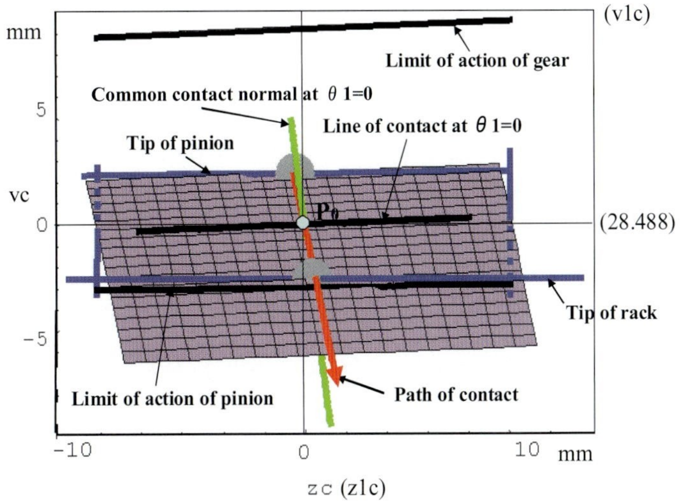

Fig.7.3-2(a) Surface of action in the coordinate system Cs seen from the negative direction of uc (u1c)

(b) Surface of action

Figures 7.3-2(a) and (b) show the surface of action drawn by Eq. (7.1-12).

Figure 7.3-2(a) is seen from the negative direction of the axis u_{1c}, where the path of contact, the line of contact and the common contact normal through P_0, the limits of action of the pinion and gear by Eq. (2.4-4) and the tip lines of the pinion and rack by Eq. (3.6-2) are drawn.

Figure 7.3-2(b) is seen from the positive direction of the axis z_{1c}, which corresponds to Fig. 2 in reference (72).

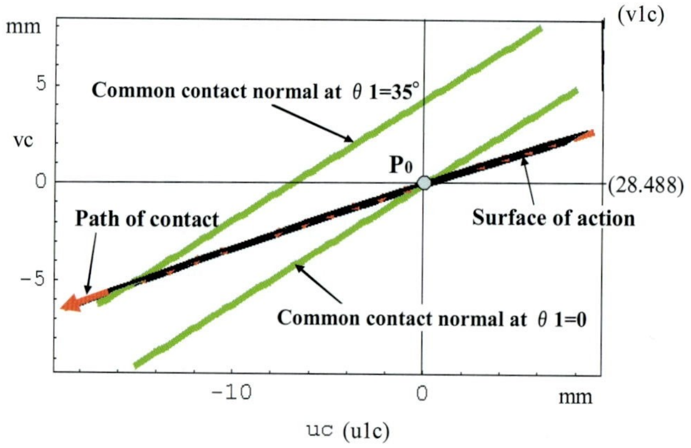

Fig.7.3-2(b) Surface of action in the coordinate system Cs seen from the direction of zc

(c) Tooth surfaces of the pinion and rack

Figure 7.3-3 shows the tooth surface of the pinion which is drawn by transforming Eq. (7.1-12) into the coordinate system C_{r1} through Eq. (7.1-5) and the plane tooth surface of the rack which is drawn by transforming Eq. (7.1-12) into the coordinate system C_{1crk} through Eq. (7.1-7), where the path of contact, the line of contact and the common contact normals, the limit of action of the pinion, the tip lines of the pinion and rack and the surface normals are drawn at the same time. The surface normal of the pinion at $\theta_1 = 35°$ exists beyond the limit of action, but there are no tooth surfaces actually at this point.

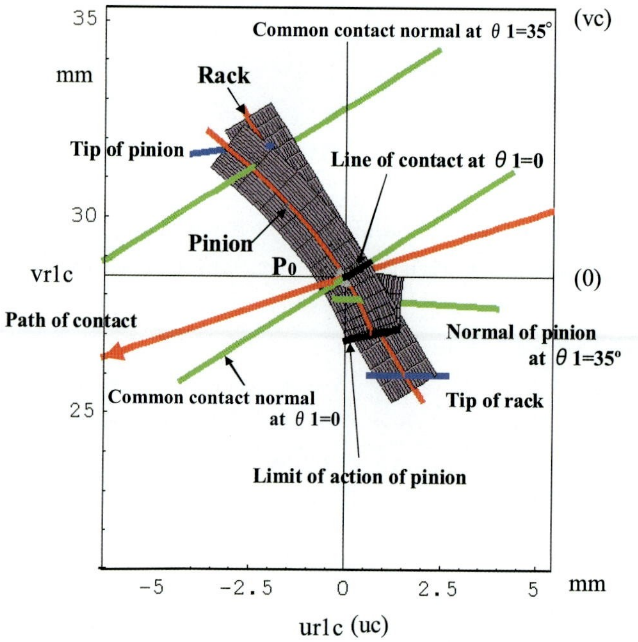

Fig.7.3-3 Tooth surfaces of pinion and rack

7.3.2 Design example (2)

The example in which gear II is replaced by a rack that is driven by a screw with constant lead L and the tooth surface of the rack is a plane mentioned in 7.2.2 is calculated here.

(1) Given conditions

(a) Lead of screw L, shaft angle and gear ratios between the pinion and the screw

$$L = 8.7315 \text{ mm},\ \Sigma = 0.00001°,\ i_{s0} = 20.50(\theta_1 = 0),\ i_{E0} = 23.60(\theta_{1E0} = 35°)$$

(b) to (d)

In order to compare one with the other, they are the same as in the design example (1).

(2) Results of calculation

(a) Base circle variation coefficient a_{1m}, radius $R_{b1}(\theta_1)$ of base cylinder and gear ratio $i_m(\theta_1)$

They are obtained by Eq. (7.2-5) as follows.

$$a_{1m} = 5.995 \text{ mm/rad.}$$
$$R_{b1}(\theta_1) = 24.217 + 5.995\theta_1$$
$$i_m(\theta_1) = 0.8465(24.217 + 5.995\theta_1)$$

Figure 7.3-4 shows the variation of the gear ratio $i_m(\theta_1)$.

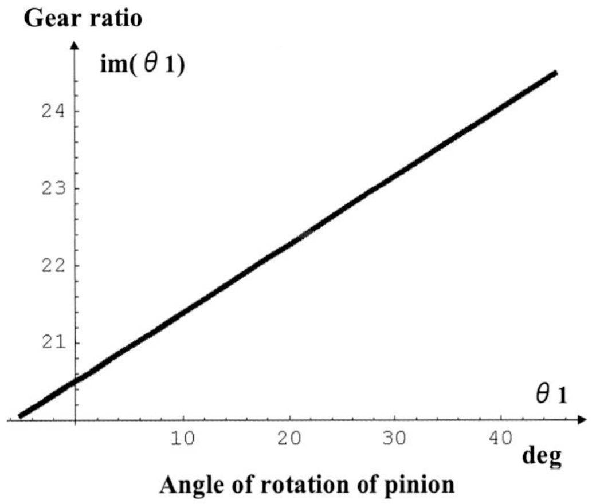

Fig.7.3-4 Variation of gear ratio

(b) Surface of action

Because Eq. (7.1-12) has $a_{1m} = a_1$ and the same $R_{b1}(\theta_1)$, the design examples (1) and (2) have the same surface of action drawn in Figs. 7.3-2(a) and (b). However, the gear ratios $i(\theta_1)$ between gear II and the pinion and $i_m(\theta_1)$ between the screw and the pinion are different as shown in Figs. 7.3-1 and 7.3-4.

(c) Tooth surfaces of the pinion and rack

Because the surface of action by Eq. (7.1-12) is the same, the tooth surfaces of the pinion and rack are the same as those in the design example (1) shown in Fig. 7.3-3.

7．3．3　Design example (3)

The example in which gear II is replaced by a rack that is driven by a screw with constant lead L and the tooth surface of a pinion is an involute helicoid mentioned in 7.2.3 is calculated here.

(1) Given conditions

(a) to (d)

In order to compare one with the other, they are the same as those in the design example (2).

(e) Radius of base cylinder of pinion

It is given as follows because the tooth surface of the pinion is an involute helicoid.

$$dR_{b1}/d\theta_1 = 0, \quad R_{b1}(\theta_1) = R_{b10} = 24.217 \text{ mm}$$

(2) Results of calculation

(a) $d\chi_1/d\theta_1 = b_{1m}$ and gear ratio $i_m(\theta_1)$

They are obtained by Eq. (7.2-8) as follows.

$$b_{1m} = 0.30354$$

$$i_m(\theta_1) = -17.4267/\cos(3.69624 + 0.30354\,\theta_1)$$

Figure 7.3-5 shows the variation of the gear ratio $i_m(\theta_1)$.

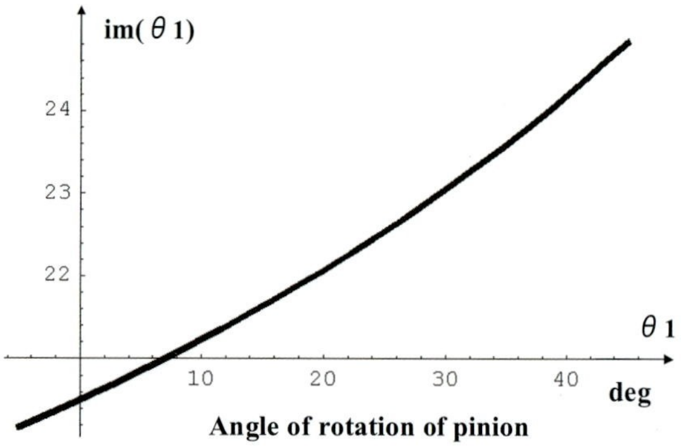

Fig.7.3-5 Variation of gear ratio

(b) Surface of action

Figures 7.3-6(a) and (b) show the surface of action drawn by Eq. (7.1-17).

(c) Tooth surfaces of the pinion and rack

Figure 7.3-7 shows the tooth surfaces of the pinion and rack which are drawn by Eqs. (7.1-5) and (7.1-7) respectively.

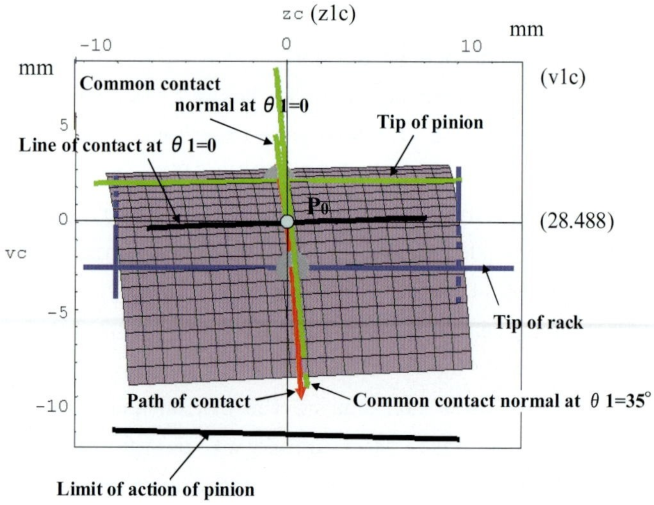

Fig.7.3-6(a) Surface of action in the coordinate system Cs seen from the negative direction of uc (u1c)

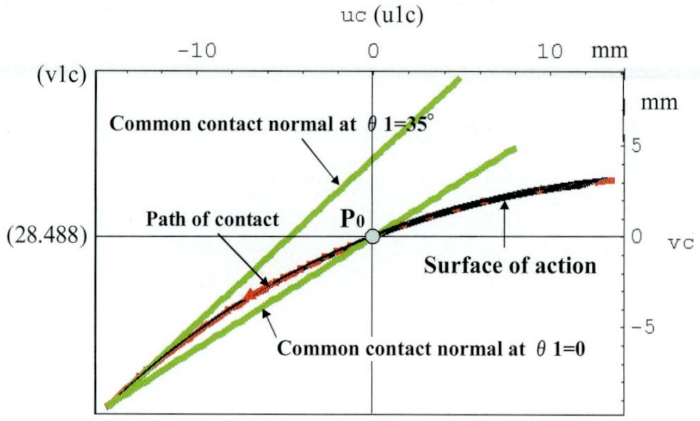

Fig.7.3-6(b) Surface of action in the coordinate system Cs seen from the direction of zc

7. 4 Summary and references

With regard to gear pairs with shaft angle $\Sigma=0$ transmitting inconstant rotational motion, the design methods based on the basic theory in Chapter 2 are clarified with the examples and it is shown that the new theory is effective for designing and analyzing gear pairs with an inconstant gear ratio, the results of which are summarized as follows.

(1) The design methods of the path of contact, the surface of action and conjugate tooth surfaces of a gear pair and its rack with shaft angle $\Sigma = 0$ transmitting inconstant rotational motion are clarified and the following applications are shown by the design examples.

(a) Design methods and examples where the tooth surface of a rack is a plane.

(b) A design method and example where the tooth surface of a pinion is an involute helicoid.

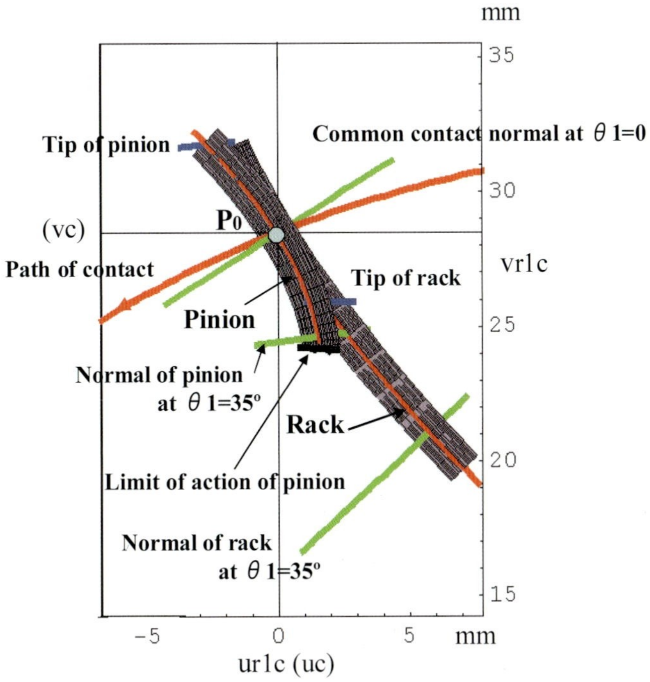

Fig.7.3-7 Tooth surfaces of pinion and rack

(2) The design example shown in reference (72) is one of the examples of item (1a) designed by another method.

References

(71) 窪田雅雄，非円形歯車に関する研究，機械試験所報告第 30 号，1959.
(72) 本多　捷，非定回転比ステアリング歯車機構，トヨタ技術，Vol.19-5，1968，608-619.

8. Summary

A new tooth geometry is proposed, which has common equations defined in common coordinate systems for almost all kinds of gears from cylindrical, worm, crossed helical to bevel and hypoid gears and makes it possible to estimate their dynamic performances at the same time, the results of which are summarized as follows.

(1) Eight coordinate systems common to all kinds of gears are defined, in which a point and the inclination angle of a normal through the point are described by five variables, which means that the point in the new theory always has a plane element. Basic equations of tooth geometry, such as the fundamental requirement for contact, paths of contact, tooth profiles and limits of action, are expressed as functions of the angle of rotation and equations of motion and bearing loads are obtained. Based on these coordinate systems and the equations, the new tooth geometry becomes common for all kinds of gears and makes it possible to deal with the dynamic rotational motion of the gears at the same time.

(2) The basic theory mentioned above is applied to the design of tooth surfaces of gears for power transmission and it is clarified that the path of contact satisfying the conditions for no variation of bearing loads is a straight line which coincides with the common contact normal and how the path of contact which realizes a pair of smooth tooth surfaces must be selected. Concepts common to all kinds of gears, such as an equivalent rack, a plane surface of action, conjugate tooth surfaces corresponding to the plane surface of action, design reference bodies of revolution (pitch hyperboloids) and contact ratios, are introduced. Based on this design method, a design example of hypoid gears is shown and compared with that designed by Wildhaber's method.

(3) Instead of giving the plane surface of action to realize no variation of bearing loads, an involute helicoid and its conjugate tooth surface having the same path of contact as that on the plane surface of action are discussed. The equations of an involute helicoid, its surface of action and its conjugate tooth surface are obtained and design examples of face gears are shown.

(4) Variations of a tooth bearing and a single flank error caused by assembly errors are discussed. They are obtained by calculating the normal amount of movement of the tooth surface of the gear along the line of contact on the fixed tooth surface of the pinion when the gear is moved from the correct position by assembly errors.

(5) The paths of contact under no loads and under load of a gear pair having involute helicoids for one member which are modified slightly and the conjugate tooth surfaces of the involute helicoids for the other are obtained. The equations of motion are solved and the dynamic increment of tooth load and the variation of bearing loads are calculated. The dynamic increment of tooth load of the gear pair having a single flank error composed of symmetrical convex curves which is practically important is algebraically obtained and verified by experiment through helical gear pairs. The dynamic increment of tooth load obtained here makes it possible to estimate the dynamic performances of noise and vibration in the design stage of gear pairs.

(6) Paths of contact, their surfaces of action and their conjugate tooth surfaces of gear pairs (rack and pinion) with shaft angle $\Sigma = 0$ transmitting inconstant rotational motion are designed and analyzed and two design examples, (a) when the tooth surface of a rack is a plane and (b) when the tooth surface of a pinion is an involute helicoid, are shown. Through these

design examples, it is shown that the new tooth geometry is effective for designing a gear pair transmitting inconstant rotational motion.

If the present definitions of pitch surface, surface of action, pressure angle, spiral angle, tooth trace, contact ratio and so on are redefined so as to be common to all kinds of gears based on the new tooth geometry, both geometric problems such as designing tooth surfaces and dynamic problems such as estimating the dynamic increment of tooth loads for all kinds of gears which have been discussed separately can be dealt with and understood in the same way using the new tooth geometry.

Postscript

The author intended to describe what are the basics of gear theories, based on which how quiet gear pairs can be designed and how the designed ones rotate and make noise and vibration, using his experiences during more than 40 years. Though about half of the contents of this book have been already published in JSME transactions or at JSME or ASME conferences, whose reputations were not so favorable, what was your impressions ?

When the vibration theory in Chapter 6 was contributed, it was severely criticized in reviews by those who had faith in the present theory. However, it meaned that the new theory was reinforced in the result because they looked at this theory from a point of view different from the author's perspective, therefore he thanks them now.

Moreover when the design theory in Chapters 2 through 4 was contributed, the former half of them were published after some twists and turns, but from an certain instant the latter half were suddenly refused to publish with the reproach that gears would never be designed by this sort of theory and that the author knows nothing about the fundamental requirement for contact. Afterward, they were resubmitted several times with modifications, but never accepted. Therefore, the author made up his mind to publish them on his own because he staked his life on this theory and would not like to leave them unfinished.

Now that this book has been written, the author believes that this theory makes it possible to discuss both the geometrical field (design) and the dynamical one (dynamic loads) of the gear theory through the common equations defined in the common coordinate systems in the same way, which have been discussed separately.

Although the people who understand this theory are limited mostly to those around the author, it is expected that if many other people accept this theory in the near future and discuss gears in the same basis with the same terms, the theory of gears will make progress further and many new gears of better performance will be developed.

Finally, the author wishes to express his appreciation to Mr. T. Akamatsu of Toyota Motor Corporation and Honorary Professor Dr. M. Katou of Tohoku University who have appreciated and supported this study which was not so favorable from the beginning. The author also wishes to thank his colleages at Toyota Motor Corporation and Toyota Central R & D Labs., Inc. and the engineers at Geason Corporation who listened to, helped and cooperated with the author in unfavorable situations. Without their cooperation, this theory would not have existed. The author thanks them heartily once again.

In publishing this book, Mr. T. Fujioka and Ms. N. Konita of SOEISHA/SANSEIDO BOOKSTORE provided helpful suggestions. The author would like to express his heartfelt gratitude to them.

The author : Sho Honda

Professional experience :

12/2005 - Present : Gear consultant.

01/1993 - 11/2005 : Toyota Central R&D Labs., Inc.
Research on a unified theory of gears.

04/1965 - 12/1992 : Toyota Motor Corporation.
Design, development and production relating to gear units.

Education :

01/1993 : D.E., Tohoku University, Japan.
Thesis: Rotational vibration of a helical gear pair with modified tooth surfaces.

03/1965 : B.E., Mechanical Engineering, The University of Tokyo, Japan.

A New Tooth Geometry and Its Applications

2016年6月11日　　初版発行

著者

Sho Honda

発行・発売

SOEISHA Co., Ltd.／SANSEIDO BOOKSTORE

株式会社創英社／三省堂書店

〒101-0051　東京都千代田区神田神保町1-1

Tel：03-3291-2295　Fax：03-3292-7687

印刷／製本

日本印刷株式会社

ISBN978-4-88142-979-2 C3053